Springer Japan KK

Constitutive Relation in High/Very High Strain Rates

Prof. Dr. Kozo Kawata
Department of Materials Science and Technology
Faculty of Industrial Science and Technology
Science University of Tokyo
2641, Yamazaki, Noda
Chiba 278, Japan

Prof. Dr. Jumpei Shioiri
Department of Mechanical Engineering
College of Engineering
Hosei University
3-7-2, Kajinocho, Koganei
Tokyo 184, Japan

ISBN 978-4-431-65949-5   ISBN 978-4-431-65947-1 (eBook)
DOI 10.1007/978-4-431-65947-1

Printed on acid-free paper

© Springer Japan 1996
Originally published by Springer-Verlag Tokyo Berlin Heidelberg New York in 1996

Typesetting: Camera ready by authors

SPIN: 10529284

PREFACE

The IUTAM Symposium on Constitutive Relation in High/Very High Strain Rates (CRHVHSR) was held October 16 ~ 19, 1995, at Seminar House, Science University of Tokyo, under the sponsorship of IUTAM, Japan Society for the Promotion of Science, The Commemorative Association for the Japan World Exposition (1970), Inoue Foundation for Science, The Japan Society for Aeronautical and Space Sciences, and Science University of Tokyo.

The proposal to hold the symposium was accepted by the General Assembly of IUTAM held in Haifa, Israel, in August 1992, and the scientists mentioned below were appointed by the Bureau of IUTAM to serve as members of the Scientific Committee. The main object of the symposium was to make a general survey of recent developments in the research of constitutive relations in high and very high strain rates and related problems in high velocity solid mechanics, and to explore further new ideas for dealing with unresolved problems of a fundamental nature as well as of practical importance. The subjects covered theoretical, experimental, and numerical fields in the above−mentioned problems in solids, covering metals, polymers, ceramics, and composites. Emphasis was given to the following fields:

1. Material characterization of solids in high velocity deformation, experimental techniques, typical data obtained by these techniques, modeling, and constitutive relations
2. Strain rate dependent elasto−visco−plastic stress waves
3. Crack initiation, propagation, and dynamic fracture toughness
4. Dynamic stress concentration
5. Structural dynamics in impact and constitutive relations of solids
6. Impact related problems

These research fields were covered in 28 invited papers, and in 16 scientific sessions chaired by the invited specialists. As indicated in the scientific program, two papers could not be presented orally by their authors. Participants, numbering 47, came from 12 countries (Austria, China, France, Germany, Japan, Korea, Poland, Romania, Russia, Sweden, UK, and USA). Discussions ranged from the microscopic level of dislocation theory to the macroscopic behavior of engineering materials and structures. From the standpoint of science, technology, and friendly exchange of opinions during social events, the aims of the IUTAM symposium were satisfactorily achieved.

With advice and suggestions provided by IUTAM, the Local Organizing Committee was responsible for planning meetings, for social activities, and for editing the proceedings. The members of the International Scientific Committee and the Local Organizing Committee are listed below.

We appreciate the contributions made by the following organizations: IUTAM, Science Council of Japan, Japan Society for the Promotion of Science, The Commemorative Association for the Japan World Exposition (1970), Inoue Foundation for Science, The Japan Society for Aeronautical and Space Sciences, and Science University of Tokyo. We express our sincere thanks to the IUTAM Officers Prof. L. van Wijngaarden, Prof. P. Germain, Prof. B.A. Boley, and Prof. F. Ziegler and to former officers Sir J. Lighthill and Prof. W. Schielen for their valuable guidance and cooperation; to the Representatives of the National Committee for Theoretical and Applied Mechanics of the Science Council of Japan, especially Prof. I. Imai, Prof. Y. Yamamoto, and Prof. T. Tatsumi, for valuable guidance and arrangements; to the members of the Scientific Committee and the Local Organizing Committee; to all participants and all supporting members for their hearty cooperation in carrying out the meeting; and to Springer–Verlag Tokyo for publishing the record of scientific progress made by the symposium.

K. Kawata

J. Shioiri

SYMPOSIUM ORGANIZATIONS

Scientific Committee

K. Kawata (Japan), Co−Chair
J. Shioiri (Japan), Co−Chair
K.B. Broberg (Sweden)
O.T. Bruhns (Germany)
D.C. Drucker (USA)

V.E. Fortov (Russia)
N. Jones (UK)
Q.S. Nguyen (France)
R. Wang (China)
F. Ziegler (Austria)

Local Organizing Committee

K. Kawata
J. Shioiri
T. Hayashi
H. Homma
M. Itabashi

K. Kishida
A. Kobayashi
T. Kurokawa
I. Maekawa
H. Matsumoto

T. Shioya
K. Takahashi
N. Takeda
K. Tanaka
S. Tanimura

Editorial Committee

K. Kawata
J. Shioiri

Symposium Sponsors

International Union for Theoretical and Applied Mechanics
Japan Society for the Promotion of Science
The Commemorative Association for the Japan World Exposition (1970)
Inoue Foundation for Science
The Japan Society for Aeronautical and Space Sciences
Science University of Tokyo

PARTICIPANTS

Asada, Kazuo, Takasago R&D Center, Mitsubishi Heavy Industries, Ltd., 2-1-1, Shinhama, Arai-cho, Takasago, Hyogo 676, Japan

Broberg, K. Bertram, Dept. of Mathematical Physics, University College Dublin, Belfield, Dublin 4, Ireland

Dioh, Ndiba N., Dept. of Mechanical Engineering, Imperial College of Science, Technology and Medicine, Exhibition Road, London SW7 2BX, UK

Ehara, Hiromitsu, Dept. of Mechanical Engineering, Faculty of Science and Technology, Science University of Tokyo, 2641, Yamazaki, Noda, Chiba 278, Japan

Enomoto, Hirotaka, Dept. of Materials Science and Technology, Faculty of Industrial Science and Technology, Science University of Tokyo, 2641, Yamazaki, Noda, Chiba 278, Japan

Gary, Gérard, Laboratoire de Mécanique des Solides, Ecole Polytechnique, 91128 Palaiseau Cedex, France

Goldsmith, Werner, Dept. of Mechanical Engineering, College of Engineering, University of California, Berkeley, Berkeley, CA 94720, USA

Hayashi, Takuo, Professor Emeritus, Dept. of Mechanical Engineering, Faculty of Engineering Science, 1-1, Machikaneyama-cho, Toyonaka, Osaka 560, Japan

Ishigami, Hideyuki, Dept. of Materials Science and Technology, Faculty of Industrial Science and Technology, Science University of Tokyo, 2641, Yamazaki, Noda, Chiba 278, Japan

Itabashi, Masaaki, Dept. of Materials Science and Technology, Faculty of Industrial Science and Technology, Science University of Tokyo, 2641, Yamazaki, Noda, Chiba 278, Japan

Jones, Norman, Impact Research Centre, Dept. of Mechanical Engineering, The University of Liverpool, P.O. Box 147, Liverpool L69 3BX, UK

Kawata, Kozo, Dept. of Materials Science and Technology, Faculty of Industrial Science and Technology, Science University of Tokyo, 2641, Yamazaki, Noda, Chiba 278, Japan

Kimura, Hiroshi, Dept. of Materials Science and Technology, Faculty of Industrial Science and Technology, Science University of Tokyo, 2641, Yamazaki, Noda, Chiba 278, Japan

Klepaczko, Janusz R., Laboratory of Physics and Mechanics of Materials, ISGMP, Metz University, Ile du Saulcy, F−57045 Metz Cedex, France

Kobayashi, Akira, Dept. of Mechanical Engineering, Faculty of Science and Technology, Science University of Tokyo, 2641, Yamazaki, Noda, Chiba 278, Japan

Kurokawa, Tomoaki, Dept. of Aeronautics and Astronautics, Kyoto University, Yoshida−Honmachi, Sakyo−ku, Kyoto, Kyoto 606−01, Japan

Li, Xin−Zen, Dept. of Precision Science and Technology, Osaka University, 2−1, Yamada−Oka, Suita, Osaka 565, Japan

Maekawa, Ichiro, Dept. of Mechanical Systems Engineering, Kanagawa Institute of Technology, 1030, Shimo−Ogino, Atsugi, Kanagawa 243−02, Japan

Maigre, Hubert, Laboratoire de Mécanique des Solides, URA CNRS 317, 91128 Palaiseau Cedex, France

Mimura, Koji, Dept. of Mechanical Systems Engineering, College of Engineering, University of Osaka Prefecture, 1−1, Gakuen−cho, Sakai, Osaka 593, Japan

Morozov, Nikita F., Math. & Mech. Faculty, St.−Petersburg University, Bibliotechnaja Pl., 2, Petrodvorets, St.−Petersburg, 198904, Russia

Nakamura, Takashige, Dept. of Materials Science and Technology, Faculty of Industrial Science and Technology, Science University of Tokyo, 2641, Yamazaki, Noda, Chiba 278, Japan

Nakano, Motohiro, Dept. of Precision Science and Technology, Osaka University, 2−1, Yamada−Oka, Suita, Osaka 565, Japan

Neimitz, Andrzej, Dept. of Mechanical Engineering, Kielce University of Technology, Al. Tysiaclecia Panstwa Polskiego 7, 25−314 Kielce, Poland

Nojima, Taketoshi, Div. of Aeronautics and Astronautics, Kyoto University, Yoshida−Honmachi, Sakyo−ku, Kyoto, Kyoto 606−01, Japan

Ogihara, Shinji, Dept. of Mechanical Engineering, Faculty of Science and Technology, Science University of Tokyo, 2641, Yamazaki, Noda, Chiba 278, Japan

Ota, Tomohito, Dept. of Materials Science and Technology, Faculty of Industrial Science and Technology, Science University of Tokyo, 2641, Yamazaki, Noda, Chiba 278, Japan

Rossmanith, Hans P., Institute of Mechanics, Technical University Vienna, Wiedner Hauptstraße 8−10/325, A−1040 Vienna, Austria

Schulze, Volker, Institut für Werkstoffkunde I, Universität Karlsruhe (TH), Kaiserstraße 12, D−76128 Karlsruhe, Germany

Shin, Hyung−Seop, Dept. of Mechanical Engineering, Andong National University, P.O. Box 203, Andong, Kyungpuk 760−600, Korea

Shioiri, Jumpei, Dept. of Mechanical Engineering, Faculty of Engineering, Hosei University, 3−7−2, Kajino−cho, Koganei, Tokyo 184, Japan

Shioya, Tadashi, Dept. of Aeronautics and Astronautics, University of Tokyo, 7−3−1, Hongo, Bunkyo−ku, Tokyo 173, Japan

Ståhle, Per, Dept. of Solid Mechanics, Luleå Institute of Technology, S−971 87 Luleå, Sweden

Suliciu, Ion, Institute of Mathematics of the Romanian Academy, P.O. Box 1−764, RO−70700 Bucharest, Romania

Takahashi, Kiyoshi, Research Institute of Applied Mechanics, Kyushu University, 6−1, Kasuga−Koen, Kasuga, Fukuoka 816, Japan

Tanimura, Shinji, Dept. of Mechanical Systems Engineering, College of Engineering, University of Osaka Prefecture, 1−1, Gakuen−cho, Sakai, Osaka 593, Japan

Torii, Shunji, Dept. of Materials Science and Technology, Faculty of Industrial Science and Technology, Science University of Tokyo, 2641, Yamazaki, Noda, Chiba 278, Japan

Uchikawa, Yukinobu, Dept. of Materials Science and Technology, Faculty of Industrial Science and Technology, Science University of Tokyo, 2641, Yamazaki, Noda, Chiba 278, Japan

Wang, Li−Lih, Ningbo University, Ningbo, Zhejiang 315211, People's Republic of China

Yamasaki, Fumitomo, Dept. of Mechanical Engineering, Faculty of Science and Technology, Science University of Tokyo, 2641, Yamazaki, Noda, Chiba 278, Japan

Yamauchi, Yoshiaki, Dept. of Precision Science and Technology, Osaka University, 2−1, Yamada−Oka, Suita, Osaka 565, Japan

Zhou, Feng−Hua, Dept. of Aeronautics and Astronautics, University of Tokyo, 7−3−1, Hongo, Bunkyo−ku, Tokyo 173, Japan

SCIENTIFIC PROGRAM

October 16 (Mon.), 1995

Opening Session [Presiding: Prof. W. Goldsmith and Prof. J. Shioiri]
Opening Address

Session 1 [Presiding: Prof. N. Jones and Prof. T. Hayashi]
Mechanical Characterization of Solids in High Strain Rate Tension
○ K. Kawata, M. Itabashi, and H. Kimura (Science University of Tokyo, Japan)

Session 2 [Presiding: Prof. G. Gary and Dr. N.N. Dioh]
Constitutive Relation on the Strain Rate and Temperature Dependency of the Flow Stress of Uniaxially Deformed Steels
○ V. Schulze and O. Vöhringer (University Karlsruhe, Germany)

A New Experimental Technique for Shear Testing at High Strain Rates Using T−Shaped Specimens *
I. Maekawa (Kanagawa Instiutute of Technology, Japan) and ○ H.−S. Shin (Andong National University, Korea)

Session 3 [Presiding: Dr. V. Schulze and Prof. I. Maekawa]
Testing Viscous Soft Materials at Medium and High Strain Rates
○ G. Gary, L. Rota, and H. Zhao (École Polytechnique, France)

High Strain Rate Constitutive Relationships for Semi−Crystalline Polymers Using a Combined Numerical and Experimental Approach
○ N.N. Dioh, A. Ivankovic, and P.S. Leevers (Imperial College, UK)

About a New Experimental Method of Identification of the Dynamic Toughness of Materials
H. Maigre (École Polytechnique, France)

Session 4 [Presiding: Prof. H.−S. Shin and Prof. T. Shioya]
Strain Rate Sensitivity of Flow Stress at Very High Rates of Strain
○ J. Shioiri (Prof. Emeritus, University of Tokyo, Japan), K. Sakino, and S. Santoh (Hosei University, Japan)

Constitutive Equations of Austenitic Stainless Steels under Uniaxial and Plastic Shearing at Very High Strain Rates
J.R. Klepaczko (Metz University, France)

A Strain—Rate and Temperature Dependent Constitutive Model with Unified Parameters for Various Materials
  ○ K. Mimura, S. Tanimura, and K. Higashi (University of Osaka Prefecture, Japan)

Session 5 [Presiding: Prof. J.R. Klepaczko and Prof. S. Tanimura]
An Energetic Characterization of the Constitutive Equations for Dynamic Rate—Dependent Plasticity
  I. Suliciu (The Romanian Academy, Romania)

October 18 (Wed.), 1995

Session 6 [Presiding: Prof. L.—L. Wang and Prof. K. Takahashi]
Loss of Localization at High Crack Speeds
  K.B. Broberg (University College Dublin, Ireland)

Crack Growth Equations
  A. Neimitz (Kielce University of Technology, Poland)

Session 7 [Presiding: Prof. K.B. Broberg and Prof. H. Maigre]
Dynamic Fracture Toughness and Crack Propagation in Brittle Material
  ○ T. Shioya and F. Zhou (University of Tokyo, Japan)

Session 8 [Presiding: Prof. I. Suliciu]
Experimental Study on Ballistic High Velocity Impact *
  A. Kobayashi, S. Ogihara, ○ F. Yamasaki, and T. Ishigure (Science University of Tokyo)

On the Behavior of Dynamic Crack Surface Ligaments
  K.—G. Sundin, P. Nilsson, and ○ P. Ståhle (Luleå Institute of Technology, Sweden)

Velocity Dependent Dynamic Fracture Toughness K_{ID} of Araldite B Simultaneously Determined by Caustic and Photoelastic Methods
  ○ K. Takahashi and M. Kido (Kyushu University, Japan)

Session 9 [Presiding: Prof. P. Ståhle]
Computer Simulation of Structural and Mechanical After—Effects of Shock Waves Induced by Impulse Laser in Solids **
  A.B. Volyntsev and A.N. Shilov (Perm State University, Russia)

Nonlinear Viscoelastic Constitutive Relations and Nonlinear Viscoelastic Wave Propagation for Polymers and Polymer—Matrix—Composites at High Strain—Rates
  ○ L.—L. Wang, D.—J. Huang, and S. Gan (University of Science and Technology of China, PROC)

Session 10 [Presiding: Prof. A. Neimitz and Prof. A. Kobayashi]
Dynamic Measurement of Elastic Moduli for Composite Materials Using Disk Specimens
  ○ Y. Yamauchi, M. Nakano, K. Kishida, T. Hashimoto (Osaka University, Japan), and Y. Sogabe (Ehime University, Japan)

Closing Session [Presiding: Prof. J. Shioiri]
    Closing remarks

○ : Speaker
  * : Paper that presented orally but not submitted as final manuscript
 * * : Although not orally presented for visa problem, Professor Kawata discussed already in another
        IUTAM Symposium at Sèvres, France the contents of the paper.
* * * : Read by Professor Li—Lih Wang

International Union of Theoretical
and Applied Mechanics

K. Kawata · J. Shioiri (Eds.)

Constitutive Relation in High/Very High Strain Rates

IUTAM Symposium, Noda, Japan
October 16–19, 1995

Springer

CONTENTS

XVI

XVIII

Mechanical Characterization of Solids in High Strain Rate Tension

Kozo Kawata, Masaaki Itabashi and Hiroshi Kimura

Department of Materials Science and Technology, Science University of Tokyo, 2641, Yamazaki, Noda, Chiba 278, Japan

Summary

Extensive studies on mechanical characterization of solids of wide categories in high strain rate <u>tension</u> are reported. A bird's−eye view of breaking strain and absorbed energy per unit volume, not only stress, is given for uniaxial tension in strain rate level of 10^{-3} s^{-1} ~10^3 s^{-1} never studied enough hitherto. In these studies, solutions of long standing questions are included. Classification of solids ranging from metals, polymers (energy elasticity and entropy elasticity) and composites to ceramics, from the standpoint of impact−resistant properties in high strain rates is also summarized.

Keywords: Characterization in high velocity tensile mechanical properties, High velocity brittleness and ductility, One bar method, Fastening technique, Specific absorbed energy per unit volume, Classification of solids as impact−resistant materials

1 Introduction

Demands for exact mechanical characterization of solids in high strain rate tension, have grown in both scientific and technological standpoint. Some experimental difficulties that prevented precise high strain rate tensile data acquisition have been overcome recently [1]. Following to high−velocity tensile characterization formulas and some experimental developments, obtained strain rate effects in tension for single phase materials: pure metals, polymers ($T < T_r$) and polymers ($T > T_r$), and for multiphase materials: steels, Al alloys (2219, 6061 and 7075), and composites (CFRP, GFRP, KFRP and particle dispersed elastomer) are reported. Finally, estimation of solids as impact−resistant materials, are stated.

2 Formulas for Mechanical Characterization in High Strain Rate Tension

For perfect determination of dynamic mechanical properties of solids and evaluation of solids as impact−resistant structural materials basing upon these data, behaviors of strain up to breaking ε_r and absorbed energy per unit volume before breaking E_{ab} should be known, not only behavior of stress. It is pointed out, here that ε_r and E_{ab} are important as measures of limit of deformability and energy absorption. In carrying out this purpose, KHKK one bar method formulas [1] and systems (Figure 1) are proved to give precise results up to final breaking of solids even if test materials are very tough and have very large breaking strain.

1

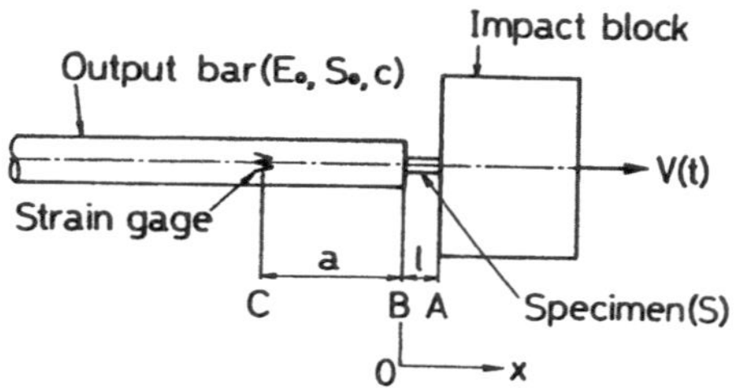

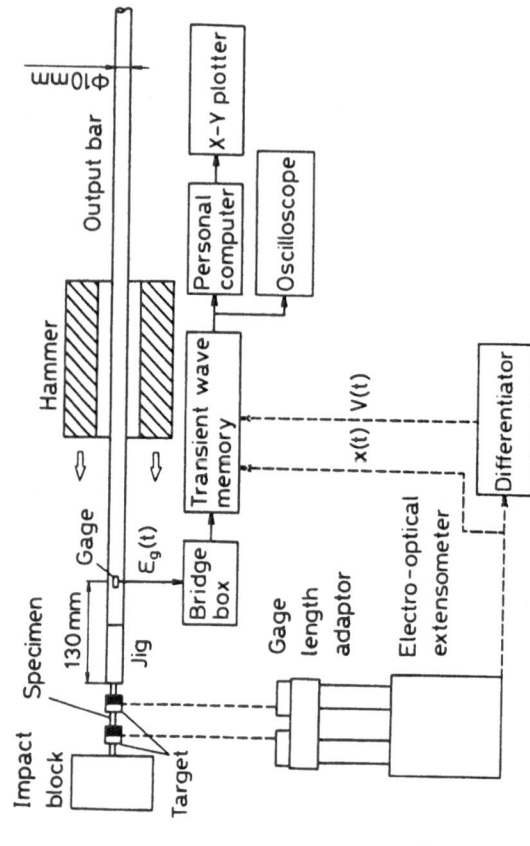

Figure 2: Two targets method

$$\sigma(t) = \frac{S_o}{S} E_o \varepsilon_g\left(t+\frac{a}{c}\right)$$

$$\varepsilon(t) = \frac{1}{l} \int_0^t \left\{ V(t) - c\varepsilon_g\left(t+\frac{a}{c}\right) \right\} dt$$

$$\dot{\varepsilon}(t) = \frac{1}{l} \left\{ V(t) - c\varepsilon_g\left(t+\frac{a}{c}\right) \right\}$$

σ : Dynamic stress
ε : Dynamic strain
$\dot{\varepsilon}$: Strain rate
t : Time after impact
S_o : Sectional area of output bar
S : Sectional area of specimen
E_o : Young's modulus of output bar
l : Length of specimen
a : Distance between point B and C
c : Elastic wave velocity
V : Velocity of impact block
ε_g : Strain at point C

Figure 1: Principle of KHKK one bar method

For precise determination of strain of strong solids, (i) it is necessary to fix completely the end parts of specimen to the both fastening parts. Another method is (ii) two targets method (Figure 2), that enables to acquire precise strain of test pieces as the difference of displacements of two ends of gage length, independently from possible pull out of end parts of specimen from the fastening areas. This method enables us complete acquisition of dynamic stress−strain diagrams in high strain rate tension up to breaking [2−4]. The strain rate ranges from 10^2 s^{-1} to 10^4 s^{-1}.

The fundamental formulas for dynamic stress $\sigma(t)$, dynamic strain $\varepsilon(t)$ and strain rate $\dot{\varepsilon}(t)$ are as follows:

$$\sigma(t) = \frac{S_0}{S} E_0 \, \varepsilon_g(t+\frac{a}{c}) \tag{1}$$

$$\varepsilon(t) = \frac{1}{\ell} \int_0^t \left\{ V(\tau) - c \, \varepsilon_g(\tau+\frac{a}{c}) \right\} d\tau \tag{2}$$

$$\dot{\varepsilon}(t) = \frac{1}{\ell} \left\{ V(t) - c \, \varepsilon_g(t+\frac{a}{c}) \right\} \tag{3}$$

where t is the time after impact. ℓ and S are the length and cross−sectional area of the specimen, and S_0, E_0 and c are the cross−sectional area, Young's modulus and longitudinal elastic wave propagation velocity of the output bar. $V(t)$ and $\varepsilon_g(t)$ are the velocity of the impact block and strain of the output bar at the distance a from the end obtained from four semiconductor strain gages. Most significant results are as follows. Test temperature is 20 ℃ in all cases.

3   Determination of Sharp Increase in Flow Stress at 10^3 s^{-1} or So in Tension for OFHC Copper

Sharp increase in flow stress at 10^3 s^{-1} or so in tension for OFHC copper at room temperature, exists. This result [5] is obtained by tension free from friction between coin−shaped compressive specimen and end planes of bars in SHPB method, and answers in the negative to Lindholm's conclusion that states no transition of rate controlling process of plastic deformation in the range of 10^3 s^{-1} and 10^4 s^{-1}, attributing to possible friction between coin−shaped specimen and end planes of the SHPB bars.

4   Characteristics of Evaluated Solids

4.1   Annealed pure metals

· Crystal lattice systems effect in high strain rate breaking strain is clearly found [6,7] as shown in Table 1.
· Pure Fe (C and N less than 3 ppm) shows remarkable high velocity brittleness [2,8]. This is an example of most remarkable high velocity brittleness similarly with PET shown later.

Table 1: Crystal lattice systems effect in breaking strain at high strain rate for annealed pure metals

Crystal lattice system	Effect in breaking strain
FCC (Al, Cu)	High−velocity ductility
BCC (Fe, Mo)	High−velocity brittleness
HCP { (Ti, c/a=1.59) (Zn, c/a=1.86) }	High−velocity ductility High−velocity brittleness

4.2 Commercial Al alloys and steel sheets for automotive bodies [2, 9]

· The ratio of maximum flow stress at 10^3 s^{-1} to that at 10^{-3} s^{-1} for steels decreases with increasing quasi−static tensile strength.
· Uniform elongation behavior for steels is dependent on microstructures, and has tendencies to decrease at 10^3 s^{-1}, compared with at 10^{-3} s^{-1} ∼10^{-1} s^{-1}. On the other hand, total (breaking) elongation (including local strain) behavior for steels shows the characteristics of strain rate insensitivity or high velocity ductility, differing from pure iron showing extremely high velocity brittleness.
· For Al alloys, uniform elongation at 10^3 s^{-1} is larger than at 10^{-3} s^{-1}.

4.3 Polymers

4.3.1 Energy elasticity (T < T$_g$) [10]

As typical examples, Figures 3 ∼ 8 are shown.
· Flow stress increases with increasing log $\dot{\varepsilon}$, as shown in Figures 3 and 4 for example.
· Dynamic strength is generally higher than static one.
· PET shows remarkable high velocity brittleness (Figure 3).
· Breaking strain decreases with increasing log $\dot{\varepsilon}$ except PC (Figure 4).
· PC is singularly of high ε_T and E_{ab} in high $\dot{\varepsilon}$. Especially E_{ab} at 10^3 s^{-1} is higher than at quasi−static tension. This is an important fact, showing high capacity of absorbing impact energy by PC.

4.3.2 Entropy elasticity (T > T$_g$) [11,12]

· Most complex behaviors ($\dot{\varepsilon}$: 10^{-3} s^{-1} ∼ 10^3 s^{-1}) as shown in Figure 9.
· In tensile strength, total (breaking) strain and absorbed energy per unit volume, three types of behavior are observed:

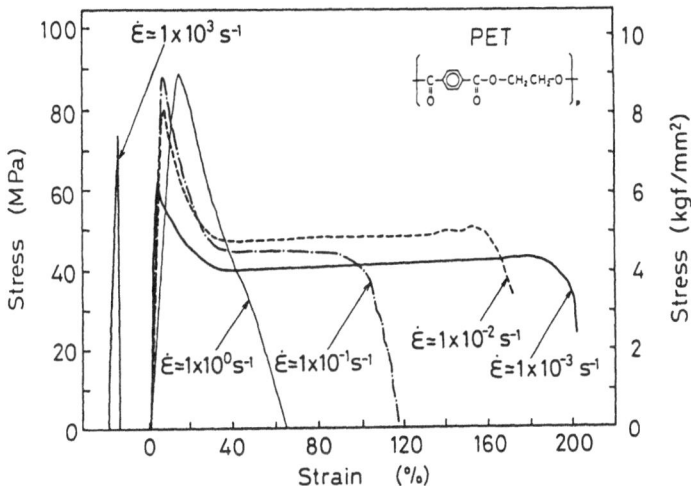

Figure 3: Tensile stress–strain relation in high and lower strain rates for PET

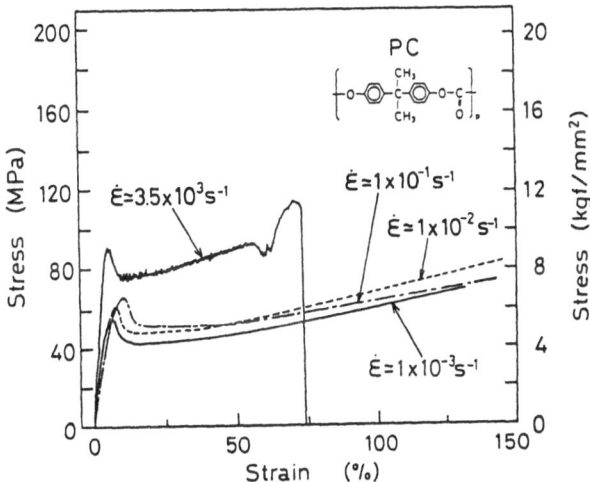

Figure 4: Tensile stress–strain relation in high and lower strain rates for PC

Type 1 Monotonously increasing with increasing strain rate.
Type 2 First increasing, next reaches to maximum value, and then decreasing with increasing strain rate.
Type 3 Monotonously decreasing with increasing strain rate.
In experimental data acquisition, special attention should be paid for the low level of stress and high level of elongation.

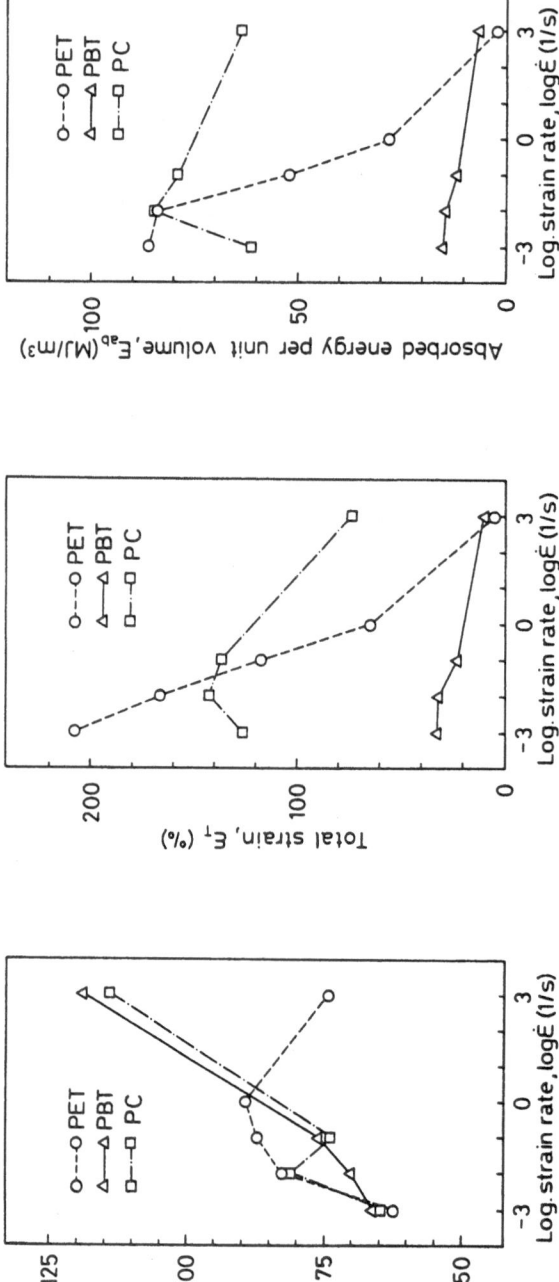

Figure 5: Strain rate effects in tensile mechanical properties of PET, PBT and PC

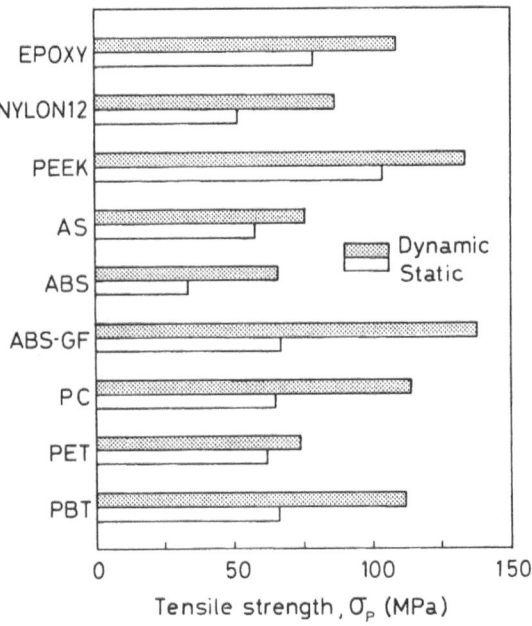

Figure 6: Comparison of dynamic and static tensile strength for polymers ($T < T_g$)

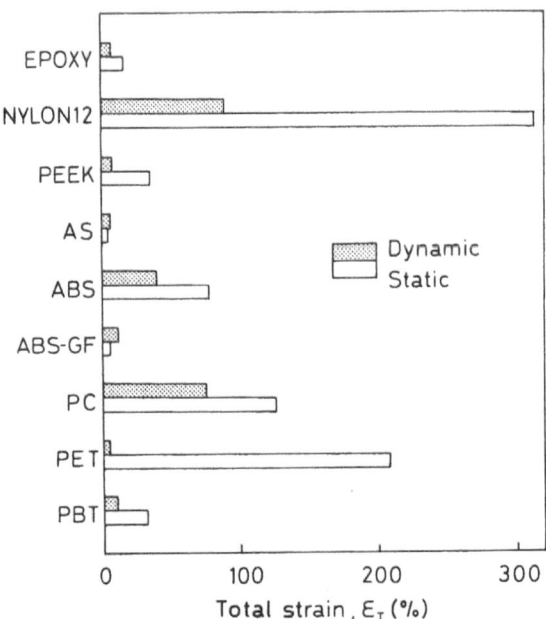

Figure 7: Comparison of dynamic and static total strain (rupture strain) for polymers ($T < T_g$)

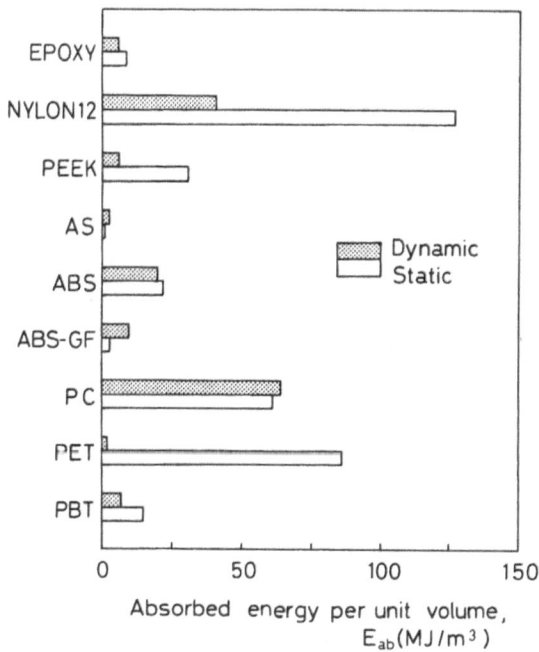

Figure 8: Comparison of dynamic and static absorbed energy per unit volume for polymers $(T < T_g)$

4.3.3 Dispersed particle strengthened elastomer (dummy solid propellant of HTPB matrix) [13]

The tendencies are as follows, using the classification shown in 4.3.2.
- Maximum flow stress: type 1
- Breaking strain: type 2
- Fracture appearances show that fracture—initiating point moves to inside of matrix in 10^3 s^{-1} tension from pure interface between matrix and dummy solid particles at 10^{-3} s^{-1}.

4.4 Hard polymer matrix FRP

In woven cloth reinforced polymers: GFRP and CFRP [14,15], high velocity ductility for the former, and high velocity brittleness to strain rate insensitivity in breaking strain for the latter are found. In these cases, specimen fastening was carried out by screw threads without pull—out from the fastening area. Recently introducing two targets method, tensile properties of the most anisotropic materials group: KFRP UD and GFRP UD [16] and CFRP UD [17] have been precisely characterized, as shown in Figures 10 ~ 12.

      KFRP UD   Strength: type 1. Breaking strain: slightly type 3.
      GFRP UD   Strength: type 1. Breaking strain: type 1 (high velocity ductility).
      CFRP UD   Strength: type 1. Breaking strain: type 4 (strain rate insensitivity).

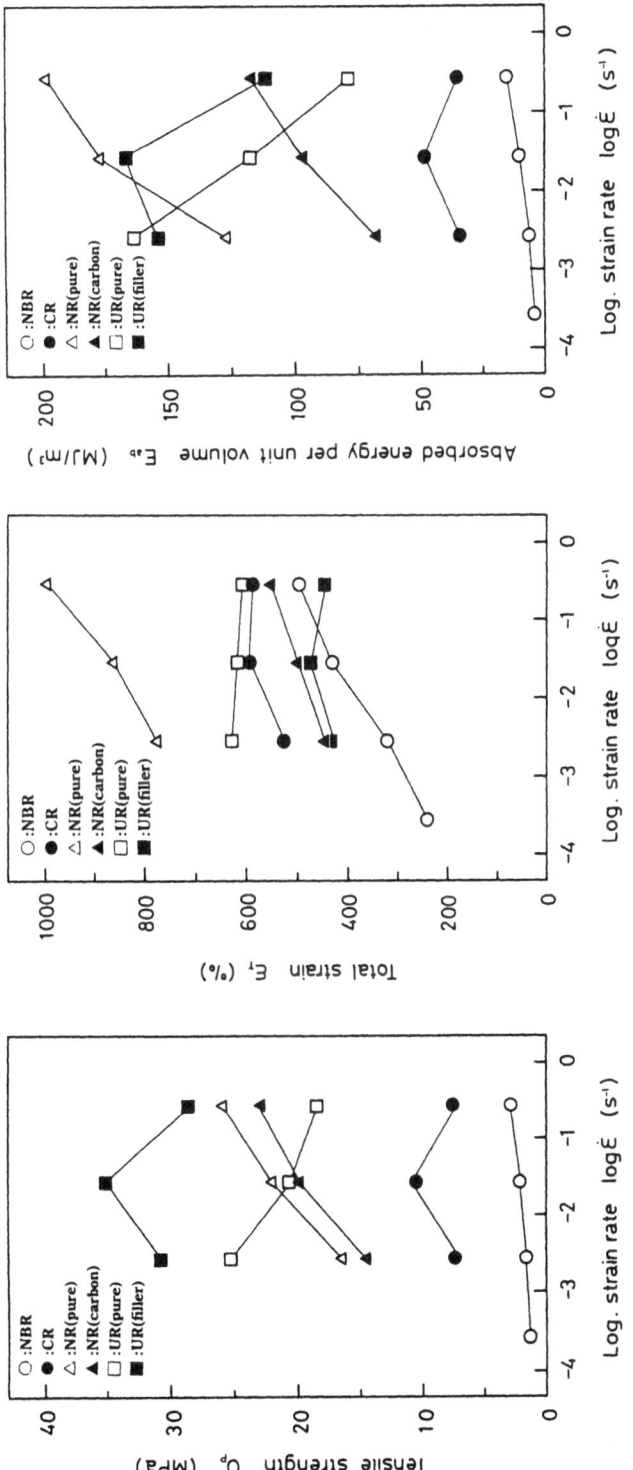

Figure 9: Strain rate effects in tensile mechanical properties of polymers ($T > T_g$)

Table 2: Key properties of KFRP UD, GFRP UD and CFRP UD in dynamic and quasi−static tension

Material	KFRP UD		
	Quasi−static	Dynamic	
Strain rate (s⁻¹)	1.82×10^{-3}	1.20×10^{2}	D/S *
Tensile strength (MPa)	969	1260	1.30
Strain at tensile strength (%)	2.35	1.39	0.59
Breaking strain (%)	2.35	2.16	0.92
Absorbed energy per unit volume (MJ/m³)	11.4	16.9	1.48
Remarks	High Velocity Brittleness ～ Strain Rate Insensitivity		

Material	GFRP UD		
	Quasi−static	Dynamic	
Strain rate (s⁻¹)	1.87×10^{-3}	0.874×10^{2}	D/S *
Tensile strength (MPa)	691	1030	1.49
Strain at tensile strength (%)	2.92	3.35	1.15
Breaking strain (%)	2.93	3.57	1.22
Absorbed energy per unit volume (MJ/m³)	10.8	22.6	2.09
Remarks	High Velocity Ductility		

Material	CFRP UD		
	Quasi−static	Dynamic	
Strain rate (s⁻¹)	5.6×10^{-5}	1.1×10^{2}	D/S *
Tensile strength (MPa)	2200	2570	1.17
Strain at tensile strength (%)	1.52	1.60	1.05
Breaking strain (%)	1.52	1.60	1.05
Absorbed energy per unit volume (MJ/m³)	16.2	23.2	1.43
Remarks	Strain Rate Insensitivity		

* : Ratio of dynamic value to quasi−static one

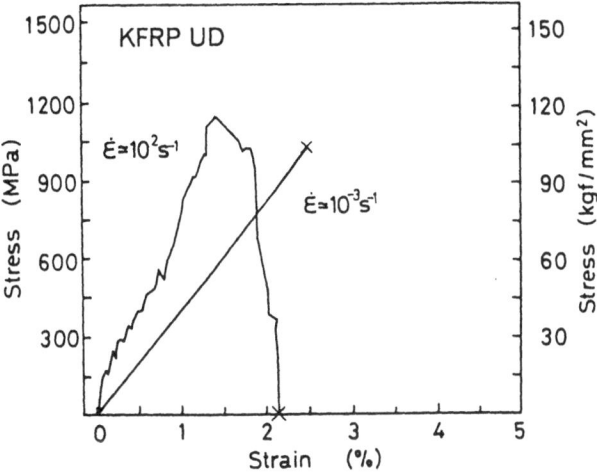

Figure 10: Tensile stress-strain relation in high and low strain rates for KFRP UD

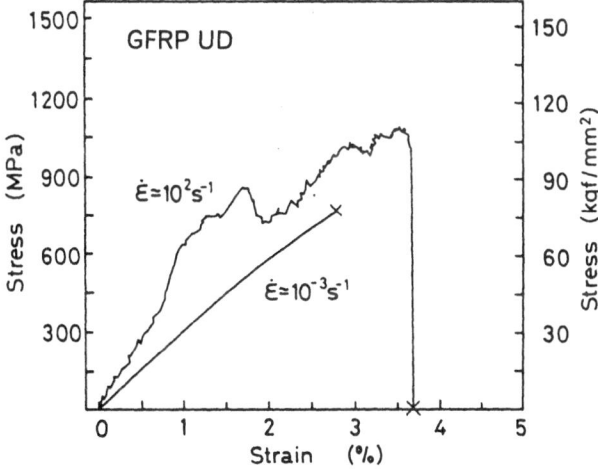

Figure 11: Tensile stress-strain relation in high and low strain rates for GFRP UD

In CFRP UD characterization, specimen fastening utilizing expansive cement is quite effective for prevention of pulling-out of specimen from fastening areas (Figure 13). The tendencies in GFRP UD and CFRP UD are similar with in woven cloth reinforced cases respectively.

The key properties in dynamic tension (D) and in quasi-static tension (S) for KFRP UD, GFRP UD and CFRP UD are summarized in Table 2.

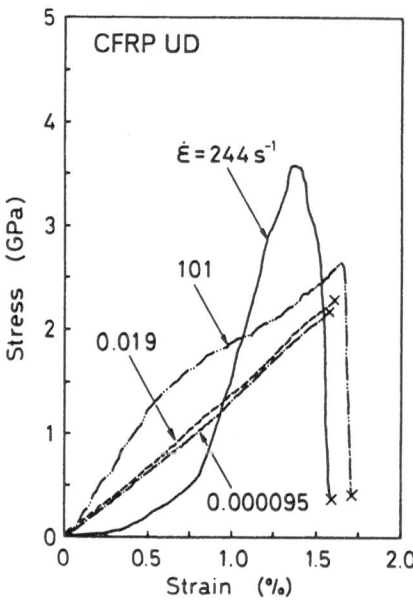

Figure 12: Tensile stress–strain relation in high and lower strain rates for CFRP UD

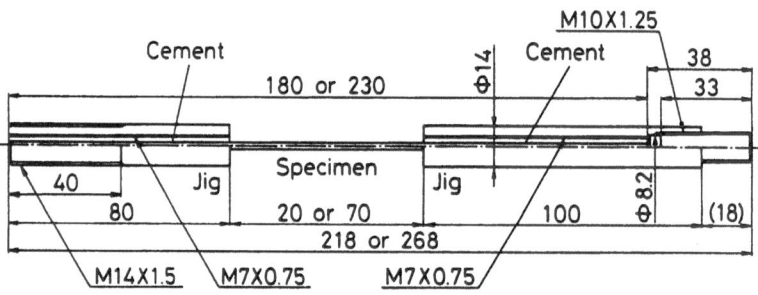

Figure 13: Specimen fastening utilizing expansive cement

5 Selective Embrittlement and Weakening of Al Alloys by Coupling of Pre–Fatigue and High Strain Rate Tension [18]

When pre–fatigued 2219 and 7075 Al alloys suffer from high strain rate (10^3 s^{-1}) tension, embrittlement and weakening occur. For low strain rate (10^{-3} s^{-1}) tension such phenomenon is not observed. For pre–fatigued 6061 Al alloy, such emprittlement is not observed even for high strain rate tension.

Strengthening mechanism of Al alloys is precipatation hardening for 2219 and 7075, on the other hand, it is solid solution hardening for 6061. To explain this newly found phenomenon DSCF (dynamic stress concentration fracture) model (Figure 14) is proposed.

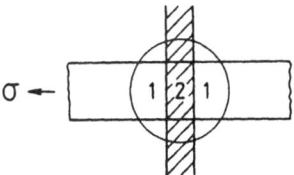

Figure 14: DSCF (dynamic stress concentration fracture) model

Table 3: Tensile properties of typical solid materials in high and low strain rates

Material		$\dot{\varepsilon}$ s^{-1}	σ_p MPa	ε_P %	ε_T %	E_{ab} MJ/m^3	E_{ab}/ρ kJ/kg
2011-F	D	1.8 ×10^3	408	6.5	19.7	63.3	23.4
	S	1.4 ×10^{-3}	389	7.3	15.2	53.3	19.7
2024-T4	D	0.69×10^3	483	10.5	15.3	66.2	24.5
	S	2.08×10^{-3}	505	10.8	15.3	70.8	26.2
7075-T651	D	0.95×10^3	593	7.3	11.6	59.6	22.1
	S	2.08×10^{-3}	601	7.5	10.5	58.2	21.6
SUS304	D	2.0 ×10^3	703	20.2	48.2	293	37.6
	S	1.38×10^{-3}	682	50.7	72.0	426	54.6
SUS347PN	D	1.30×10^3	746	24.7	47.9	315	40.4
	S	2.09×10^{-3}	671	44.7	61.9	371	47.6
Pure Iron B	D	1.04×10^3	578	0.7	2.3	6.4	0.82
	S	2.08×10^{-3}	223	23.4	58.6	108	13.8
GFRP 1	D	1.81×10^3	555	6.8	9.1	35.2	16.8
	S	1.04×10^{-3}	325		4.2	7.7	3.7
GFRP 6	D	0.58×10^3	773	9.7	14.3	78.9	37.6
	S	0.87×10^{-3}	467		3.5	9.6	4.6
CFRP 1	D	1.86×10^3	658	2.2	3.3	11.0	6.9
	S	0.99×10^{-3}	554		3.3	9.8	6.1
CF/GF	D	0.85×10^3	1039	4.4	8.2	44.9	23.6
Hybrid 1	S	1.56×10^{-3}	940		2.7	15.5	8.2
PET	D	0.76×10^3	74.3	4.3	5.2	2.1	1.3
	S	1.05×10^{-3}	62.3	3.9	207	86.2	53.9
PC	D	1.14×10^3	114	72.5	75.5	63.9	53.2
	S	1.05×10^{-3}	64.6		126	61.4	51.2
PYREX	D	2.4 ×10^2	392	0.64	0.69	1.76	0.70
	S	1 ×10^{-2}	255		0.34	0.44	0.18

σ_P : Tensile strength

ε_P : Strain at σ_P

ε_T : Total strain (Rupture strain)

E_{ab} : Absorbed energy per unit volume

ρ : density

D : in dynamic tension

S : in quasi−static tension

6 Classification of Solids Basing upon High Strain Rate Tensile Mechanical Characterization

Typical index values tensile strength σ_P , strain at tensile strength ε_P , total strain ε_T , and absorbed energy per unit volume E_{ab} and specific absorbed energy per unit volume E_{ab}/ρ (ρ : density) in high and low strain rate tension are summarized in Table 3.

The materials are selected from four groups: 1) metallic crystalline materials: pure iron, structural steels and Al alloys, 2) composite materials: GFRP, CFRP and CF/GF hybrids, 3) polymeric materials: PET and PC, and 4) inorganic brittle material: Pyrex glass.

The highest and lowest values of E_{ab} and E_{ab}/ρ of solid materials of four categories are summarized in Table 4. In E_{ab}/ρ , all categories of 1) metals and alloys, 2) polymer matrix composites, and 3) polymers (T < T_g) contain superior candidates for impact−resistant solid materials.

In Table 4, the highest value of 53.2 for E_{ab}/ρ corresponding to PC (polycarbonate) should be paid attention, although the value of 63.9 for E_{ab} itself is lower than 78.9 and 315.

Table 4: E_{ab} and E_{ab}/ρ in high strain rate tension of 10^3 s^{-1} for solid materials of four categories

Solid materials	E_{ab} (MJ/m³)	E_{ab}/ρ (kJ/kg)
1) Metals, Alloys	6.4 ~ 315	0.82 * ~ 40.4
2) Polymer matrix composites	11.0 ~ 78.9	6.9 * ~ 37.6
3) Polymers (T<T_g)	2.1 ~ 63.9	1.3 * ~ 53.2
4) Inorganic brittle material	1.76	0.70

* : corresponds to the cases of most remarkable "high velocity brittle" materials

7 Conclusions

1. Using KHKK one bar method formulas, high strain rate tensile mechanical properties of solids of five categories are characterized up to breaking.

2. To secure precise strain determination especially for materials of high modulus and low strain, 1) extremely complete chucking or 2) two targets method are necessary.

3. Sharp increase in flow stress at strain rates in the order of 10^3 s^{-1} in tension for OFHC copper at room temperature, exists. This result answers in the negative to Lindholm's conclusion that states no such sharp increase.

4. Single phase materials

4.1. In pure metals, crystal lattice systems effect in high velocity brittleness and high velocity ductility is clarified. Pure iron shows the most remarkable high velocity brittleness.

4.2. In polymers (T < T_g), that is, glassy polymers, strain rate dependence of σ_P , ε_T and

E_{ab} is classified into three types. In σ_P , D/S ratio (Dynamic value ($\sim 10^3$ s^{-1}) vs. quasi–Static value ($\sim 10^{-3}$ s^{-1})) is generally higher than 1. On the other hand, in ε_T , D/S ratio is generally lower than 1. PET shows remarkable high velocity brittleness. On the other hand, PC shows D/S ratio of E_{ab} larger than 1, differing from other many polymers in glassy state.

4.3. In polymer (T > T_g), that is, elastomers, strain rate dependence of σ_P , ε_T and E_{ab} is also classified into three types.

5. Multiphase materials

5.1. A new fact that the coupling of pre–fatigue and high strain rate tension decreases the strengths of Al alloys of precipitation hardening type, such 2219 and 7075, is found. This phenomenon is understood by developing DSCF model.

5.2. In HTPB dummy solid propellant, tensile mechanical properties in 10^3 s^{-1} range, is first clarified. Fractured surface in high strain rate tension is clearly different from in static tension. This is also understood by DSCF model.

5.3. CFRP and KFRP show high velocity brittleness or strain rate insensitivity in breaking strain. In measuring these properties, two targets method is effective. GFRP shows remarkable high velocity ductility.

6. Estimation of solids as impact–resistant materials is made, basing upon extended tensile data acquisition up to breaking by KHKK one bar method. The estimation may be made in two ways, that is, by E_{ab} itself and by E_{ab} / ρ . The highest value of specific energy absorption shown by polycarbonate should be paid attention.

7. In next stage, the research of high strain rate tensile properties should be extended to multiphase materials more and more, not limited in single phase materials.

8 Acknowledgments

The authors wish to express their hearty thanks to the long standing supports by the Ministry of Education, Science and Culture, Science University of Tokyo, University of Tokyo, Japan Society for Promotion of Sciences, the Commemorative Association for the Japan World Exposition (1970), Toray Science Foundation, many companies and to the students of Kawata Laboratory who assisted eagerly research works.

9 References

[1] K. Kawata, S. Hashimoto, K. Kurokawa and N. Kanayama, A New Testing Method for the Characterization of Materials in High Velocity Tension, *Mechanical Properties at High Rates of Strain 1979*, Institute of Physics, Bristol and London, 1979, pp. 71–80.

[2] K. Kawata, S. Hashimoto, S. Sekino, and N. Takeda, Macro– and Micro–Mechanics of High–Velocity Brittleness and High–Velocity Ductility of Solids, *Macro– and Micro–Mechanics of High Velocity Deformation and Fracture*, Springer–Verlag, Berlin, 1987, pp. 1–25.

[3] K. Kawata, I. Miyamoto, M. Itabashi and S. Sekino, On the Effects of Alloy Components in the High Velocity Tensile Properties, *Impact Loading and Dynamic Behaviour of Materials*, DGM Informationsgesellschaft Verlag, Oberursel, Vol. 1, 1988, pp. 349–356.

[4] K. Kawata, *High Velocity Brittleness and High Velocity Ductility of Structural Solids*, Report of Research Project (No. 61420023), Grant–in–Aid for Scientific Research of the Ministry of Education, Science and Culture, 1989.

[5] K. Kawata, T. Kawagoe and K. Saino, Mechanical Properties in Dynamic Fracture of OFHC Copper and Some Cu—Zn Alloys in 10^3/sec Tensile Strain Rate Range, *Dynamic Fracture*, Chuo Technical Drawing, Toyohashi, 1990, pp. 287—296.

[6] K. Kawata, S. Fukui, J. Seino and N. Takada, Some Analytical and Experimental Investigations on High Velocity Elongation of Sheet Materials by Tensile Shock, *Behaviour of Dense Media under High Dynamic Pressure*, Paris, Dunod, 1968, pp. 313—323.

[7] K. Kawata, S. Hashimoto and K. Kurokawa, Analyses of High Velocity Tension of Bars of Finite Length of BCC and FCC Metals with Their Own Constitutive Equations, *High Velocity Deformation of Solids*, Springer—Verlag, Berlin, 1979, pp. 1—15.

[8] K. Kawata, S. Hashimoto, N. Takeda and S. Sekino, On High—Velocity Brittleness and Ductility of Dual—Phase Steel and Some Hybrid Fiber Reinforced Plastics, *Recent Advances in Composites in the United States and Japan*, ASTM STP 864, 1983, pp. 700—711.

[9] E. Nakanishi, M. Itabashi and K. Kawata, The Fast and Slow Deformation Behaviour in Commercial Aluminum Alloys and Steel Sheets for Automotive Bodies, *Structural Failure, Product Liability and Technical Insurance*, IV, Elsevier Science Publishers, 1993, pp. 423—430.

[10] K. Kawata, S. Ninomiya and S. Fujitsuka, High Velocity Deformation for Polymeric Materials in 10^3/s Region, *Proceedings of the 19th Aircraft/Aerospace Materials Symposium*, University of Tokyo, 1989, pp. 51—56, (in Japanese).

[11] K. Kawata, H.—L. Chung and M. Itabashi, Mechanical Characterization and High Velocity Ductility of HTPB Propellant Binder, *Proceedings of the 6th Japan—U.S. Conference on Composite Materials*, Technomic Publishing, Lancaster, 1993, pp. 771—781.

[12] K. Kawata, M. Itabashi and T. Ota, Strain Rate Effects on Mechanical Properties for Rubber Materials, *Proceedings of the 37th JSASS/JSME Structures Conference*, 1995, pp. 105—108 (in Japanese).

[13] H.—L. Chung, K. Kawata and M. Itabashi, Tensile Strain Rate Effect in Mechanical Properties of Dummy HTPB Propellants, *Journal of Applied Polymer Science*, Vol. 50, 1993, pp. 57—66.

[14] K. Kawata, A. Hondo, S. Hashimoto, N. Takeda and H.—L. Chung, Dynamic Behaviour Analysis of Composite Materials, *Composite Materials; Mechanics, Mechanical Properties and Fabrication*, Japan Society for Composite Materials, Tokyo, 1981, pp. 2—11.

[15] K. Kawata, S. Hashimoto and N. Takeda, Mechanical Behaviours in High Velocity Tension of Composites, *Progress in Science and Engineering of Composites*, North—Holland, Vol. 1, 1982, pp. 829—836.

[16] K. Kawata, M. Itabashi and M. Tomi, Dynamic Tensile Mechanical Properties of One—Dimensional Continuous—Fiber Reinforced KFRP and GFRP, *Proceedings of the 4th Zairyo—no Shogeki Mondai Symposium*, Society of Materials Science, Japan, 1993, pp. 1—4, (in Japanese).

[17] H. Kimura, K. Kawata and M. Itabashi, High Velocity Tensile Mechanical Behavior of Unidirectional CFRP Thin Strip, *Proceedings of the 37th JSASS/JSME Structures Conference*, 1995, pp. 21—24 (in Japanese).

[18] K. Kawata, M. Itabashi and S. Kusaka, Behaviour Analysis of Pre—Fatigue Damaged Aluminum Alloys under High—Velocity and Quasi—Static Tension, Proceedings of IUTAM Symposium on Micromechanics of Plasticity and Damage of Multiphase Materials, (August 29 ~ September 1, 1995), Sèvres, France, (in printing).

Constitutive Relation on the Strain Rate and Temperature Dependence of the Flow Stress of Uniaxially Deformed Steels

V. Schulze and O. Vöhringer

Institut für Werkstoffkunde I, Universität Karlsruhe (TH), D-76128 Karlsruhe, FRG

Summary

After a short explanation of the theoretical background of thermally activated glide of dislocations, results of experimental investigations with specimens of normalized SAE 1045at 81K $\leq$ T $\leq$ 398K and $5 \cdot 10^{-5} s^{-1} \leq \dot{\epsilon} \leq 1 \cdot 10^{-2} s^{-1}$ will be presented and described by a quantitative constitutive model. Extrapolations to high strain rates $2.5 \cdot 10^{-1} s^{-1} \leq \dot{\epsilon} \leq 1 \cdot 10^{+3} s^{-1}$ will be compared with further experimental results. The influence of soluted alloying elements on the constitutive parameters will also be discussed using investigations at a HSLA steel, low alloyed steels and a tool steel.

Keywords: Steels, Strain-rate and Temperature Dependence of Flow Stress, Activation Enthalpy, Pseudo Solid Solution Softening

1 Introduction

At temperatures T $\leq$ 0.3 T_m (T_m = melting or solidification temperature in K), elastic-plastically deformed metallic polycrystals show a characteristic dependence of flow stress σ on temperature T and strain rate $\dot{\epsilon}$ according to thermally activated glide of mobile dislocations across short-range obstacles. The quantitative description of this correlation [1] provides that at temperatures lower than a strain-rate depending critical temperature T_0 the flow stress σ is additively composed of an athermal component σ_G (depending on temperature only in the same manner as the shear modulus G or Young's modulus E) and a thermal component σ^*

$$\sigma = \sigma_G + \sigma^* = \sigma_{G,o} \frac{G(T)}{G(0K)} + \sigma^*(T,\dot{\epsilon}) \tag{1}$$

The first component is due to the effect of long range dislocation obstacles, the second one is caused by short-range obstacles which may be overcome by slip dislocations due to thermal fluctuations. σ^* increases with decreasing temperature and increasing strain rate. At T = 0K, where the probability that thermal fluctuations occur becomes zero, σ^* shows its maximum, strain-rate independent value.

2 Theoretical Background

The free activation enthalpy ΔG for thermally activated glide of mobile dislocations across short-range obstacles is correlated to the thermal flow stress σ^*, the strain rate $\dot{\epsilon}$, the temperature T and the Boltzmann-constant k by

$$\dot{\epsilon}(T,\sigma^*) = \dot{\epsilon}_o \exp\left[-\Delta G(\sigma^*)/kT\right] \tag{2}$$

with

$$\dot{\epsilon}_o = \rho_m \, b \, L \, v / M_T$$

where ρ_m = density of mobile slip dislocations, b = amount of burgers vector, L = average distance of short range obstacles, v = part of the Debye-frequency v_D and M_T = Taylor-factor. $\Delta G(\sigma^*)$ is determined by the interaction energy of slip dislocations and the short range obstacles, which can be

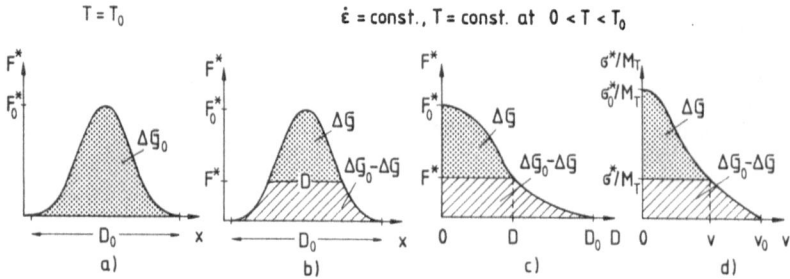

Fig. 1: Force-distance-curve at $T = T_0$ (a) and $0 < T < T_0$ (b) as well as transformation to thermal flow stress vs. activation volume (b-d) (schematically).

described by force-distance-curves (F^*,x-curves) as shown in Fig. 1. The area below the curve characterizes the free activation enthalpy of the whole obstacle, ΔG_0. If $\Delta G = \Delta G_0$, no external force F^* and therefore no thermal flow stress is neccessary for slip. Then eq. 2 leads to

$$T_0 = \frac{\Delta G_0}{k \ln (\dot{\epsilon}_0 / \dot{\epsilon})} \qquad (3)$$

as the critical temperature above which $\sigma^* = 0$. At temperatures $0K < T < T_0$ the thermally activated part of ΔG decreases thus requiring a mechanical force F^* which raises up to its maximum F_0^* at $T = 0K$. Figure 1 shows that the $F^*(x)$-curves may be transformed to $\sigma^*(v)$-curves with the activation volume $v = b\, L\, D$ (D = effective thickness of the short range obstacle) and $\sigma^*/M_T = F^*/bL$. Both, $F^*(x)$ and $\sigma^*(v)$, allow to determine the dependence of the activation enthalpy on the thermal force or stress using the relationship

$$\Delta G (\sigma^*) = \Delta G_0 - \int_0^{F^*} D(F^*)\, dF^* = \Delta G_0 - \frac{1}{M_T} \int_0^{\sigma^*} v (\sigma^*)\, d\sigma^* \qquad (4)$$

If $\Delta G(\sigma^*)$ is approached by a power law function [2-4] with two exponents m and n which are determined by the shape of the short range obstacle, eq. 2 and 3 lead to

$$\sigma^* = \sigma_0^* \left[1 - \left(\frac{T}{T_0} \right)^n \right]^m \qquad (5)$$

A quantitative analysis of the influence of temperature and strain-rate on the flow stress at $T \leq T_0$ requires stress-strain-measurements at different strain-rates $\dot{\epsilon}$ and strain-rate-jump tests in the interesting region of temperatures. After separation of the athermal stress component by relating the values of flow stress to the temperature-dependent shear modulus the thermal components of flow stress can be evaluated as a function of temperature and strain-rate. Differentiation of eq. 4 at constant temperature with respect to σ^* and eq. 2 show, that the activation volume

$$v(T,\dot{\epsilon}) = M_T\, k\, T\, \left. \frac{\Delta \ln \dot{\epsilon}}{\Delta \sigma^*} \right|_{T=const.} \qquad (6)$$

may be determined by strain-rate-jump tests if the changes in flow stress $\Delta\sigma^*(T,\dot{\epsilon})|_T$ caused by changes of strain rate $\Delta\ln\dot{\epsilon}$ are measured. All $\sigma^*(T,\dot{\epsilon})$- and $v(T,\dot{\epsilon})$-values lead to the correlation $\sigma^*(v)$ which is expected to be independent of the strain-rate. Fitting a hyperbola

$$\frac{\sigma^*}{M_T} = \frac{A}{(B \cdot v - C)^B} - D \qquad (7)$$

to this curve by a nonlinear least-squares-algorithm allows for analytical integration according to eq.

4 [5]. This leads to $\Delta G_o - \Delta G(T)|_k$-values which show a straight line with the intercepts ΔG_o and $T_o(\dot\epsilon)$. Using eq. 3, $\dot\epsilon_o$ can be determined. All $\sigma^*(T,\dot\epsilon)$- and $\Delta\sigma^*(T,\dot\epsilon)$-values available were put into a nonlinear least-squares algorithm in order to evaluate σ_o^*, m and n.

3 Materials investigated and Experimental Procedures

Most experiments were carried out with SAE 1045 (german grade Ck 45, 0.47 C, 0.30 Si, 0.81 Mn, 0.04 Cr, 0.01 Mo, 0.03 Cu, 0.03 Ni and rest Fe, all in wt.-%) in a normalized heat treatment condition. The specimen had a measuring area of 60 mm in length and 3.8 mm in diameter. Investigations with a HSLA-steel, low alloyed steels and a tool steel (see Tab. 1) were carried out with a different specimen geometry with $l_o = 25$mm and $d_o = 5$ mm.
Tensile tests were performed in a 500kN-universal testing machine and a tempering chamber which allowed tests between 81K and 423K. Besides simple tensile tests, strain-rate-jump tests with jumps in $\dot\epsilon$ of a factor of 10 were carried out at distinct values of ϵ_p and evaluated according to [6]. Tests at high strain-rates $2.5 \cdot 10^{-1}s^{-1} \le \dot\epsilon \le 1 \cdot 10^{+3}s^{-1}$ were performed with a Split-Hopkinson-Pressure-bar [7].

4 Experimental Results

Figure 2a displays the dependence of the 1.5%-proof stress ($R_{p1.5}$) on the temperature at both strain rates investigated. At temperatures lower than 300K, the curves separate and show higher $R_{p1.5}$-values at $\dot\epsilon = 1 \cdot 10^{-3}s^{-1}$ than at $\dot\epsilon = 1 \cdot 10^{-4}s^{-1}$. Further tensile tests were carried out at the temperatures 173, 193, 223 and 243K and at strain-rates between $5 \cdot 10^{-5}s^{-1}$ and $1 \cdot 10^{-2}s^{-1}$, resulting in the $R_{p1.5}$-values in Fig. 2b which increase nonlinearly with logarithmically increasing strain-rate.

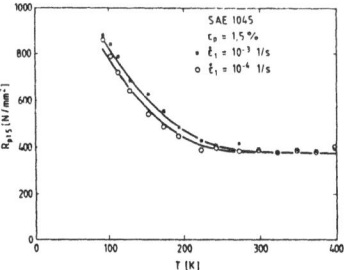 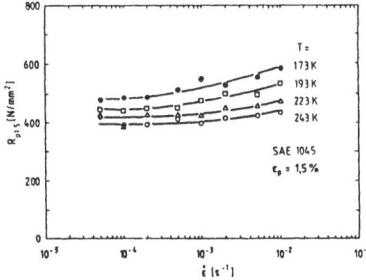

Fig. 2: Influence of strain-rate and temperature on 1.5%-proof stresses (lines= calculation).

Strain-rate-jumps at $\epsilon_p = 1.5\%$ from $\dot\epsilon_1 = 1 \cdot 10^{-4}s^{-1}$ to $\dot\epsilon_2 = 1 \cdot 10^{-3}s^{-1}$ and $\dot\epsilon_1 = 1 \cdot 10^{-3}s^{-1}$ to $\dot\epsilon_2 = 1 \cdot 10^{-2}s^{-1}$ cause changes of the thermal flow stress $\Delta\sigma^*$ which depend on T and $\dot\epsilon_1$ (Fig. 3a). They increase with decreasing temperature, reach maxima at about 140K and slightly decrease at even lower temperatures. Jumps at $\dot\epsilon_1 = 1 \cdot 10^{-3}s^{-1}$ always show higher $\Delta\sigma^*$-values than those at $\dot\epsilon_1 = 1 \cdot 10^{-4}s^{-1}$.
Thermal stress values σ^* and activation volumes v, which are determined from $\Delta\sigma^*$-values according to eq. 6, were always investigated in pairs at the same temperatures and strain-rates, so that they could be plotted in $\sigma^*/M_T(v)$-diagrams. As can be seen from Fig. 3b, this curve does not reveal any significant dependence on the strain-rate and the hyperbola (eq. 7) leads to a good agreement with the experimental data. Integration of the hyperbola at distinct points of $v(T,\dot\epsilon)$ leads to ΔG_o-ΔG-values, which decrease linearly with increasing temperature (Fig. 4). This decrease turns out to be lower at the higher strain-rate. Linear regression of this dependence, eq. 3 and averaging of the values determined at $\dot\epsilon = 1 \cdot 10^{-4}s^{-1}$ and $1 \cdot 10^{-3}s^{-1}$ result in $\Delta G_o = 0.64$eV and $\dot\epsilon_o = 4.4 \cdot 10^{+7}s^{-1}$. With these constants the iteration of $\sigma^*(T,\dot\epsilon)$- and $\Delta\sigma^*(T,\dot\epsilon)$-values could be carried out, leading to $\sigma_o^* = 2000$N/mm^2, m = 1.78 and n = 0.50. Fig. 2 shows that measured and calculated values of $R_{p1.5}$ agree very well.

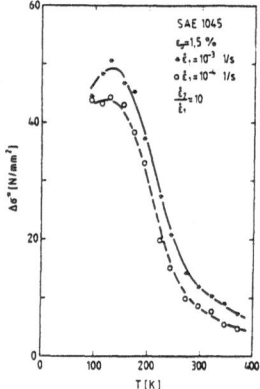
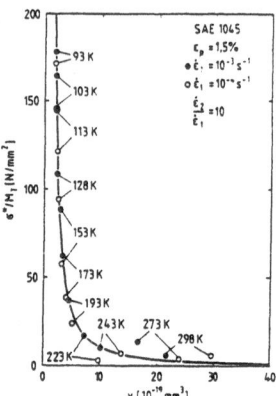

Fig. 3: Influence of strain-rate on thermal flow stress jumps observed at $\dot{\epsilon}_1/\dot{\epsilon}_2 = 10$ and $\epsilon_p = 1.5\%$ vs. temperature (a) and correlation of thermal flow stress related to M_T and activation volume at different starting strain-rates (b).

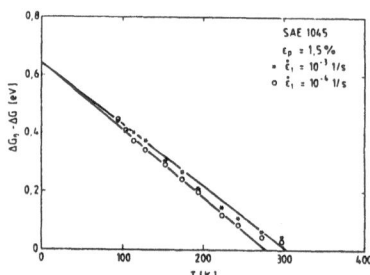

Fig. 4: Influence of strain-rate on $\Delta G_o - \Delta G(T)$ vs. temperature.

The same procedure was used to determine the constitutive constants at other values of ϵ_p between 0.2% and 5% and leads to the values shown in Fig. 5a-c as a function of ϵ_p. While ΔG_o and $\dot{\epsilon}_o$ are independent of plastic strain for 2.5% $\leq \epsilon_p \leq$ 5% they increase with decreasing $\epsilon_p <$ 2.5%. The values of σ_o^*, m and n are not influenced significantly by ϵ_p over the whole ϵ_p-range investigated. States of other steels without Lüders deformations were investigated with the same procedure and lead to constitutive parameters which are independent of plastic deformation.

Tab. 1 illustrates in addition to the parameters of SAE 1045 those of several other steels investigated in [8] in the same way as here. While m and n are nearly independent of the material and the material state, the ΔG_o-values lie between 0.42 and 0.78 eV, the $\dot{\epsilon}_o$-values between $7.29 \cdot 10^5$ and $9.77 \cdot 10^8 \text{s}^{-1}$ and the σ_o^*-values between 1056 and 2000 N/mm^2.

5 Discussion

As could be seen from section 4, the iterative algorithm leads to a good description of the temperature and strain-rate dependence of the flow stress of normalized SAE1045. This is emphasized by Fig. 6 which shows an extrapolation of the calculated flow stresses up to $\dot{\epsilon} = 1 \cdot 10^{+4} \text{s}^{-1}$. The measured compressive flow stress values determined in a Hopkinson-bar agree well with the calculated curve,

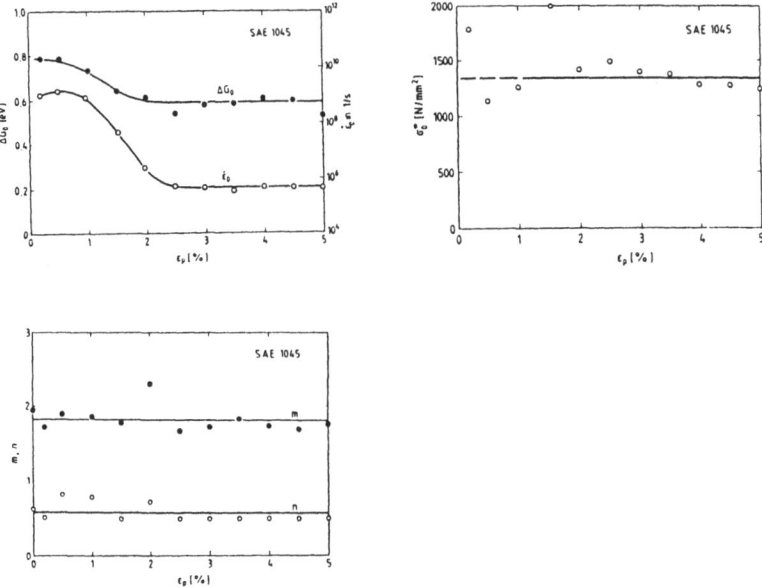

Fig. 5: Constitutive parameters ΔG_0, $\dot{\epsilon}_0$, σ_0^*, m and n vs. plastic strain.

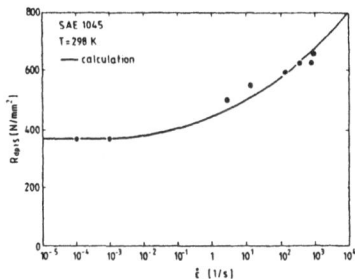

Fig. 6: Extrapolation of the constitutive equation at ϵ_p = 1.5% to tests at very high strain-rates for SAE 1045.

though σ_0^* = 2000N/mm^2 is very high. This also holds for the low alloyed steel 1¼Cr-Mo-3Ni and the tool steel 9Cr-3Si in the same manner (see Fig. 7a and b). It is evident that the measured values of the 0.2%-proof stress at room temperature and T = 213 and 223K, respectively, can be well described by the calculated curves given by the constitutive equation.

The observed values of m and n are within the range observed empirically by [2,3,9-11]and within m = 1-2 and n= 0.5-1, which were theoretically determined by [12-14] assuming that the double kink mechanism of [15] is the strain-rate-determining thermally activable process. This is believed to hold for the steels investigated.

At SAE1045, the determination of ΔG_0 and $\dot{\epsilon}_0$ at different plastic strains leads to constant values for $\epsilon_p \geq$ 2.5% (Fig. 5). In this case, ΔG_0 = 0.6eV and $\dot{\epsilon}_0$ = 4.3·10^{+5}s^{-1} lie within the interval of these values determined by other authors with pure iron or low carbon steel specimens [16-20]. However, with decreasing $\epsilon_p \leq$ 2.5% increasing values of ΔG_0 and $\dot{\epsilon}_0$ were observed. This could be caused by the inhomogeneous deformations (Lüders deformations) which increase up to 2.5% at T ≈ 100K. They

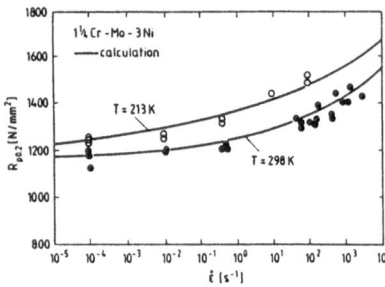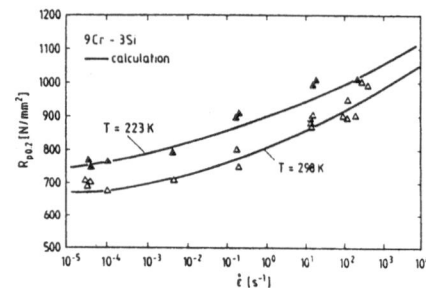

Fig. 7: Extrapolation of the constitutive equation at ϵ_p = 0.2% to tests at very high strain-rates for the low alloyed steel 1¼Cr-Mo-3Ni (a) and the tool steel 9Cr-3Si (b).

cause local strain-rates which can be some orders of magnitude higher than the macroscopic strain-rate. This effect can produce a microscopic decrease of the probability of thermally activated slip and can therefore cause an increase of ΔG_o and $\dot{\epsilon}_o$ observed. In order to separate the influence of the Lüders-deformation, flow stresses at $\epsilon_p < \epsilon_{Lo}$ were corrected to the values which should occur at homogeneous deformation. Using a Ludwik-equation [21]

$$\sigma = \sigma_o + A\, \epsilon_p^{n^*} \qquad (8)$$

with the properties σ_o, A and n^*, which depend from material state and/or temperature, the stress-strain curves at $\epsilon_p > \epsilon_{Lo}$ were fitted and thus, at $\epsilon_p < \epsilon_{Lo}$ lowered flow stress values could be extrapolated. At ϵ_p = 0.2%, the evaluated constitutive parameters using these corrected flow stresses are similar to the average parameters of the uncorrected flow stresses at $\epsilon_p \geq 2.5\%$ (m = 1.75, n = 0.52, ΔG_o = 0.48eV, ϵ_o = $1.34\cdot10^{+3}$ s^{-1}, σ_o^* = 1301 N/mm^2). This indicates that the Lüders behaviour causes higher values of the constitutive parameters at small plastic strains. The results at the steels mentioned in Tab. 1 [8], emphasize this because they are characterized by homogeneous deformation at the onset of plasticity and their constitutive parameters remain nearly independent of plastic strain.

Tab. 1: Constitutive parameters and 0.2%-proof-stress of different steels, $^*)$ = corrected values.

steel American grade	German grade	ϵ_p [%]	$R_{p0.2,\,298K}$ [N/mm^2]	m	n	ΔG_o [eV]	$\dot{\epsilon}_o$ [s^{-1}]	σ_o^* [N/mm^2]
SAE1045	Ck 45	0.2	330	1.72	0.52	0.78	$9.77\cdot10^8$	1783
		1.5		1.78	0.50	0.64	$4.4\cdot10^7$	2000
		≥2.5		1.78	0.53	0.58	$7.29\cdot10^5$	1352
		0.2 $^*)$		1.72	0.52	0.48	$1.34\cdot10^3$	1301
ISO 4950	StE690	0.2-5	830	1.76	0.50	0.42	$2.19\cdot10^7$	1313
1¼Cr-Mo-1¼Ni	30CrNiMo 5 5	0.2-5	1380	1.58	0.50	0.40	$2.96\cdot10^6$	1289
1¼Cr-Mo-3Ni	35 NiCrMo12 5	0.2-5	1220	1.84	0.50	0.69	$7.28\cdot10^7$	1194
1½Cr-1¾Mo	45 CrMo 6 7	0.2-5	1420	1.65	0.50	0.47	$8.06\cdot10^5$	1355
9Cr-3Si	X 45 CrSi 9 3	0.2-5	670	1.98	0.54	0.72	$1.47\cdot10^7$	1056

The values of σ_0^* listed in Tab. 1 indicate that alloying of steels leads to decreasing values of σ_0^* and therefore to decreasing dependences of flow stress on temperature and strain-rate. This is often described in literature as pseudo solid solution softening [e.g. 22-25]. The influence of soluted alloying elements on σ_0^* was studied at different metals and is supposed to be caused by a decrease of Peierls-potential or kink energy. The steels investigated here have different contents of the alloying elements Cr, Mo, Si, Ni and Mn. As only the solid soluted parts of these elements can influence the value of σ_0^* and Cr and Mo will be nearly totally precipitated as carbides, only the influence of Ni, Si and Mn - which are assumed to be totally soluted - is regarded more closely. Therefore σ_0^* is determined under the conditions $\bar{m} = 1.75$ and $\bar{n} = 0.5$ for all steels, since the small differences of the actual values of m and n have a weak effect on the extrapolated σ_0^*-values. Assuming linear and non-interacting dependences of the soluted contents of Ni, Si and Mn on σ_0^*, a regression analysis was performed which yields to

$$\sigma_0^* = 1653 \ N/mm^2 - 121 \ N/(mm^2 \, wt-\%)\left([wt\%-Ni] + 2.43 \, [wt\%-Si] + 1.70 \, [wt\%-Mn] \right) \qquad (9)$$

Figure 8 shows the very good description of the σ_0^*-values of the steels achieved by this equation. In accordance with literature [22-25] the influence of Si and Mn is much higher than that of Ni.

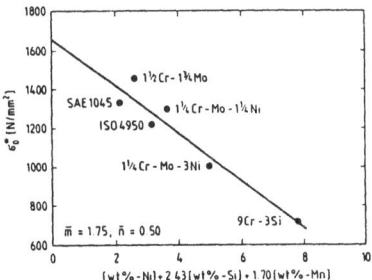

Fig. 8: Influence of Nickel, Silicium and Manganese on the constitutive parameter σ_0^* assuming $\bar{m} = 1.75$ and $\bar{n} = 0.5$.

6 Conclusion

After a short overview of the theoretical background of thermally activated glide of dislocations a constitutive, power-law equation for the flow stress is presented. An algorithm developed for the determination of the material constants used in this equation was applied to experimental data of normalized SAE1045 evaluated at 81 K $\leq$ T $\leq$ 398 K and $\dot{\epsilon} = 1 \cdot 10^{-4} s^{-1}$ as well as $1 \cdot 10^{-3} s^{-1}$. The agreement of calculated and experimental data is good, even at extrapolations to high strain-rates $\dot{\epsilon} \leq 1 \cdot 10^4 s^{-1}$. This is also valid for a HSLA steel, low alloyed steels and a tool steel. Due to pseudo solid solution softening effects, the values of σ_0^* of these steels are decreased by contents of Ni, Si and Mn in different ways. The dependence of the constitutive constants of SAE1045 on plastic strain in the interval 0.2% $\leq \epsilon_p \leq$ 2.5% correlates with the Lüders-deformations observed at this steel. The exponents m and n of the constitutive equation show that the strain-rate-controlling process for dislocation slip is the double kink mechanism.

24

7 Literature

[1]   A. Seeger: Theorie der Kristallplastizität, III. Die Temperatur- und Geschwindigkeitsabhängigkeit der Kristallplastizität, Z. Naturforschung 9a (1954), 870/881

[2]   U.F. Kocks, A.S. Argon, M.F. Ashby: Thermodynamics and Kinetics of Slip, Prog. Mat. Sci, Vol. 19, Pergamon Press, Oxford, (1975)

[3]   O. Vöhringer: Die strukturmechanischen Grundlagen der plastischen Verformung von viel-kristallinen α-Kupfer-Legierungen, Habilitationsschrift, Universität Karlsruhe (1972)

[4]   K. Ono: Temperature dependence of dispersed barrier hardening, J. Appl. Physics 39 (1968), 1803/1806

[5]   J. Stoer: Einführung in die numerische Mathematik I, Springer-Verlag, Berlin (1983)

[6]   D. Munz, E. Macherauch: Geschwindigkeitswechselversuche an α-Messing, Z. Metallkde. 6 (1966), 442/451

[7]   L. W. Meyer: Werkstoffverhalten hochfester Stähle unter einsinnig dynamischer Beanspruchung, Dr.-Ing.-Dissertation, Universität Dortmund (1982)

[8]   F. Burgahn: Einsinniges Verformungsverhalten und Mikrostruktur ausgewählter Stähle in Ab-hängigkeit von Temperatur und Verformungsgeschwindigkeit, Dr.-Ing. Dissertation, Universität Karlsruhe (1991)

[9]   G. Wielke: Thermally Activated Glide of Model Crystals in Strain-Rate Change and Stress Relaxation Experiments, Z. Metallkde. 68 (1977), 199/204

[10] F.A. Smidt: An analysis of thermally activated flow in α-iron based on T-τ^*-considerations, Acta Met. 17 (1969), 381/392

[11] K. Hüttebräucker: Das zügige Verformungsverhalten normalisierter untereutektoider Kohlenstoffstähle, Dr.-Ing.-Dissertation, Universität Karlsruhe (1977)

[12] J. Friedel: Dislocations, Gauthier-Villars, Paris (1956)

[13] J.E. Dorn, S. Rajnak: Nucleation of Kink Pairs and the Peierls Mechanism of Plastic Deformation, Trans. Met. Soc. AIME 230 (1964), 1052/1064

[14] R.L. Fleischer: Rapid Solution Hardening, Dislocation Mobility and the Flow Stress of Crystals, J. Appl. Phys. 33 (1962), 3504-3508

[15] A. Seeger: The temperature and strain-rate dependence of the flow stress of body centered cubic metals: a theory based on Kink-Kink interactions, Z. Metallkde. 72 (1981), 369/380

[16] B.L. Mordike, P. Haasen: The influence of temperature and strain rate on the flow stress of α-iron single crystals, Phil. Mag. 7 (1962), 459/474

[17] Z.S. Basinski, J.W. Christian: The influence of temperature and strain rate on the flow stress of annealed and decarburized iron at subatmospheric temperatures, Aust. J. Phys. 13 (1960), 299/308

[18] P. Wynblatt, A. Rosen, J.E. Dorn: The effect of temperature, strain rate and structure on the flow stress of Fe-2 pct. Mn alloy, Trans. Met. Soc. AIME 233 (1965), 651/655

[19] H. Conrad, S. Frederick: The effect of temperature and strain rate on the flow stress of iron, Acta Met. 10 (1962), 1013/1020

[20] K. Okazaki, M Kagawa, Y. Aono: An analysis of thermally activated flow in an Fe-0.056 at.-% Ti alloy using stress relaxation, Z. Metallkde. 67 (1967), 47/56

[21] P. Ludwik: Über den Einfluß der Deformationsgeschwindigkeit bei bleibenden Deformationen mit besonderer Berücksichtigung der Nachwirkungserscheinungen, Phys. Z. 10(1909), 411/417

[22] W.A. Spitzig, W.C. Leslie: Solid-solution softening and thermally activated flow in alloys of Fe with 3 at.-% Co, Ni or Si, Acta Met. 129 (1971),1143/1152

[23] T. Tanaka, S. Watanabe: The Temperature Dependence of the Yield Stress and Solid Solution Softening in Fe-Ni and Fe-Si Alloys, Acta Met. 129 (1971), 991/1000

[24] W.C. Leslie, R.J. Sober, S.G. Babcock, S.J. Green: Plastic Flow in Binary Substitutional Alloys of bcc Iron - Effects of Strain-rate, Temperature and Alloy Content, Trans. ASM, 62(1969), 690/710

[25] E. Pink, R.J. Arsenault: Low-Temperature Softening in Body-centered Cubic Alloys, Progress in Materials Science, 24 (1980), 1/50

Testing viscous soft materials at medium and high strain rates

G.Gary, L.Rota and H.Zhao

Laboratoire de Mécanique des Solides, Ecole Polytechnique, 91128 PALAISEAU CEDEX, France

Summary

This paper presents a review of recent results found in our Laboratory and new techniques allowing for a valuable analysis of strain rate effects on the behaviour of soft viscous materials. Those results concern the use of Hopkinson bars: -*a)* the waves' dispersion correction for viscoelastic 3D bars; -*b)* inverse calculation methods that are needed for the analysis of the results of transient tests when the hypothesis of homogeneity in the specimen is not valid.

Keywords: Hopkinson bars, waves' dispersion correction, viscoelastic bars, inverse methods, low impedance materials, viscoplastic materials.

1. Introduction

The Split Hopkinson Pressure Bar (SHPB), or Kolsky's apparatus, is a commonly used apparatus for testing materials at high strain rates, in the range of approximately 300 to 3000 s^{-1}. Historically, the first use of a long thin bar to measure the pulse shape induced by an impact is considered due to Hopkinson [1]. This method has been well established after the critical work of Davies [2]. The experimental set-up with two long bars and a short specimen has been introduced by Kolsky [3]. The Split Hopkinson bar technique, which has been initially used in compression, has been extended to the tension (Harding et al., [4]) and to the torsion (Duffy et al., [5]).

Kolsky's original SHPB analysis is based on some basic assumptions. (i) The waves propagating in the bars can be described by the one-dimensional wave propagation theory. (ii) The stress and strain fields in the specimen are uniform in its axial direction. (iii) The specimen inertia effect is negligible. (iv) The friction effect in the compression test is also negligible.

Those assumptions have been extensively studied, and often overcomed, in past decades.

This paper will focus on some particular points related to two different kinds of problems. In section 2 we consider measurement problems that are directly related to the SHPB arrangement. In section 3, we consider the identification problems relating material behaviour to experimental measurements.

2. SHPB measuring techniques. Viscoelastic dispersion correction

2.1 SHPB measuring technique

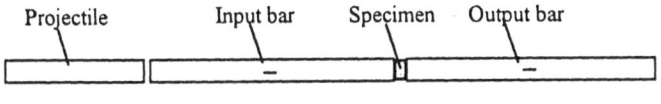

Fig. 1. SHPB test set-up

A typical SHPB set-up is outlined in Fig. 1. It is composed of long input and output bars with a short specimen placed between them. The impact of the projectile at the free end of the input bar develops a compressive longitudinal incident wave $\varepsilon_i(t)$. Once this wave reaches the bar specimen interface, a part of it, $\varepsilon_r(t)$, is reflected, whereas another part goes through the specimen and develops in the output bar the transmitted wave $\varepsilon_t(t)$. Those three basic waves recorded by the gages cemented on the input and output bars allow for the measurement of forces and velocities at the two faces of the specimen.

This measurement technique is based on the wave propagation theory and on the superposition principle. According to the wave propagation theory, the stress and the particle velocity associated with a single wave can be calculated from the associated strain measured by the strain gages. Using the superposition principle, the stress and the particle velocity in one section are calculated from the two waves propagating in opposite directions in this section. When the waves are known at bar-specimen interfaces, the forces and the velocities at both faces of the specimen are given by the following equations (1).

$$F_{input}(t) = S_B\, E\,(\varepsilon_i(t) + \varepsilon_r(t)) \qquad V_{input}(t) = C_0\,(\varepsilon_i(t) - \varepsilon_r(t))$$
$$F_{output}(t) = S_B\, E\, \varepsilon_t(t) \qquad V_{output}(t) = C_0\, \varepsilon_t(t) \qquad (1)$$

where S_B, E and C_0 are respectively the bar's cross-sectional area, Young's modulus, and the elastic wave speed.

As the three waves are not measured at bar-specimen interfaces in order to avoid their superposition, they have to be shifted from the position of the strain gages to the specimen faces, in time and distance. This shifting leads to two main perturbations. First, waves change in their shapes on propagating along the bar. Second, it is very difficult to find an exact delay in the time shifting to ensure that the beginnings of the three waves correspond to the same instant. Those perturbations, if not controlled, can introduce errors in the final result, especially in the range of small strains.

The input force is proportional to the sum of the incident and the reflected wave recorded in the input bar. A correct measure of this force needs that the reflected wave is significantly different of the opposite value of the incident wave. The impedance of the specimen cannot then be too small compared to the one of the bars. This condition leads to the use of low impedance bars which are more often made of a viscoelastic material.

2.2 Correction for wave dispersion

The wave dispersion effects on longitudinal elastic waves propagating in cylindrical bars have been studied experimentally by Davies [6]. On the basis of the longitudinal wave solution for an infinite cylindrical elastic bar given by Pochhammer [7], and Chree [8], a dispersion correction has been proposed and verified by experimental data. Even though the Pochhammer-Chree solution is not exact for a finite bar, it is found easily applicable and sufficiently accurate [6].

In the case of elastic bars, using the assumptions of the propagation of harmonic waves (Eqn. 2) in an infinite cylindrical rod leads to a Pochhammer-Chree's frequency equation that gives a relation between the wave number ξ and the frequency ω ([7],[8]). Using the same assumptions, this approach is generalised to the case of bars made of any linear viscoelastic material [9].

The displacement $\underline{u}(\underline{X}, t)$ is written in the following form :

$$\underline{u}(\underline{X}, t) = \frac{1}{2\pi} \int_{-\infty}^{+\infty} \underline{u}^*(\underline{X}, \omega) e^{-i\omega t} d\omega \quad \text{with } \underline{u}^*(\underline{X}, \omega) = \underline{u}^*(r, \theta, \omega) e^{i\xi(\omega)z} \qquad (2)$$

where $\underline{u}(\underline{X}, t)$, $\underline{u}^*(\underline{X}, \omega)$ are respectively displacement as a function of time and of the frequency. $\underline{X}$ denotes the space vector the components of which are r, θ, z in cylindrical coordinates.

Considering a linear viscoelastic media, the constitutive law can be written in the frequency domain as follows [10]:

$$\underline{\sigma}^{*}(\omega) = \lambda^{*}(\omega)tr(\underline{\varepsilon}^{*}(\omega))\underline{1} + 2\mu^{*}(\omega)\underline{\varepsilon}^{*}(\omega) \qquad (3)$$

where $\underline{\sigma}^{*}(\omega), \underline{\varepsilon}^{*}(\omega), \lambda^{*}(\omega), \mu^{*}(\omega)$ are respectively the stress tensor, the strain tensor, and two material coefficients.

As it is presented in [9], the complete solution of the governing equation, with boundary conditions on the external surface of the bars, leads to a frequency equation that gives a relation between the wave number ξ and the frequency ω (Eqn 4).

$$f(\xi) = (2\alpha / r_{0})(\beta^{2} + \xi^{2})J_{1}(\alpha.r_{0})J_{1}(\beta.r_{0}) - (\beta^{2} - \xi^{2})^{2}J_{0}(\alpha.r_{0})J_{1}(\beta.r_{0})$$
$$- 4\xi^{2}\alpha.\beta.J_{1}(\alpha.r_{0})J_{0}(\beta.r_{0}) = 0 \qquad (4)$$

with $\alpha^{2} = \dfrac{\rho\omega^{2}}{\lambda + 2\mu} - \xi^{2}$; $\beta^{2} = \dfrac{\rho\omega^{2}}{\mu} - \xi^{2}$; where J_{0}, J_{1} are the Bessel functions, λ and μ are the elastic coefficients, r_{0} is the radius of the bar.

A "viscoelastic frequency equation" is then obtained. This equation takes the same form as the classical one obtained in elasticity. However, in the present case, the argument ξ in the equation is a complex number. [11]

In this equation, ξ represents the complex change in phase function of the frequency ω. Its real part gives the relation between the frequency and the associated phase velocity and its imaginary part gives the relation between the frequency and the associated attenuation coefficient.

The explicit relation $\xi(\omega)$ is found numerically from equation (4) by solving a non-linear bi-dimensional optimisation problem respectively dealing with the real and the imaginary parts of the complex number ξ.

The shifting function is carried out numerically using the Fast Fourier Transformation (FFT) technique. The associated dispersion correction is more important than the correction for an equivalent purely elastic system, in particular because of a significant dumping effect. An example of the influence of this correction is shown in fig. 2a and 2b.

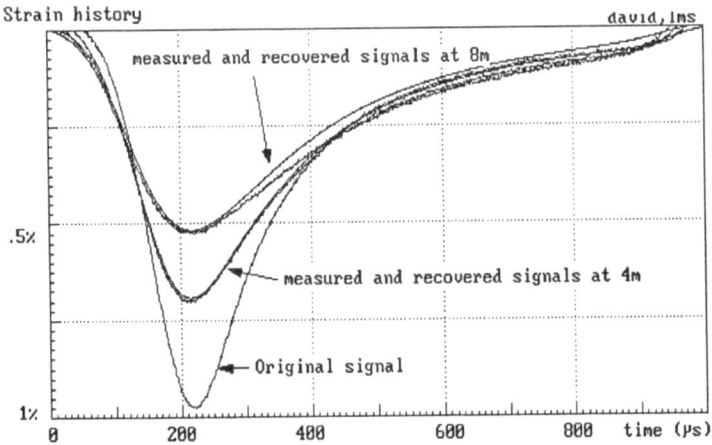

Fig. 2a Example of dispersion effects in a 40 mm diameter Nylon bar.

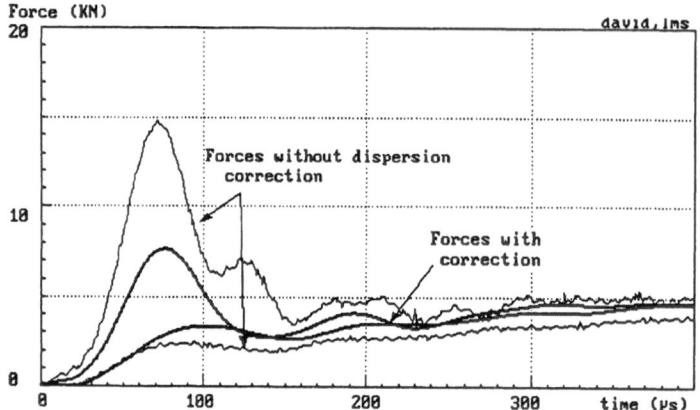

Fig. 2b Influence of dispersion correction on forces measurement
(for a 3m long input bar)

3. Material behaviour identification

3.1 Classical SHPB analysis

The Split Hopkinson pressure bar arrangement can give very accurate measurements of forces and velocities at both sample faces if the data processing is carefully performed. There remains the second kind of problems of SHPB mentioned in the introduction, which consist of relating material properties to measured forces and velocities at the two specimen faces. The classical analysis assumes the axial uniformity of stress and strain fields in the specimen. An average stress strain curve can be obtained from Eqn. 5a and 5b, deduced of Eqn. 1, which leads to the so-called two-waves analysis.

$$\dot{\varepsilon}_s(t) = \frac{V_{output}(t) - V_{input}(t)}{l_s} = \frac{-2C_0\varepsilon_r(t)}{l_s} \quad (5.a) \qquad \sigma_s(t) = \frac{F_{output}(t)}{S_s} \quad (5.b)$$

This assumption is obviously not correct at the early stage of the test because of the transient effects: the loading starts at one face of the specimen whereas the other face remains at rest. A three-waves analysis has been then proposed to use the average of the two forces to calculate the stress (Eqn. 5.d) instead of Eqn. 5.b [12].

$$\dot{\varepsilon}_s(t) = \frac{V_{output}(t) - V_{input}(t)}{l_s} \quad (5.c) \qquad \sigma_s(t) = \frac{F_{input}(t) + F_{output}(t)}{2S_s} \quad (5.d)$$

This classical analysis has been significantly improved by various authors to take account of radial inertia and eventually friction effects. In order to minimise friction effects, Davies & Hunter [13] have recommended an optimal length/diameter ratio of the specimen. In this case, radial and longitudinal inertia effects should be taken into account. The correction, based on the assumption of the axial uniformity of fields, is proposed. Other propositions for the correction of inertia and of friction effects can be found in later works [14,15,16]. Most of those corrections have been analysed and proved by the numerical simulation work of Bertholf & Karnes [17]. Those aspects are disregarded in this paper, and following the work reported by other authors [18,19,20], a one-dimensional simulation of the wave propagation in the specimen is used to investigate the assumption of axial uniformity of stress and strain fields.

3.2 SHPB analysis by inverse calculation

3.2.1. Introduction.
Looking at input and output forces in fig. 2b indicates, at least in the first third of the test, that the assumption of homogeneity of stresses is obviously not verified. It can be deduced that strain-rate and strain fields will be also non-homogeneous as it is observed in fig. 3.

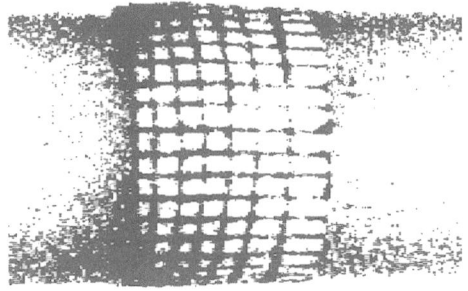

Fig. 3. High speed camera frame of a foam specimen under dynamic loading.
(black lines are equidistant in the unloaded state)

It is then necessary to develop a method which permits relating the material behaviour to the measured forces and velocities without assumption of uniformity.

3.2.2. Theoretical approach
A SHPB test provides superabundant measurements that are forces and velocities at both ends of the specimen. Accordingly, an identification technique based on an inverse calculation method [21] can be introduced. We assume that an appropriate form of the material behaviour is known, with parameters to be determined. Using a part of data - two velocities, for example, (V_i, V_o) - as input data, another part of data - the two forces (F_i, F_o) - is used to determine the best fit between them and the calculated forces, using a transient calculation. A set of parameters which gives the calculated forces in good agreement with the measured ones can be then found. One could also use an inverse method with SHPB waves used as measurements (as done in [11]). The reason of our choice is only to be more general in order to use data obtained with other devices.
To model this transient behaviour, one has to choose a constitutive equation, written as a relation between mechanical fields and parameters $\{p_i\}$. Each constitutive equation defines a class of behaviour. Inside this class, each particular behaviour corresponds to a set of parameters. For a given constitutive equation, one looks for the set of parameters that enables the description of measured forces *and* velocities. Let A be the operator associated with a given constitutive equation that enables to calculate (F_i, F_o, V_i, V_o) knowing the parameters $\{p_i\}$ and boundary conditions (BC) being either forces (F_i, F_o) or speeds (V_i, V_o). This calculation is called the *direct problem*. Our aim is to solve the associated *inverse problem* for experimental data $(F_i^d, F_o^d, V_i^d, V_o^d)$:

$$\text{find } \{p_i\} \text{ so that } (F_i^d, F_o^d, V_i^d, V_o^d) = A(\{p_i\}, BC) \qquad (6)$$

This inversion is an ill-posed problem because experimental measurements are not *a priori* in the range of *ImA* of *A* (i.e. all possible responses for all set of parameters). Consequently, this ill-posed problem is solved through an inverse calculation. The complete theoretical process is presented in [22] and only key points are underlined here.

3.2.3 Key points of the method

The choice of the constitutive equation is obviously very important. One could indeed argue that if the constitutive law of the material is known, there is no need for inverse calculations. For this reason, it is emphasised that the goal of this method is not to identify the constitutive law of the material using only one test. The purpose is to simulate the stress-strain field from forces and velocities measured at the boundary, in order to optimise the analysis of each test. Consequently, for visco-plastic materials, a quite general elasto-visco-plastic model of Sololovsy [23] and Malvern [24] type is used (Equ. 7)

$$\frac{\partial \varepsilon}{\partial t} = \frac{1}{E}\frac{\partial \sigma}{\partial t} \quad if \ \sigma \le \sigma_s \ ; \quad \frac{\partial \varepsilon}{\partial t} = \frac{1}{E}\frac{\partial \sigma}{\partial t} + g(\sigma,\varepsilon) \quad if \ \sigma > \sigma_s \qquad (7)$$

where

$$g(\sigma,\varepsilon) = \frac{(1+\frac{E_t}{E})\sigma - \sigma_s - E_t\varepsilon}{\eta}$$

Such a type of model allows for fast direct calculation. It gives good results for a large class of materials, as it is shown in Fig. 4a.

One has then to find a set of parameters that verifies, *as well as possible*, the compatibility between measured quantities and the constitutive equation. It leads to a minimisation problem. Looking for the *best* set of parameters is then not a trivial question and needs a special physical attention to the chosen mathematical *distance* used in the minimisation process. If, for instance, speeds are used as input values, the *distance* is calculated between real and simulated forces (using a quadratic distance, for example) but forces could also be used as input values and the *distance* calculated between real and simulated speeds. The chosen method should also be able to take account of complementary information about the test: - information on the displacement field obtained by high speed camera frames, - different quality of the data : when low impedance materials are tested, the input speed is known with a better precision than the input force, as it has been seen before.

3.2.4. Type of subsequent results

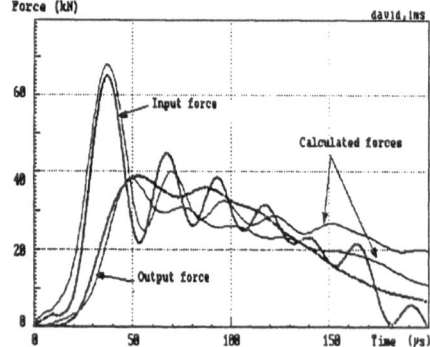

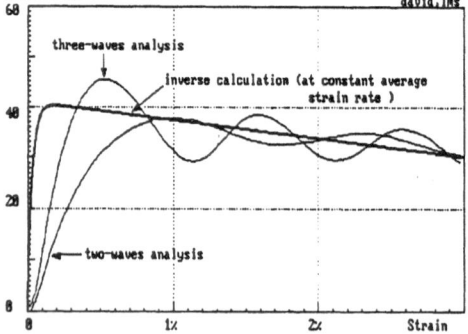

Fig. 4a. Simulated forces and real forces for salt

Fig. 4b Stress-strain curves from different analyses

In Fig. 4a, comparison for salt-rock between the measured forces and the calculated ones is shown. The chosen model, with the set of parameters that give the best agreement with experimental data, can be considered as the representative model of the specimen in this test. As a result, the stress and strain fields in the specimen are known so that a stress-strain curve is found and can be given for a constant strain rate through the identified model. In our example, the stress-strain curve obtained with this method is compared with those of *two-waves* or *three-waves* classical SHPB analysis. In the case of salt, those two curves are quite far from that of the present method, especially in the range of small strains (Fig. 4b). As a result, the inverse calculation technique is the only way to obtain accurate results for this type of materials.

4. Conclusion

Some foam materials used in automotive industry dummies are low impedance materials that can support large strains and show a strong strain-rate effect. Testing and modelling such materials at high strain rates leads to solve special difficulties.
Because of low impedance, Hopkinson bars made of low impedance materials are used. Such bars are not purely elastic but viscoelastic and need the use of a special waves' correction dispersion theory. The associated technique is successfully applied in our Laboratory with 3m long Nylon bars with the signal processing is included in the software for SHPB analysis.
Because of the strong strain-rate effect, measured input and output forces are sometimes significantly different and SHPB analysis must be done using an inverse method. This method in still under study to find the mathematical process in agreement with Hopkinson bars' data and material properties. Current results using such an approach already give satisfying results.

References

[1] B. Hopkinson, A method of measuring the pressure in the deformation of high explosives by the impact of bullets, *Phil. Trans. Roy. Soc.*, 1914, Vol. A213, pp. 437-452.
[2] R.M. Davies, A critical study of Hopkinson pressure bar, *Phil. Trans. Roy. Soc.*, 1948, Vol. A240, pp.375-457.
[3] H. Kolsky, An investigation of the mechanical properties of materials at very high rates of loading, *Proc. Phys. Soc.*, 1949, Vol. B62, pp. 676-700.
[4] J. Harding, E.D. Wood, J.D. Campbell, Tensile testing of materials at impact rates of strain. *J. Mech. Engng. Sci.*, 1960, Vol. 2, pp. 88-96
[5] J. Duffy, J.D. Campbell, R.H. Hawley, On the use of a torsional split Hopkinson bar to study rate effects in 1100-0 aluminium, *J. Appl. Mech.*, 1971, Vol. 38, pp. 83-91
[6] R.M. Davies, A critical study of Hopkinson pressure bar. *Phil. Trans. Roy. Soc.*, 1948, Vol. A240, pp. 375-457.
[7] L. Pochhammer, Uber die fortpflanzungsgeschwindigkeiten kleiner schwingungen in einem unbergrenzten isotropen kreiszylinder, *J. für die Reine und Angewandte Mathematik*, 1876, Vol. 81, pp. 324-336.
[8] C. Chree, The equations of an isotropic elastic solid in polar and cylindrical coordinates, their solutions and applications, *Cambridge Phil. Soc. Trans.* 1889, Vol. 14, pp. 250-369.
[9] H. Zhao G. Gary, A three dimensional longitudinal wave propagation in an infinite linear viscoelastic cylindercal bar. Application to experimental techniques, *J. Mech. Phys. Solid*, Vol 43, 1335-1348, 1995
[10] D.R. Bland, *The Theory of Linear Viscoelasticity*, Oxford Univ. Press., Oxford, 1960
[11] H. Zhao, Analyse de l'essai aux barres de Hopkinson, Application à la mesure du comportement dynamique des matériaux. *Ph. D. Thesis*, ENPC, Paris, 1992.

[12] U.S. Lindholm, Some experiments with the split Hopkinson pressure bar. *J. Mech. Phys. Solids*, 1964, Vol. 12, pp. 317-335.

[13] E.D.H. Davies, S.C. Hunter, The dynamic compression testing of solids by the method of the split Hopkinson pressure bar, 1963, *J. Mech. Phys. Solids*, Vol. 11, pp.155-179.

[14] J.R. Klepaczko, Lateral inertia effects in the compression impact experiments, *Reports of Inst. Fund. Technical Research*, 1969, n° 17, Warsaw.

[15] C.K.H. Dharan, F.E. Hauser, Determination of stress-strain characteristics at very high strain rates, *Exp. Mech.*, 1970, Vol. 10, pp. 370-376.

[16] J.Z. Malinowski, J.R. Klepaczko, Dynamic frictional effects as measured from the split Hopkinson bar, *Int. J. Mech. Sci.*, 1986, Vol. 28, pp. 381-391.

[17 L.D. Bertholf, J. Karnes, Two-dimensional analysis of the split Hopkinson pressure bar system, *J. Mech. Phys. Solids*, 1975, Vol. 23, pp. 1-19.

[18] A.F. Conn, On the use of thin wafers to study dynamic properties of metals, *J. Mech. Phys. Solids*, 1965, Vol. 13, pp. 311-327.

[19] F.E. Hauser, Techniques for measuring stress-strain relations at high strain rates, *Exp. Mech.*, 1966, Vol. 6, pp. 395-402.

[20] W.E. Jahsman, Reexamination of the Kolsky technique for measuring dynamic Material behavior. *J. Appl. Mech.*, 1971, pp. 77-82.

[21] H.D. Bui, *Introduction aux problèmes inverses en mécanique des matériaux*. Editions Eyrolles, Paris 1993 // English translation, *Inverse problems in the mechanics of materials: an introduction*, CRC Press, Boca Raton, 1994.

[22] L. Rota, An inverse approach for identification of dynamic constitutuve equations, *Proceedings of the 2nd international symposium on inverse problems - ISIP'94*, A.A.Balkema, Rotterdam, Brookfield, 1994.

[23] V.V. Sokolovsky, The propagation of elastic-viscous-plastic waves in bars. *Prikl. Mat. Mekh.*, 1948, Vol. 12, pp. 261-280.

[24] L.E. Malvern, Propagation of longitudinal waves of plastic deformation. *J. Appl. Mech.*, 1951, Vol. 18, pp. 203-208.

High strain rate constitutive relationships for semi-crystalline polymers using a combined numerical and experimental approach.

N. N. Dioh, A. Ivankovic and P. S. Leevers

Department of mechanical engineering.
Imperial College of Science, Technology and Medicine
Exhibition Road, London SW7 2BX

Summary

Thermo-mechanical experimental data is presented for a pipe grade medium density polyethylene commercially used in gas and water pipelines, which must be resistant to rapid crack propagation. The data which includes stress/strain results over a strain rate range 10^{-4} to 10^4 s^{-1} and a temperature range -20 to 20°C are fairly well described by the Eyring relationship, even though wave propagation effects should be accounted for at high strain rates. In addition, corresponding infrared measurements show that at high strain rates where the testing conditions are adiabatic, temperature increases of the order of 2°C are recorded in compression tests and values in excess of 60°C are recorded in tensile tests due to localised necking within a small portion of the specimen.

A thermo-mechanical constitutive model based on the Prandtl-Reuss elastic-plastic relationships has been developed for this material using a finite volume numerical technique. Comparisons between the numerical model and experimental stress, strain and strain rate results are presented in order to assess the applicability of the model. A brief formulation of the finite volume numerical code is also presented. The importance of adiabatic heating in sustaining rapid crack propagation in polymers will be discussed in the context of a recently developed thermal decohesion model for semi-crystalline polymers.

Keywords: polyethylene, high strain rate, finite volume numerical technique, necking, constitutive relationships.

1 Introduction

The now widespread use of thermoplastics in pressurised pipelines calls for ever increasing knowledge of the behaviour of these materials with respect to their susceptibility to rapid crack propagation (RCP); a phenomenon which is known to occur in these pipelines under critical conditions of temperature and pressure. Recent work on a thermal decohesion model by Leevers [1,2] indicates that it is now possible to predict the dynamic fracture resistance of pipe-grade polymers, simply from their bulk properties such as yield stress, weight-average chain contour length, density, specific heat capacity, latent heat of fusion and melting temperature. In this model, the crack tip is represented by a Dugdale cohesive zone where crack propagation occurs when the rise in temperature at the crack tip, associated with fibril formation and subsequent drawing, is sufficiently high and localised for chain separation to take place, thereby leading to failure of the entire craze. It is therefore important that the thermo-mechanical behaviour of the material, (in terms of a closed form constitutive relationship) is known at strain rates and temperatures relevant to RCP conditions. The results presented in this paper will be restricted to medium density polyethylene (MDPE).

As a consequence of the interest in RCP, studying the tensile behaviour of these materials would seem more appropriate but since compressive tests at high strain rates are easier to perform, these were chosen instead. The effect of temperature and strain rate on the mechanical behaviour of polymers in compression will be similar to that in tension especially for semi-crystalline materials whose properties do not depend as strongly on hydrostatic stresses as those of say amorphous polymers. However, tensile tests at high strain rates were performed in order to study the adiabatic conditions associated with the neck propagation and failure. As mentioned before, this is thought to be closely related to the processes taking place at the tip of a running crack.

The strategy adopted here is to first of all create a thermo-mechanical stress strain data-base for MDPE under a wide range of strain rates and temperatures on which the development of a closed form constitutive relationship will be based. The strain rates and temperatures chosen in the test represent those encountered in-service in pipelines. The J-2 theory has been chosen as a starting point for the constitutive relationship since it has been used to describe with some success the post yield behaviour of some semi-crystalline polymers.

Measurements of the temperature rise associated with the high strain rate compressive and tensile tests are presented. A brief formulation of the numerical model used to assess the validity of the J-2 flow theory for describing the behaviour of MDPE is also presented and comparisons between the numerical and experimental results are made.

2 Thermo-mechanical experimental data

2.1 Materials

A MDPE, Rigidex 002/50, manufactured by BP Chemicals was chosen for study in the present work. It was supplied in the form of compression moulded sheets. The compressive specimens were machined into 12.7 mm diameter discs of thickness 4.4 mm or 1.5 mm. Before testing, each specimen was coated with thin layer of petroleum jelly, which has been shown to be a good lubricant for testing polymers at high strain rates [3]. For the tensile tests, the specimens were manufactured according to the ASTM standard D1822 and were of gauge length 20 mm, width 3 mm and thickness 3 mm.

2.2 Compression tests

The low and intermediate rate tests were carried out on an Instron and a Mayes testing machine respectively. Due to the small specimen thicknesses used, a linear voltage differential transducer (LVDT) was used to monitor specimen deformation. Strain rate was calculated from the LVDT readings and checked against the chosen crosshead velocity of the machine. All specimens were tested to an optimum strain of 30%.

The high strain rate tests were carried out in a split Hopkinson pressure bar (SHPB) apparatus with bars made from 15.8 mm diameter high strength aluminium [4,5]. In all test set-ups, lower temperatures were achieved by surrounding the specimen with a cooling chamber and in the case of the SHPB, by using a cooling sleeve. Strain rates were altered by varying the crosshead velocities in the case of the conventional test machines and by varying impact velocity in the case of the SHPB. During high strain rate deformation, the temperature rise associated with the specimen was monitored using an infrared camera of line frequency 3500 Hz.

2.2.1 Results

Typical stress strain results obtained at room temperature from tests at low, intermediate and high strain rate regimes are shown in figure 1. The tests were then repeated at 0°C and -20°C and the complete set of results presented in the so-called Eyring plot of yield stress against logarithmic strain rate. From this, the strain rate sensitivity of MDPE Rigidex 002/50 can be assessed. Since MDPE does not exhibit a yield point in compression, the results are presented in terms of a percentage flow stress against the corresponding logarithmic strain rate (figure 2). It is clear from figure 2 that at high strain rates, MDPE appears to show an increase in strain rate sensitivity and my not be described by the one-parameter Eyring model [6]. However, further experimentation, numerical and analytical modelling has shown that this apparent increase in sensitivity at high strain rates can be explained simply in terms of wave propagation effects[7]. The impact velocity employed in the SHPB test as well as the choice of specimen dimensions may in some cases account for this upturn. Figure 3 shows that the apparent increase in strain rate sensitivity disappears when the specimen thickness is reduced from 4.4 mm to 1.5 mm. For this material therefore, the results indicate that the simple Eyring model adequately describes its behaviour in the strain rate range studied.

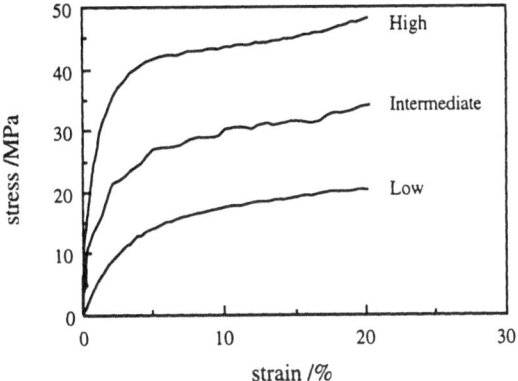

Figure 1: Influence of strain rate on the stress strain behaviour of MDPE at 23°C. Low, intermediate and high strain rate tests corresponds to strain rates of 3.8×10⁻⁴, 10 and 4.5×10³s⁻¹ respectively.

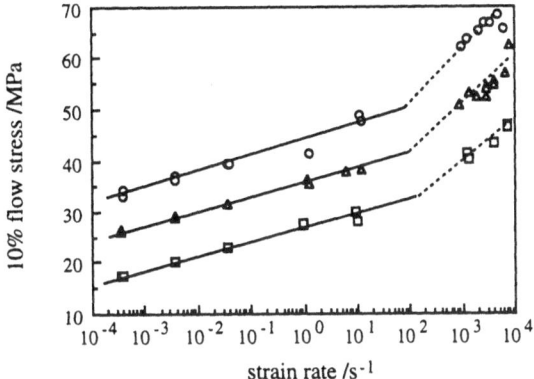

Figure 2: Effect of temperature and strain rate on the measured flow stress of MDPE. Symbols: □, 23°C; ▲, 0°C; ○, -20°C.

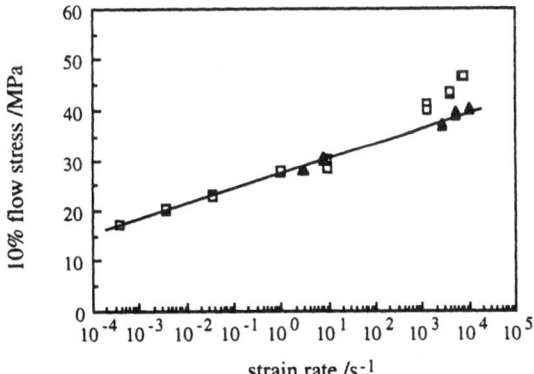

Figure 3: Effect of specimen thickness on the measured flow stress of MDPE at 23°C. Symbols: □, 4.4 mm thick specimens; ▲, 1.5 mm thick specimen.

Figure 4 shows infrared measurements of the temperature rise during deformation of a MDPE specimen tested in a SHPB. The results correspond to a strain of approximately 30%. As is evident from figure 4, the increase in temperature during deformation is of the order of 2°C and will therefore not have a significant effect on the measured flow properties presented in figures 1 to 3.

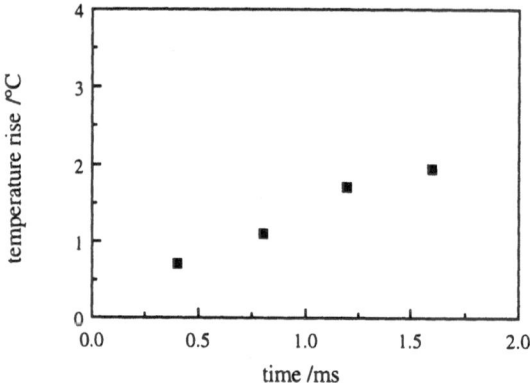

Figure 4: Adiabatic temperature rise in a MDPE specimen tested in the SHPB.

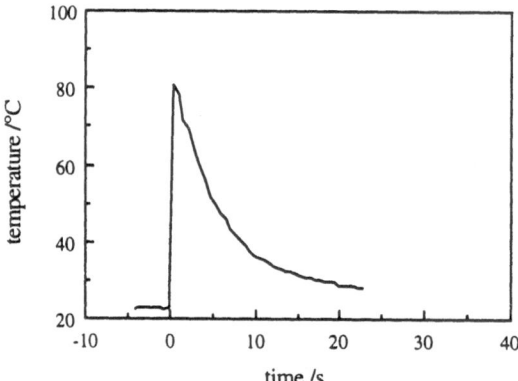

Figure 5: Adiabatic temperature rise associated with the necking
of a MDPE specimen during tensile testing.

2.3 Tensile tests

Figure 5 shows the infrared temperature measurements obtained from a tensile specimen deforming under adiabatic conditions at high strain rates. Measurements have shown that the draw ratio within the necked region of the specimen is of the order of 5. As seen from the figure, the temperature rise associated with the process of neck formation is of the order of 60°C. This value is sufficiently high to cause substantial softening within the specimen. However, in order to study how this increase in temperature affects the failure mechanisms within the neck, additional experiments are necessary. Nevertheless, the results indicate that the temperature rise associated with high strain rate adiabatic deformation is sufficiently high to govern the failure mechanisms at high loading rates.

3 Numerical simulation of experimental data

Numerical simulations of both the SHPB test and the high strain rate tensile test have been performed using the finite volume numerical technique. The numerical work presented here assumes that the material behaves in a non-linear rate independent manner and is represented by a bi-linear stress strain relationship whose characteristics are depicted in table 1. In simulating the SHPB test, the material is assumed to be non-linear elastic with the same primary and secondary moduli as that of table 1. So long as unloading is not considered this model should yield similar results to an elastic-plastic one. In simulating the high strain rate tensile test, the material is assumed to behave in a non-linear thermo-elastic plastic manner such that temperature predictions can be performed. The purpose of the modelling is two-fold; firstly, to show that the Prandtl-Reuss thermo elastic plastic model yields reasonable results for semi-crystalline polymers and secondly, to illustrate that stress wave propagation effects may be responsible for the upturn seen in the SHPB results .

3.1 Formulation of the finite volume numerical model

As a starting point, the equation of motion:

$$\frac{\partial}{\partial t}\int_V \rho \frac{\partial u_i}{\partial t}dV \;=\; \int_S t_i \, dS \;+\; \int_V \rho f_i dV, \tag{1}$$

and the equation for thermal energy balance:

$$\frac{\partial}{\partial t}\int_V \rho c T dV \;=\; \int_S q_j n_j \, dS + \int_V S_T dV, \tag{2}$$

in integral form have to be solved at any instant in time t for the given solid of volume V, bounded by a surface S. In the above equations, u_i is the displacement vector, $t_i = \sigma_{ij} n_j$ is the traction vector, σ_{ij} is the stress tensor, n_j is the unit vector of outward normal to the surfaces S, f_i is the body force, ρ is the mass density, T is the temperature, q_j is the heat flux, c is the specific heat capacity and S_T is the heat source. In order to solve the above system of equations, the following constitutive relations are used [8]:

$$d\sigma_{ij} \;=\; 2Gd\varepsilon_{ij} + \lambda d\varepsilon_{kk}\delta_{ij} - (3\lambda + 2G)\alpha dT\delta_{ij} - \left\{ \frac{3G\sigma_{ij}^d \sigma_{kl}^d d\varepsilon_{kl}}{\overline{\sigma}^2\left(\dfrac{H'}{3G}+1\right)} \right\} \tag{3}$$

$$q_j = k\frac{\partial T}{\partial x_j} \tag{4}$$

where

$$\varepsilon_{ij} \;=\; \frac{1}{2}\left(\frac{\partial u_i}{\partial x_j} + \frac{\partial u_j}{\partial x_i}\right) \tag{5}$$

is the strain tensor,

$$\sigma_{ij}^d = \sigma_{ij} - \frac{1}{3}\sigma_{kk}\delta_{ij} \tag{6}$$

is the stress deviator and

$$\overline{\sigma} = \left(\frac{3}{2}\sigma_{ij}^d \sigma_{ij}^d\right)^{1/2} \tag{7}$$

is the effective stress as defined by the Von Mises yield criterion. G and λ are Lame's constants, defined as $G = E/2(1 + v)$ and $\lambda = vE/(1 + v)(1 - 2v)$ respectively, where E is Young's Modulus and

v is Poisson's ratio. H' is the plastic modulus, δ_{ij} is the Kronecker delta, k is the heat conductivity of the solid and α is the thermal expansion coefficient. Neglecting the thermal elastic strains, the heat source S_T may be approximated as [9]:

$$\int_V S_T dV = \int_V f\sigma_{ij}\dot{\varepsilon}_{ij}^p dV \tag{8}$$

where $\dot{\varepsilon}_{ij}^p$ is the plastic strain rate, and f is defined as the work rate to heat rate conversion fraction which for the present work will be taken as 1.0. The plastic strain rate term is calculated from the thermal elastic plastic constitutive relationship used [8]. Under elastic conditions, the plastic term in the brackets { } disappears and the constitutive relationship reduces to that of an incremental Hooke's law. By neglecting body forces and combining equations (1) and (2) with the respective constitutive relations, a closed system of equations can be obtained which are then solved by choosing an appropriate coordinate system and boundary conditions. These are usually in terms of displacements and/or forces for the momentum equations and temperature and/or heat fluxes for the heat equation.

The finite volume solution of the above set of equations involves either a two or three dimensional discretisation of the solution domain. In the scheme, the time domain is divided into a number of time steps δt, and the space domain into a number of cells. For the two-dimensional case, each cell is bounded by four faces whereas for the three-dimensional case, each cell has six neighbours. A linear distribution of displacements between neighbouring points is assumed and in the present work, the deformation of the cell is neglected. The inertia force is approximated using the mean value theorem. A detailed formulation and implementation of the method in stress analysis can be found elsewhere [7,10]. The material properties used in the numerical modelling are shown in table 1.

Table 1: Material properties used for numerical modelling

Elastic modulus E (GPa)	1.7
Plastic Modulus H' (GPa)	0.136
Density ρ (kgm^{-3})	950
Yield stress σ_y (MPa)	25.2
Poisson's ratio v	0.4
Thermal expansion coefficient α (/K)	1.5×10^{-4}
Specific heat capacity c (J/kg/K)	1945
Conductivity k (J/s m K)	0.5

The bi-linear stress strain properties are approximated from the high strain rate experimental stress strain results at room temperature. The numerical model of the tensile specimen has dimensions similar to that used in the experiments. The solution domain is made up of 240 cells. With one end of the specimen fixed, the other end is displaced at a constant rate of 5 m s^{-1} (similar to that of the experiment) and the stress, strain and temperature profiles are monitored within the deforming specimen. A comparison between the temperature predicted numerically (assuming a draw ratio of 5) and that obtained experimentally is shown in figure 6. Even though the results are in good agreement, it is recognised that a more appropriate model would be required if explanations of the micro-mechanisms associated with flow in these materials are of the essence.

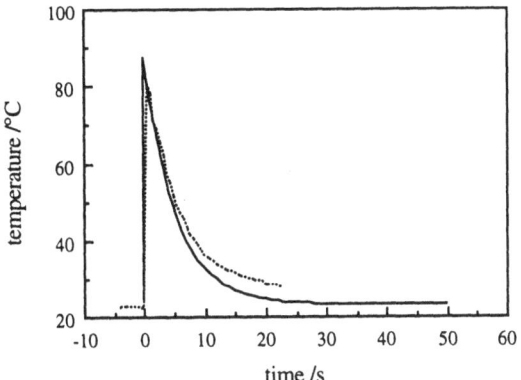

Figure 6: Comparison between numerical and experimental temperature
rise associated with necking of a MDPE specimen.

In simulating the SHPB test, a 3-D model was created. Care was taken to ensure that the dimensions of the real geometry were accurately represented. A uniform mesh of 35000 elements was used so as to avoid the generation of spurious wave reflections. The mesh size was dictated by the thickness of the sample (4.4 mm compared to 750 mm long bars). The specimen is again modelled as having the primary and secondary moduli listed in table 1. The simulation is carried out for two impact velocities and the results are shown in figure 7. As mentioned before, in simulating the SHPB test, a non-linear elastic version the code is used due to the complexities associated with a corresponding three-dimensional formulation of a thermo-elastic plastic version such as that used for the two-dimensional solution presented above. This is simply achieved by omitting the bracketed terms in equation 3, neglecting thermal effects and using either the primary or the secondary moduli in equation (3).

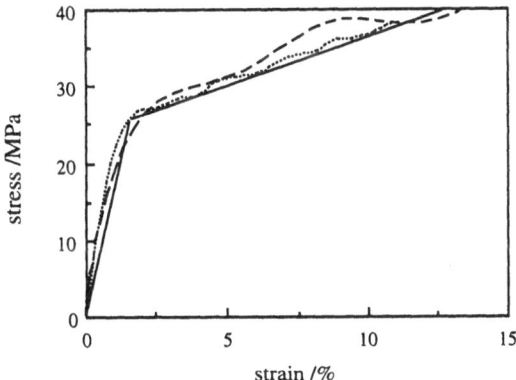

Figure 7: Numerically determined stress strain curve at impact velocities of 12 and 30 m s^{-1} .

Input bi-liear material model, ———; 12 m s^{-1}, - - - - - 30 m s^{-1}, - - - -.

It is clear from figure 7 that even though the material behaviour is rate independent, the regenerated stress strain curves (obtained in the same manner as in the experiments [4]) do show an apparent rate effect with higher flow stresses predicted at the higher impact velocity. This apparent increase in sensitivity arises due to stress wave propagation effects in the deforming specimen [7]. It has also been shown that it manifests itself more strongly at higher impact velocities since the conditions of stress and strain uniformity (on which the classical SHPB analysis is based) are less likely to be satisfied at higher impact velocities due to the presence of slow moving plastic wave fronts within the specimen.

4   Conclusions

Thermo-mechanical experimental data has been presented for a medium density polyethylene in the strain rate range 10^{-4} to 10^4 s^{-1} and the temperature range -20°C to 20°C. The results indicate that within this region, the behaviour of the material is fairly well described by the Eyring relationship and at the higher strain rates, care must be taken in interpreting the experimental results due to stress wave propagation effects. It has also been shown that measurements of the temperature rise associated with the high strain rate adiabatic deformation of this medium density polyethylene yield low values, of the order 2°C, in compression tests and high values, of the order 60°C, in tension tests. The experimental results have been validated by using a finite volume numerical technique in which the material is modelled as one whose behaviour is governed by the J-2 flow theory.

5   References

[1]   P. S. Leevers. *Impact and dynamic fracture of tough polymers by thermal decohesion in a Dugdale zone.* To be published in International Journal of Fracture.

[2]   P. S. Leevers and R. E. Morgan. *Impact fracture of Polyethylene, a non-linear elastic thermal decohesion model.* To be published in Engineering Fracture Mechanics

[3]   S. M. Walley, J. E. Field, P. H. Hope and N. A. Safford. *Phil. Trans. R. Soc. Lond,* A328, 1989, pp. 1.

[4]   U. S. Lindholm,*J. Mech. Phys. Solids* 12, 1964, pp. 317

[5]   N. N. Dioh, P. S. Leevers, and J. G. Williams, *Polymer* 34, 1993, pp. 4230

[6]   I. M. Ward and D.W. Hadley. *Mechanical properties of solid polymers,* John Willey and Sons Ltd, Chichester, 1993

[7]   N. N. Dioh, A. Ivankovic, P. S. Leevers and J. G. Williams. *Proc. R. Soc. Lond,* A449, 1995, pp. 187

[8]   T. Muraki, J. J. Bryan, and K. Masubuchi. *Trans. ASME: Journal of Engineering Materials and Technology,* 1975, pp. 81

[9]   A. B. Boley and H. J. Weiner. *Theory of Thermal Stresses,* Wiley, New York, 1962.

[10] A. Ivankovic, I. Demirdzic, J. G. Williams and P. S. Leevers, *Int. J. Fracture.* 66, 1994, pp. 357

About a new experimental method of identification of the dynamic toughness of materials

Hubert Maigre[1] and Daniel Rittel[2]

1 Laboratoire de Mécanique des Solides, URA CNRS 317, 91128 Palaiseau Cedex, FRANCE
2 Faculty of Mechanical Engineering, Technion, 32000 Haïfa, ISRAEL

Summary

A new experimental method has been developped to evalutate dynamic toughness of brittle materials. This approach is based on invariance properties of an integral and is applied to a Kolsky Bar apparatus. This method allows also mixed mode loading to investigate dynamic fracture criteria including mixity. Moreover a technique of dual treatment gives precious information to validate all the assumptions made to analyse the experiments. Some results on PMMA and commercial glass are presented to illustrate this method.

Keywords: Dynamic Fracture, Dynamic Toughness, Stress Intensity Factor, Kolsky Bars

1 Introduction

Dynamic properties of materials are now essential to the designers to compute structures subjected to transient loadings or shocks. These properties are not always well-known because standard procedures are missing and the complexity of existing procedures makes impossible systematic tests on materials. This the case of the toughness K_{IC} which is assessed in static whereas the dynamic toughness K_{IdC} is still evaluated using various experimental procedures [1] more or less accurate or easy to perform. Some sophisticated techniques are based on optical properties of the material [2] and can not be applied to opaque materials like metals. Recently, we introduced the dynamic H-integral [3] and we used the path-independent property of this integral to relate the dynamic loading on a structure to the evolution of the dynamic stress intensity factor at the crack-tip. This theoretical result has been applied to Kolsky Bars in conjunction with the Compact Compression Specimen (CCS). Knowing precisely the instant of the onset of the crack propagation we evaluate the dynamic toughness as the dynamic stress intensity factors at this instant. We present in this paper results on PMMA and glass. First we recall the principles of the method. Next we discuss the experimental results obtained on these materials.

2 Path-independent H-Integral

The H-integral is a path-independent contour integral which relates forces (F) and displacements (u) applied on the external surface S of a cracked solid (fixed crack length a) to the stress intensity factors K_I^u and K_{II}^u at the crack-tip [3]:

$$H(t) := \frac{1}{2} \int_S \left\{ F \overset{*}{\underset{t}{}} \frac{\partial v}{\partial a} - \frac{\partial T}{\partial a} \overset{*}{\underset{t}{}} u \right\} dS = \frac{1-v^2}{E} \left\{ K_I^u \overset{*}{\underset{t}{}} K_I^v + K_{II}^u \overset{*}{\underset{t}{}} K_{II}^v \right\} \tag{1}$$

This expression stands for 2D linear elactic media in plane deformation. The time dependence of the dynamic effects is explicitely taken into account through the time convolution product $\overset{*}{\underset{t}{}}$. T, v,

41

K_I^v and K_{II}^v refer to an adjoint dynamic displacement field v which is supposed known (calculated numerically in our case). Practically, $K_I^u(t)$ and $K_{II}^u(t)$ are solution of a linear convolution equation, since the adjoint field has been calculated and H has been estimated from the experimental forces and displacements.

2.1 Separation of mixed modes

H-integral includes combined opening and shearing modes [4] when the loading or the solid are not symmetrical with respect to the initial crack. To get separately $K_I^u(t)$ and $K_{II}^u(t)$ we choose respectively two appropriate adjoint fields: the first one in pure mode I ($K_{II}^v := 0$) and the second one in pure mode II ($K_I^v := 0$).

2.2 Dual analysis

The evaluation of H(t) requires both the knowledge of the forces and the displacements applied to the solid. In fact, if the adjoint field is choosen such that its displacements on S are constant with respect to the crack variation ($\partial v/\partial a := 0$), $K_I^u(t)$ and $K_{II}^u(t)$ depend only on the experimental displacements u. Similarly, we can define an other adjoint field with the applied forces constant with respect to the crack variation ($\partial T/\partial a := 0$). In this case, the dynamic stress intensity factors are evaluated only from the experimental forces and we note them $K_I^F(t)$ and $K_{II}^F(t)$. These two ways to analyse the experimental data give us a precious tool to assess the quality of the experiments and the validation of the theoritical assumptions. If every thing goes right, the stress intensity factors from the displacements are identical to the stress intensity factors from the forces. At the begining of the crack propagation ($t = t_{frac}$) the two families of stress intensity factors must begin to diverge because the hypothesis of a fixed crack is no longer valid:

$$t < t_{frac} \quad => \quad K_I^u(t) = K_I^F(t) \quad \text{and} \quad K_{II}^u(t) = K_{II}^F(t)$$

$$t > t_{frac} \quad => \quad K_I^u(t) \neq K_I^F(t) \quad \text{and} \quad K_{II}^u(t) \neq K_{II}^F(t)$$

This divergence can be used to detect the onset of the crack propagation.

3 Experimental setup

We implement the preceding theoretical results to the Kolsky Bar apparatus. This device provides a mean to apply dynamic impact and to measure forces and displacements at the interface between the bars and a specimen. We designed the special specimen CCS to be inserted between the bars without extra devices to turn compression into tension at the crack-tip (fig. 1).

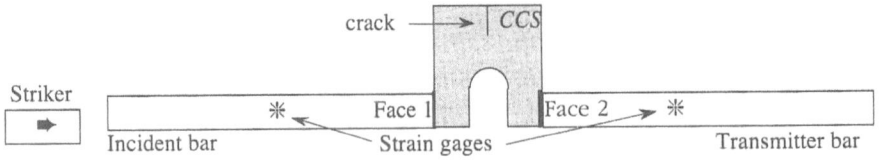

Fig.1: Experimental setup. The CCS is inserted between two instrumented bars (Kolsky Bars). Transient signals are recorded on two strain gages. Forces and displacements are determined at the specimen interfaces.

It can be noted that the geometry of the CCS is symmetrical, but loads are applied on one side only. Consequently, both mode I and mode II are present but it is obvious that the opening mode should be predominant. Kolsky apparatus giving the resultant forces (F_1 and F_2) and the mean displacements (u_1 and u_2) at the interfaces, H-integral at the boundary reduces to the sum of four convolution products between scalar quantities:

$$H(t) : = \frac{1}{2}\left\{F_1 *_t \frac{\partial v_1}{\partial a} - \frac{\partial T_1}{\partial a} *_t u_1 + F_2 *_t \frac{\partial v_2}{\partial a} - \frac{\partial T_2}{\partial a} *_t u_2\right\} \tag{2}$$

For a given series of experiments on a material, one has to generate four adjoint fields corresponding to the mixed mode and dual analysis. Since the set of adjoint fields must be calculated once, a large sample size can easily be tested.

4 Experimental results

Tests were performed on 16.5 mm diameter Kolsky Bars. Acquisition sampling was carried out at 1 MHz. Furthermore, a single wire fracture gage was glued at the vicinity of the notch tip to detect the onset of the crack propagation at the surface.

4.1 PMMA dynamic toughness

The present result concerns commercial PMMA whose properties are listed in table 1.

Young's modulus E [GPa]	Poisson ratio v	Density ρ [kg m^{-3}]
static: 3.70 dynamic: 5.76	0.42	1182

table 1: mechanical properties of the PMMA used in this study

Due to the viscosity of the PMMA, a dynamic modulus was determined by the time required for waves to travel in a rod specimen. The initial notches were machined mechanically with an average root radius 0.2 mm. Typical experimental signals (incident, transmitter pulses) are shown in fig. 2.

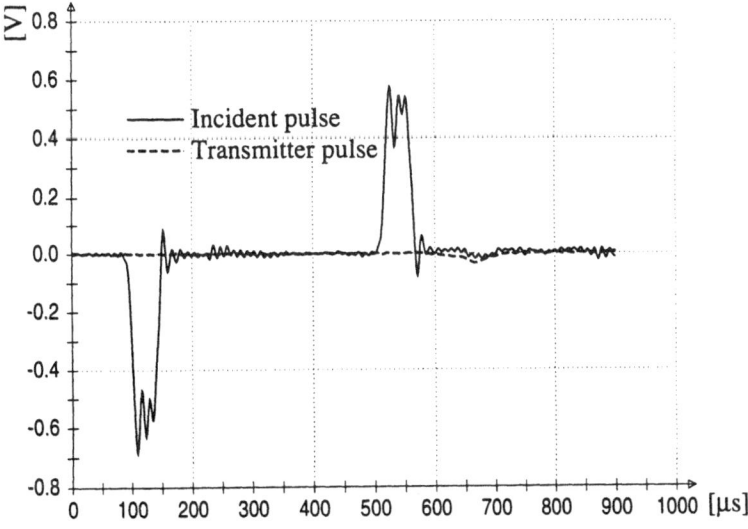

Fig.2: Typical incident and transmitter pulses

For this example this incident pulse is 70µs long. One can see that the transmitter signal is very small because of the very short time needed to fracture the whole specimen. After treatment of these gages signals, we get the interfacial forces and velocities applied to the specimen (fig. 3).

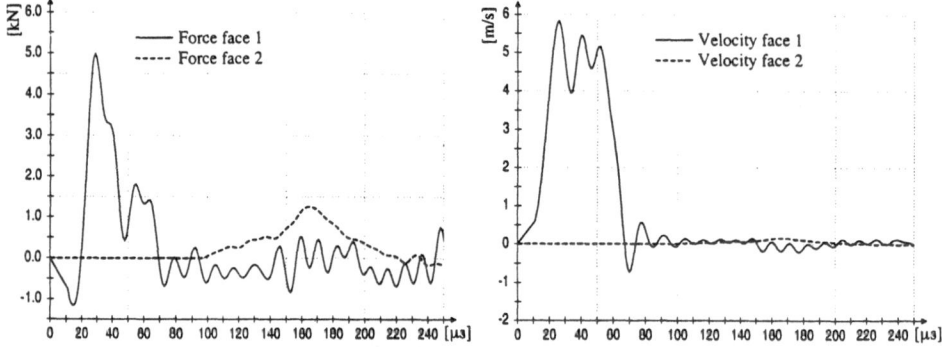

Fig.3: Interfacial forces and velocities applied on the CCS. Note that the forces are not in equilibrium.

Following equation (2), we then calculated the four H-integral with the four adjoint fields. By deconvolution we obtain four evolutions of stress intensity factors (three of them are shown fig. 4).

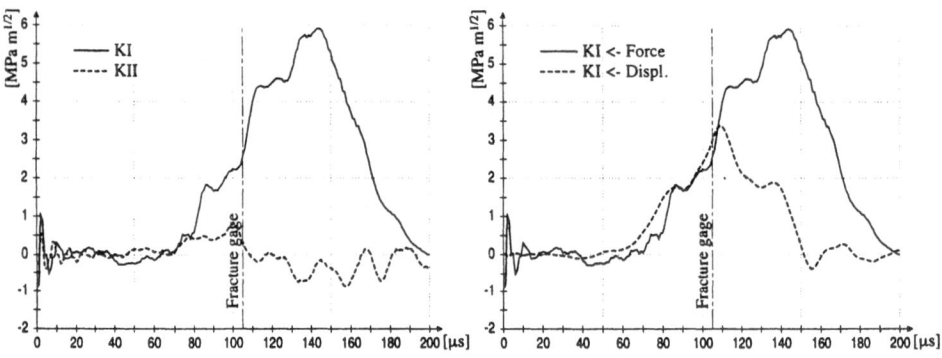

Fig.4: Dynamic stress intensity factor evolutions. The time of fracture given by the fracture gage glued on surface is also reported.

Signals take typically 60µs to reach the crack and force its opening. Fracture occurs soon after. Mode II exists because of the non symmetrical loading but very soon the mode I is dominant. This predominance of the opening is confirm by the observation of the broken specimen (fig.5). The crack path is not straight but the begining of the propagation is along the direction of the initial notch. The dual analysis gives identical evolutions until 110µs which is nearly the instant given by the fracture gage. We suppose that the crack propagation begins at this time and we obtain a critical stress intensity factor of 2.4 MPa m$^{1/2}$.

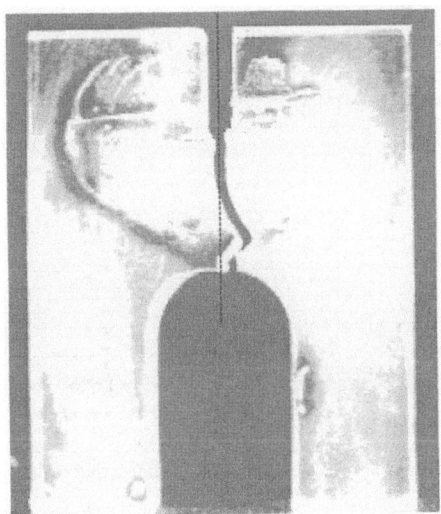

Fig.5: PMMA CCS after dynamic loading. The crack does not deviate at the onset of the propagation which confirms the existence of dominant mode I at this time.

Using this technique on 16 specimens we have shown that for this material the fracture toughness increases markedly with stress intensity rate [5] and this was already observed [6]. This variation correlates qualitatively with the visual aspect of the fracture surfaces. Indeed at higher rates some hackles can be observed in the vicinity of the crack-tip whereas a mirror aspect is observed at lower rates.

4.2 Glass dynamic toughness

We have also tested glass with this technique combining Kolsky Bars and Compact Compression Specimen. The material properties of the studied glass are listed in table 2.

Young's modulus E [GPa]	Poisson ratio ν	Density ρ [kg m^{-3}]
dynamic: 80	0.23	2503

table 2: mechanical property of the glass used in this study

We prepared 10 specimens. The initial notches were machined mechanically with an average root radius 1.0 mm. Four adjoint fields have been also calculated. We do not present again the experimental signals and the interfacial loads which are very similar to those on PMMA. We present directly after deconvolution results of a typical experiments (fig. 6). Like PMMA, the non-symmetrical loading does not generate significant mode II. There is also a good agreement between the analysis from forces and the analysis from displacements until 90μs which the time given by the fracture gages glued on both side of the specimen. What is more surprising is the very high value of the toughness around 12 MPa m$^{1/2}$ which is 10 times much higher than the static value 1 MPa m$^{1/2}$. In fact we have to observe the specimen after testing (fig. 7).

46

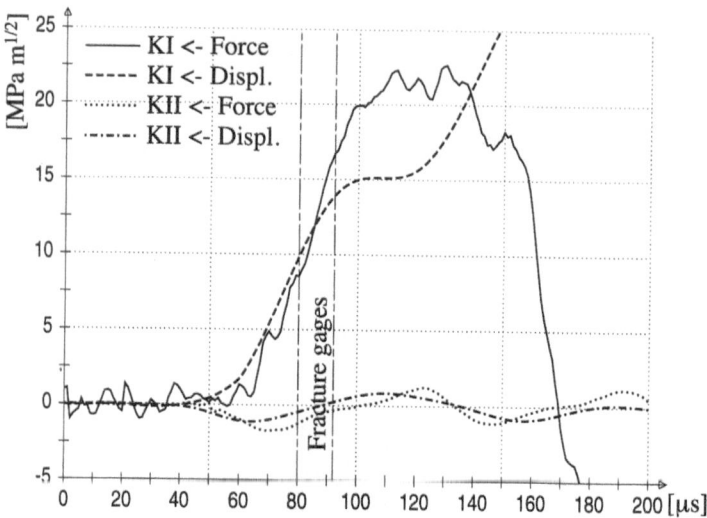

Fig.6 : Evolutions of the dynamic stress intensity factors for glass. Mode I is soon predominant. Analysis from experimental forces or displacements are very similar until the time of fracture recorded on the faces of the specimen.

Fig.7 : Glass specimen after dynamic testing. Note the 40° kink of the crack at the begining of the propagation.

The initial deviation of the crack around 40° from the axis of the notch is not compatible with a fracture occuring during dominant mode I. A possible explanation is that fracture initiates much earlier than supposed until now when mode II is similar to mode I. On fig. 8 we plot the evolutions of mode I and mode II stress intensity factors until 80μs. We present only the results obtained from experimental displacements because of the noise on those from forces.

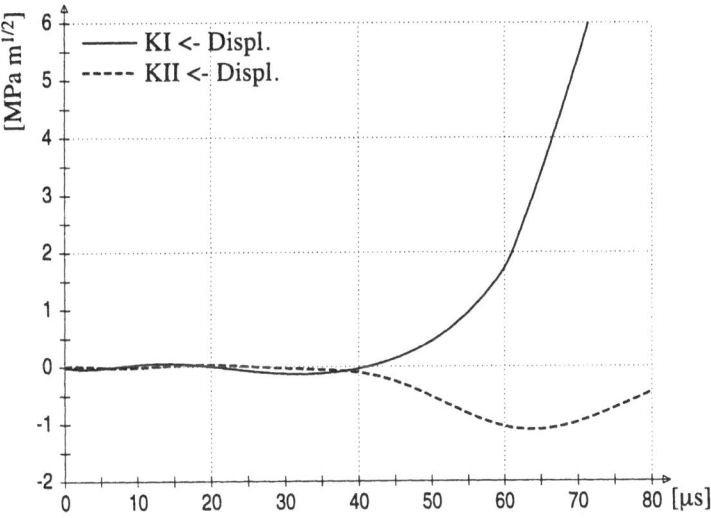

Fig. 8: Dynamic stress intensity factors at the very begining of the sollicitation at the crack-tip.

Until 65μs mode II is equivalent to mode I. During this time stress intensity factors do not exceed 2 MPa m$^{1/2}$ which is closer to the static value. So it seems that frature occurs very soon and it can not be detected either by the fracture gages or the separation in the dual analysis. Defining a criteria including mode I, mode II and the angle of kinking one can imagine to find when the fracture initiates.

5   Conclusions

We have presented a new convenient method to evaluate dynamic toughness of brittle materials. It is based on the H-integral and it is applied to Split Hopkinson Bars with CCS specimens. This method allows identification of mode I and mode II contributions. The dual analysis using independently experimental forces or displacements gives indications about the quality of the experiments and can also be used to detect the crack inititation. Experiments on PMMA have shown the good applicability of this method. Moreover, the results have confirmed the already observed increase of toughness with the the loading rate. Experiments on glass work as well but it seems much more difficult to detect the exact instant of the crack initiation. Experiments on very brittle materials have to be done very carefully and the use of the observed kink angle could provide a mean to reveal the initiation and the mode of fracture.

References

[1] A.S. Kobayashi, *Handbook of experimental Mechanics*, Prentice Hall, Englewood Cliffs, N.J.
[2] J. Beinert, J.F. Khalthoff, Experimental determination of dynamic stress intensity factors by shadow pattrerns, *Experimental evaluation of stress concentration and stress intensity factors*, 1981, pp. 281-330

[3] H.D. Bui, H. Maigre, D. Rittel, An new approach to the experimental determination of the dynamic stress intensity factor, *International Journal of Solids and Structures*, Vol. 29, 1992, pp.2881-2895

[4] D. Rittel, H. Maigre, Mixed-mode quantification for dynamic fracture initiation: application to the compact compression specimen, *International Journal of Solids and Structures*, Vol. 30, 1993, pp.3233-3244

[5] D. Rittel, H. Maigre, Dynamic fracture toughness determination using the CCS technique: apllication to PMMA, *Proceedings of ICF8*, Kiev, 1993

[6] H. Wada, C.A. Calder, T.C. Kennedy, M. Seika, Measurement of impact fracture toughness with single point bending using air gun, *Proceedings of International Symposium on Impact Engineering*, 1992, pp. 569-574

Strain Rate Sensitivity of Flow Stress at Very High Rates of Strain

J. Shioiri, K. Sakino and S. Santoh

College of Eng., Hosei Univ., Koganei-shi, Tokyo 184, JAPAN

Summary

In order to clarify the mechanism of the steep rise in the flow stress, widely seen in metallic materials at strain rates above about 5×10^3/sec, two strain rate change tests were conducted. The strain rate reduction test was made for polycrystalline aluminium, copper, iron and niobium at strain rates up to about 2×10^4/sec. The measurement of the attenuation of the ultrasonic pulse superimposed upon the dynamic plastic deformation, a kind of strain rate change test using a high frequency perturbation in the strain rate, was made for polycrystalline aluminium and copper at strain rates up to 1×10^4/sec. The results of the above two experiments show that the instantaneous strain rate plays a more important role than the strain rate history in the steep rise in the flow stress. By comparing the results of the above experiments with the results of theoretical analyses made for fcc and bcc metals on the basis of the dislocation kinetics, it is shown that, in fcc metals, the steep rise in the flow stress is caused by the phonon drag upon dislocations while, in bcc metals, the rise is a little gentler and governed by the stress dependency of the activation energy for the kink pair generation on screw dislocations.

Keywords: Plastic deformation, high strain rate, steep rise in flow stress, strain rate change test, ultrasonic attenuation test, dislocation kinetics.

1 Introduction

It is known that in metallic materials the flow stress plotted against the logarithm of the strain rate shows a steep rise at strain rates above about 5×10^3/sec. However, the mechanism of the steep rise has not yet been clarified and this has been a big obstacle to consistent understanding of the dynamic properties. For this phenomenon two types of interpretation have been given. One type of interpretation attributes it to some particular rate-controlling mechanism of deformation. Ferguson et al. paid attention to the phonon drag upon the moving dislocation [1]. The other type of interpretation ascribes this phenomenon to the strain rate history dependency of the internal structure formed in the course of deformation. Regarding the "Threshold stress", the yield stress at $0°K$, as the measure of the internal structure, Follansbee et al. showed that the flow stress measured at a given threshold stress, instead of a given strain, exhibited no steep rise in the flow stress and they emphasized the role of the internal structure reflecting the strain rate history [2]. If the former type of interpretation is the case, the flow stress in the steep rise region should depend upon the strain rate at the instant, while if the latter type is the case, the flow stress should basically depend upon the strain rate history. Therefore, in order to clarify the mechanism of the steep rise in the flow stress, the strain rate change test will provide important experimental information. However, for this purpose, the strain rate change test must be done at very high strain rates and a testing technique with a high time-resolving capability is required. In this paper, two strain rate change tests conducted by the present author's group are reported. One is the strain rate reduction test using a newly devised apparatus with a high time-resolving capability. The other is the attenuation measurement of the ultrasonic pulse superimposed upon the dynamic plastic deformation. The latter can be regarded as a strain rate change test using a high frequency perturbation in the strain rate.

2 Strain Rate Reduction Test

The strain rate change test, the test in which the strain rate is suddenly increased or decreased, has been used to evaluate the role of the strain rate history in the flow stress. However, this type of test has been done at strain rates below about 2×10^3/sec, because of the limitation in the time-resolving capability of the test system. Probably, only one strain rate change test done at very high strain rates is the pressure shear test for foil specimens with the flying plate technique [3]. Recently, Sakino and Shioiri developed a new system for the strain rate change test [4], and conducted a series of decremental strain rate tests for aluminium, copper, iron and niobium at strain rates above 5×10^3/sec [5,6,7,8,9]. The newly developed test system is shown in Fig.1 schematically. In this system, the impact bar compresses directly the specimen attached at the end of the output bar with grease, and then by collision with the decelerator its end surface velocity is decreased suddenly. The flow stress of the specimen is detected by the strain gauge attached on the output bar and recorded in the digital memory with a sampling time of 0.1μsec. From the velocity of the impact bar just before the contact with the specimen, V_0, and the time series data of the nominal stress of the specimen stored in the digital memory, σ_n, the strain rate of the specimen, $\dot{\varepsilon}$, is obtained in the following form by using the one-dimensional bar wave approximation:

(a) before the strain rate reduction

$$\dot{\varepsilon} = (1/L_0)[V_0-(A\sigma_n/a_1c_1\rho_1)-(A\sigma_n/a_3c_3\rho_3)] \qquad (1)$$

(b) after the strain rate reduction

$$\dot{\varepsilon} = (1/L_0)[\{V_0-(A\sigma_n/a_1c_1\rho_1)\}/\{1+(a_2c_2\rho_2/a_1c_1\rho_1)\}-(A\sigma_n/a_3c_3\rho_3)] \qquad (2)$$

where ρ, c and a are the density, bar wave velocity and cross sectional area, respectively; subscripts 1, 2 and 3 indicate the impact bar, decelerator and pressure bar, respectively; A and L are the initial cross sectional area and length of the specimen, respectively. The mass of the specimen is neglected.

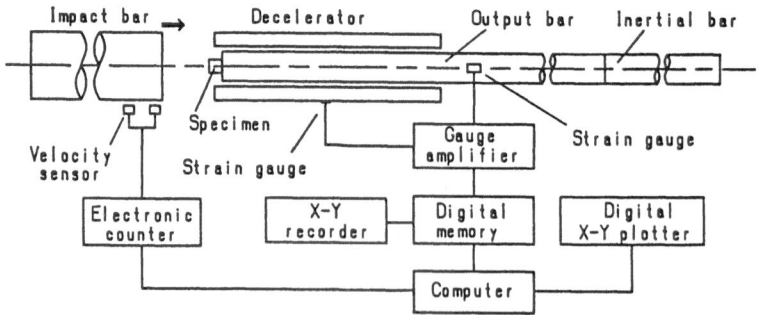

Fig.1. Devised system for strain rate reduction test

The impact bar and decelerator are made of Ti-15Mo-5Zr-3Al alloy and have the same diameter, 10mm or 13mm. The ratio a_2/a_1 is taken around 0.75. The output bar is 4mm in diameter and 300mm in length. Since for the output bar a sufficiently high mechanical impedance is required, for aluminium specimens a maraging steel bar and for copper, iron and niobium specimens a tungsten bar are used. In order to diminish the effect of the inertia very small specimens are used, i.e. in aluminium and niobium specimens the diameter and length are both 2mm and in copper and iron specimens are 1.5mm. Removing the specimen and directly hitting the output bar, the reduction of the velocity of the impact bar end obtained by collision with the decelerator can be tested. The result shows a well-controlled stepwise deceleration. The amount of the decrease in the velocity is in close agreement with the prediction based upon the one-dimensional bar wave approximation. The fall time is 0.3μsec or less.

To obtain a high time-resolving capability, in the stress measuring system, an inverse analysis is applied to the stress data recorded in the digital memory using the experimentally determined transfer function of the stress measuring system. The real flow stress response of the specimen is obtained from the recorded stress data in the form of the Fourier transform as follows:

$$X(\omega)=Y(\omega)/G(\omega) \tag{3}$$

where $X(\omega)$, $Y(\omega)$ and $G(\omega)$ are the Fourier transforms of the real flow stress, of the recorded flow stress and of the transfer function (impulse response) of the stress measuring system, respectively. In the transformation and inverse transformation of the numerical data, the FFT algorithm is utilised together with an appropriate window function. The transfer function of the system was determined experimentally from the recorded output for a known input imposed upon the specimen side end of the output bar utilising Eq.(3). As the known input a stepwise input caused by the collision of the bar of the same diameter and material as the output bar was used. The response to the unit step input thus obtained was reported in Ref.[4]. Although the experimentally determined response would include the response of the electronic apparatus besides the effect of the elastic wave dispersion in the pressure bar, the obtained response was fairly similar to the approximate solution for the elastic wave dispersion given in terms of the Airy integral [10].

Some of the experimental results necessary for the discussion in this paper are shown below. Strain rate reduction tests have been made by reducing the strain rate by 40~45% from 1×10^4~2×10^4/sec. Therefore, the strain rate after reduction is still in the steep rise region, and, accordingly, the results can give experimental information directly related to the steep rise in the flow stress. Tests have been conducted for polycrystalline specimens of aluminium (5N, cold-worked, machined, and 300°C 3hr vacuum-annealed), copper (OFHC, cold-worked, machined, and 600°C 3hr vacuum-annealed), iron (0.01%C, cold-worked, machined, and 800°C 3hr vacuum-annealed) and niobium (99.8% purity, cold-worked, machined, and 980°C 3hr high-vacuum-annealed). Constant strain rate tests have been also conducted for the same specimens using the same apparatus and an Instron type machine at strain rates from about 1×10^{-3}/sec to 2×10^4~3×10^4/sec. A result of the strain rate reduction test for aluminium is shown in Fig.2.

In the strain rate reduction tests, the role of the instantaneous strain rate in the flow stress is evaluated with the stress fall ratio (SFR) defined by

$$SFR = \Delta\sigma_f/\Delta\sigma_d \tag{4}$$

where, as indicated in Fig.2, $\Delta\sigma_f$ is the amount of the fall of the stress in the strain rate reduction test and $\Delta\sigma_d$ is the flow stress difference between

52

two constant strain rate deformations at the strain rates before and after the strain rate reduction, respectively. If the flow stress depends fully upon the instantaneous strain rate SFR should be unity, while if it depends fully upon the strain rate history SFR should be zero.   In Fig. 3, the values of SFR for aluminium are shown. The average values of SFR for the tested four metals are

Aluminium	Copper	Iron	Niobium
0.82	0.76	0.96	0.95

Those values are near to unity, although the values for two fcc metals are a little lower than those for two bcc metals. The above results indicate that in the remarkable rise in the flow stress at very high strain rates the instantaneous strain rate plays more important role than the strain rate history.

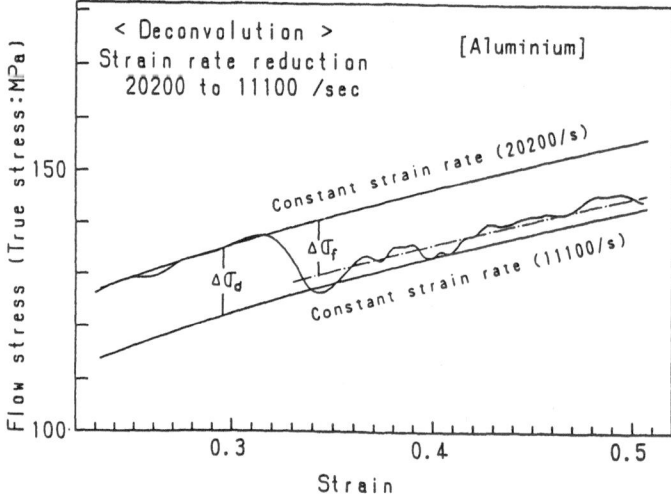

Fig. 2. Flow stress response to a sudden strain rate reduction (Aluminium)

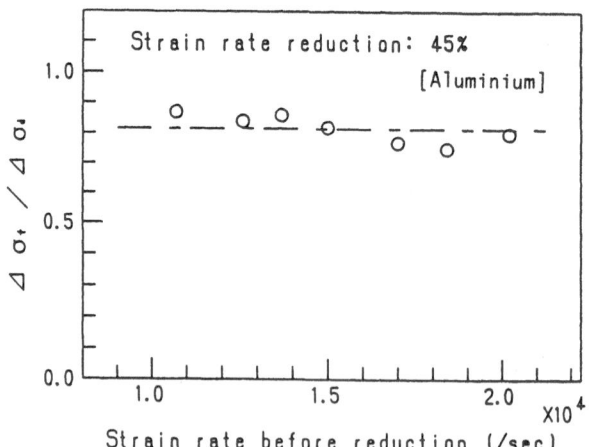

Fig. 3. Summary of results of strain rate reduction tests
in terms of SFR$= \Delta \sigma_f / \Delta \sigma_d$ (Aluminium)

2 Ultrasonic Attenuation Test

In order to obtain experimental information on the behaviour of dislocations at high strain rates, Shioiri and Satoh tried the measurement of the attenuation of the ultrasonic pulse superimposed upon the dynamic plastic deformation [11]. Continuum mechanically, this method is a strain rate change test using a small amplitude high frequency perturbation in the strain rate. Since the principle of this method as a strain rate change test and the recent developments in the experimental technique were reported in Ref.[12], in this paper only the results of the recent measurements will be described together with a necessary outline of the principle.

Using the small perturbation approximation, the attenuation of the ultrasonic wave superimposed upon the dynamic plastic deformation can be related to the dependency of the flow stress upon the instantaneous strain rate as

$$(d\sigma/d\ln\dot{\varepsilon}) = 2\dot{\varepsilon}\,\Omega\,G\,/[\,(\Delta\lambda)\,f] \tag{5}$$

where σ is the flow stress, $\dot{\varepsilon}$ the strain rate, Ω the orientation factor, G the shearing modulus, $(\Delta\lambda)$ the attenuation caused by the dynamic plastic deformation upon which the ultrasonic pulse is superimposed, f is the ultrasonic frequency. In the present work, the dynamic plastic deformation is imposed by the compressional split Hopkinson pressure bar method, and the ultrasonic pulse is sent at a right angle to the axis of the dynamic compression. Assuming the Schmid factor of the active slip system of the dynamic compression to be 0.5, Ω is given as

$$\Omega = (3E/32G)(1-2\nu)/(1-\nu^2) \tag{6}$$

where E is the longitudinal modulus and ν is Poisson's ratio.

Measurements were conducted for polycrystalline aluminium and copper at strain rates up to about 1×10^4/sec. The materials and their heat treatments are the same as those used in the strain rate reduction test. The results were reported in Ref.[9]. In Fig.4, the strain rate dependency of the flow stress $(d\sigma/d\log\dot{\varepsilon})$ calculated from the attenuation data using Eqs.(5,6) is compared

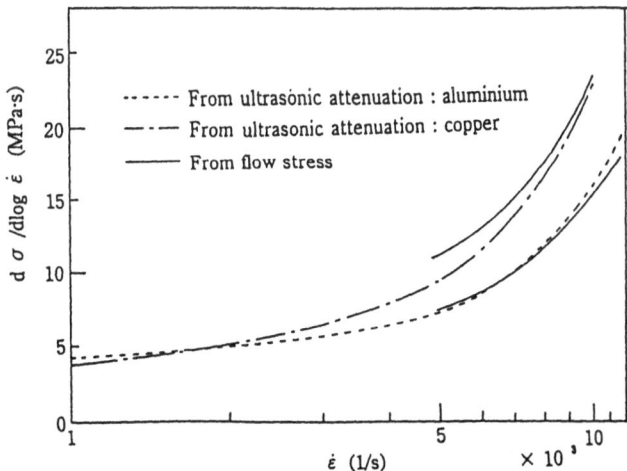

Fig.4. Strain rate dependency of flow stress calculated from attenuation data and flow stress data

with $(d\sigma/d\log\dot{\varepsilon})$ obtained directly from the flow stress measurement. Close agreements are seen in the cases of both aluminium and copper. If it is considered that, as pointed out in Ref. [12], $(d\sigma/d\log\dot{\varepsilon})$ obtained from the attenuation data does not include the component due to the strain rate history, it may be concluded that in the flow stress at very high strain rates, where the steep rise is seen, the instantaneous strain rate plays the primary role. This conclusion is the same as the conclusion of the strain rate reduction test.

4 Discussion

The results of the strain rate reduction test and the ultrasonic attenuation test, conducted at very high strain rates, show that the flow stress depends primarily upon the instantaneous strain rate and the effect of the strain rate history is secondary. This means that the strain rate dependency of the flow stress is caused mainly by the rate-controlling mechanism of the motion of the dislocation. In the following, analysis is made on the basis of the dislocation kinetics. Since there is a big difference between fcc metals and bcc metals in the mechanism of the drag upon the moving dislocation, the analysis is made for fcc and bcc metals separately.

4.1 Fcc metals

In fcc metals, since the Peierls force is negligibly small, the phonon drag is the dominant intrinsic drag upon dislocation motion. As the extrinsic drag, point obstacles to be cut through with the aid of the thermal activation should be considered. In pure metals, the forest dislocation is the dominant point obstacle to be considered. As shown schematically in Fig.5, the mobile dislocation segments move repeating the thermally assisted cutting of the forest dislocations and the phonon-drag-controlled jump motion alternately. Since the thermally assisted cutting requires a waiting time, the motion of the mobile segment in the above process is considered to be jerky. The rate of the shear strain due to the above jerky motion of the dislocation segments can be given by the following Kumar-Kumble type equation [13, 11]:

$$\dot{\gamma} = \frac{NL^2b}{t_t+t_v} = \frac{NL^2b}{\nu^{-1}\exp\{[U-Lb^2(\tau-\tau_a)]/kT\}+LB/(\tau-\tau_a)b} \tag{7}$$

where t_t is the waiting time for the thermally assisted cutting, t_v the time required by one jump motion under the control of the phonon drag, N the number of the moving segments per unit volume, L the distance between the adjacent forest dislocations, b the Burgers vector, U the activation energy of cutting, ν the frequency factor, k the Boltzmann constant, T the absolute temperature, τ the resolved shear stress, τ_a the long range athermal stress, and B is the

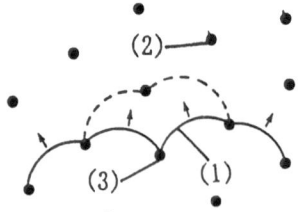

(1) moving dislocation segment

(2) forest dislocation

(3) point thermally assisted cutting occurred

Fig.5. Jerky motion of mobile dislocation segments

phonon drag coefficient (damping factor). At lower strain rates (i.e. at lower stresses) $t_i \gg t_v$ and Eq.(7) can be approximated as

$$\dot{\gamma} \fallingdotseq NL^2 b \nu \exp\{-[U-Lb^2(\tau-\tau_a)]/kT\} \tag{7'}$$

This is the thermal activation flow rate-controlled by the thermally assisted cutting of the forest dislocations, in which the flow stress depends linearly upon the logarithm of the strain rate. At very high strain rates (i.e. at higher stresses) $t_i \ll t_v$ and Eq.(7) is approximated as

$$\dot{\gamma} \fallingdotseq NLb^2 \tau/B \tag{7''}$$

This is the phonon-drag-controlled viscous flow in which the flow stress depends linearly upon the strain rate itself. As seen above, Eq.(7) can cover a wide strain rate range including the thermal activation flow range and the phonon-drag-controlled flow range together with the transition range between them.

Nowadays, it is possible to use the reliable values for the most of the physical constants and quantities in Eq.(7). In the following, fitting of Eq.(7) to the results of the constant strain rate test covering a wide range of strain rates mentioned in section 2 is made. The used values of the physical constants and quantities are as follows:

	Aluminium	Copper	Unit
G	2.67×10^{11}	4.4×10^{11}	dyn/cm^2
b	2.8×10^{-8}	2.5×10^{-8}	cm
ν	1×10^{13}	1×10^{13}	/sec
B	4.9×10^{-5} [14]	1.35×10^{-4} [14]	dyn sec/cm^2
U	$Gb^3/5$ [15]	$Gb^3/5$ [15]	

The unknown quantities in Eq.(7) are the distance between the adjacent forest dislocations, L, the density of moving dislocations, NL, and the long range athermal stress, τ_a. Those quantities will depend upon the structure formed during the deformation, and in the present treatment the values of those quantities are determined so that a good fit of the predicted curves to the experimental data is obtained. The results of the strain rate reduction tests and the ultrasonic attenuation tests show that the strain rate dependency of the flow stress is mainly due to the instantaneous strain rate and the effect of the strain rate history is secondary. This means that L, NL and τ_a should be basically independent of the strain rate, or more exactly the strain rate history, though, of course, it may depend upon the strain. Taking account of the above point, L is determined from the gradient of the σ vs. $\log \dot{\varepsilon}$ plot of the experimental flow stress in a relatively low strain rate range where σ is linearly dependent upon $\log \dot{\varepsilon}$, NL is determined from the strain rate at which the flow stress begins to rise steeply, and τ_a is determined to obtain a good fit in the stress level. The Taylor factor is assumed to be 3.07. Thus determined L, NL and τ_a are as follows:

	Aluminium			Copper			Unit
	$\varepsilon=0.1$	$\varepsilon=0.2$	$\varepsilon=0.3$	$\varepsilon=0.1$	$\varepsilon=0.2$	$\varepsilon=0.3$	
L	8.6×10^{-6}	6.7×10^{-6}	5.7×10^{-6}	6.6×10^{-6}	4.0×10^{-6}	2.9×10^{-6}	cm
NL	1.8×10^{7}	1.5×10^{7}	1.3×10^{7}	4.1×10^{7}	3.1×10^{7}	2.1×10^{7}	cm^{-2}
τ_a	9.7	13.1	15.5	47.0	64.3	70.3	MPa

The curves predicted by Eq.(7) using the above numerical values are shown in Fig.6 together with the flow stress data obtained from the constant strain

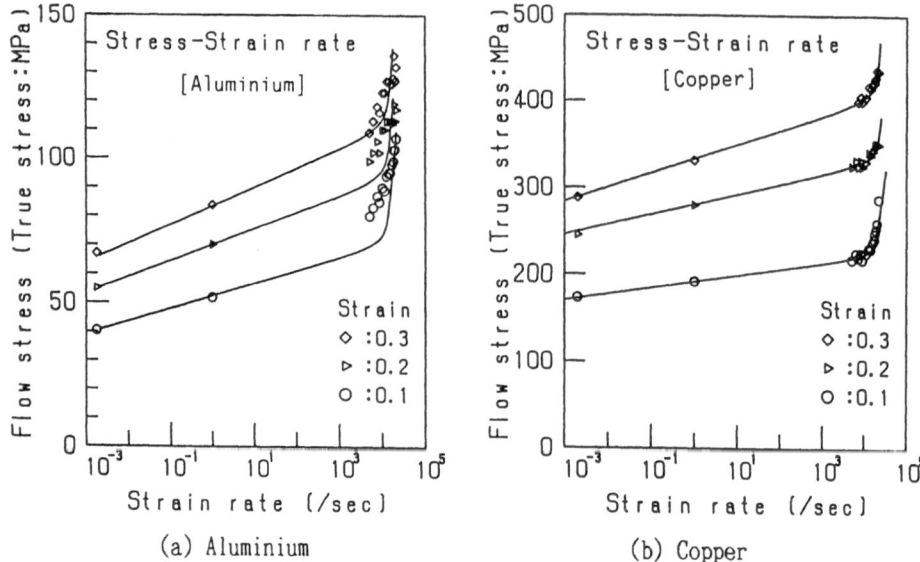

Fig.6. Strain rate dependency of flow stress:
comparison of calculated curves with experimental results

rate tests mentioned in Section 2. A fairly good agreement is seen between the predicted curves and experimental results. The steep rise seen in the predicted curves is due to the transition from the thermal activation flow to the phonon-drag-controlled flow. The agreements between the predicted curves and experimental data at this strain rate range indicates that the steep rise in the flow stress observed in the experimental data is due to the phonon drag upon dislocation motion and not due to the strain rate history dependence of the structure evolution. This conclusion is also supported by the results of the strain rate reduction test and the ultrasonic attenuation test.

4.2 Bcc metals

In bcc metals, since the Peierls resistance acting upon the screw dislocation is very large, at the beginning of the deformation only the edge component moves leaving the screw component, but, as the edge component can not multiply by itself, the movable edge component is soon exhausted. The main deformation proceeds by the screw dislocation, which is possible to multiply, overcoming the high Peierls potential barrier by generating the kink pair with the aid of the thermal activation. In bcc metals, as the stress necessary for the above process is very large compared with the stress needed by the other kinds of drag upon the dislocation, through a wide range of the strain rate, the strain rate may be given in the form of the Arrhenius equation as

$$\dot{\gamma} = \dot{\gamma}_0 \exp[-E(\tau^*)/kT] \tag{8}$$

where $\dot{\gamma}_0$ is the pre-exponential term, $E(\ )$ the apparent activation energy of kink pair generation, τ^* the effective stress $= \tau - \tau_u$, τ the applied resolved shear stress, τ_u the long range athermal stress. Aono et al. conducted the measurements of $E(\ )$ for extra pure iron and 150atppmC iron over a wide range of τ at low strain rates but by changing the temperature [16]. Since the result of the above measurements by Aono et al. shows that the effect of the carbon content upon the activation energy is small, in the

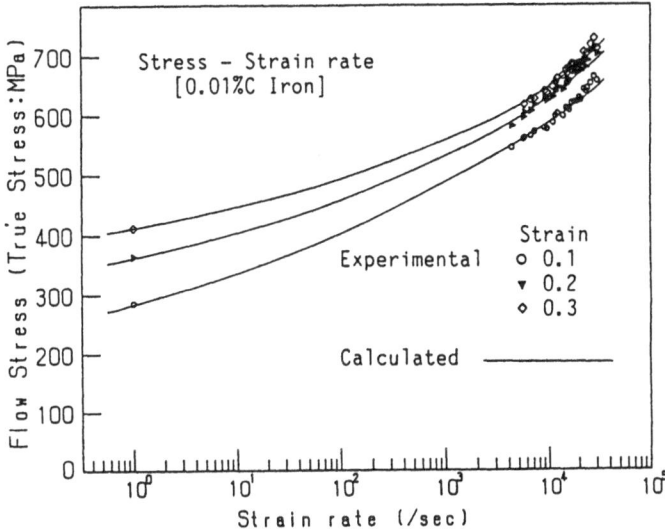

Fig. 7. Strain rate dependency of flow stress of 0.01%C Iron:
comparison of calculated curves with experimental results

comparison of Eq. (8) with the results of the present work for 0.01%C iron, the value of the activation energy obtained by Aono et al. for 150atppmC iron is used. As the flow stress of bcc metals is sensitive to the temperature, in the dynamic deformation range the correction for temperature is made assuming an adiabatic deformation. Further, confirming that the volume of the twin is about 3% at the highest strain rate, the effect of the twin is neglected. The calculated σ vs. $\log \dot{\varepsilon}$ curves are shown in Fig. 7 and compared with the results of the constant strain rate test mentioned in Section 2. In this calculation τ_a is taken to be 47.3MPa, 76.0 MPa and 93.5 MPa at ε =0.1, 0.2 and 0.3, respectively, $\dot{\gamma}_0$ to be 1.51×10^8/sec and the Taylor factor to be 2.75 ({110}<111> + {112}<111> + {123}<111> slip is assumed). As seen in this figure, a fairly good agreement is seen. The athermal stress τ_a is caused by the structure evolution in course of the deformation, but, assuming that τ_a is independent of the strain rate history, the above agreement is obtained. The rise in the flow stress at very high strain rates, being less steep than in fcc metals, is governed by the stress dependency of the activation energy of the kink pair generation. The rise is, therefore, not of structure-evolutional origin.

The experimental results obtained in the present work for niobium are very similar to the results for 0.01%C iron, and probably the same interpretation as given above for iron may be applicable.

5 Conclusion

Two strain rate change tests, the strain rate reduction test and the ultra-sonic attenuation test, were conducted at very high strain rates. The results indicate that the steep rise in the flow stress seen at strain rates above about 5×10^3/sec is caused by the rate controlling mechanism of the dislocation motion and the effect of the structure evolution depending upon the history of the strain rate is small. By analysing the above results on the basis of the dislocation kinetics, it is concluded that, in fcc metals, the steep rise is

due to the phonon drag upon moving dislocations while, in bcc metals, the rise
is a little gentler and is governed by the stress dependency of the activation
energy of the kink pair generation on screw dislocations.

References

[1] W. G. Ferguson, A. Kumar, J. E. Dorn, Dislocation Damping in Aluminum at High
    Strain Rates, *J. Appl. Phys.*, Vol. 38, 1967, pp. 1836-1869.
[2] P. S. Follansbee, U. F. Kocks, A Constitutive Description of the deformation
    of Copper Based on the Mechanical Threshold Stress as an Internal State
    Variable, *Acta Metallurgica*, Vol. 36, 1988, pp. 81-93.
[3] W. Tong, R. J. Clifton, Pressure-Shear Impact Investigation of Strain Rate
    History Effects in Oxygen-Free High-Conductivity Copper, *J. Mech. Phys.
    Solids*, Vol. 40, 1992, pp. 1251-1294.
[4] K. Sakino, J. Shioiri, Dynamic Flow Stress Response of Aluminium to Sudden
    Reduction in Strain Rate at Very High Strain Rates, *J. Physique* IV,
    Colloque C3, supplement au J. Physique III, Vol. 1, 1991, pp. C3-35-42.
[5] K. Sakino, J. Shioiri, Effect of Steep Reduction in Strain Rate on the
    Dynamic Flow Stress of Aluminium at Very High Strain Rates, *Trans. Jap.
    Soc. Mech. Eng.*, Vol. 58, 1992, No. 553 A, pp. 1703-1709. (In Japanese)
[6] K. Sakino, J. Shioiri, Effect of Steep Reduction in Strain Rate on the
    Dynamic Flow Stress of Copper at Very High Strain Rates, *Trans. Jap. Soc.
    Mech. Eng.*, Vol. 59, 1993, No. 566 A, pp. 2317-2322. (In Japanese)
[7] K. Sakino, J. Shioiri, Effect of Steep Reduction in Strain Rate on Dynamic
    Flow Stress of BCC Metals at Very High Strain Rates, *Trans. Jap. Soc.
    Mech. Eng.*, Vol. 60, 1994, No. 579 A, pp. 2561-2566. (In Japanese)
[8] J. Shioiri, K. Sakino, Roles of Instantaneous Strain Rate and Strain Rate
    History in Flow Stress at Very High Rates of Strain, *Proc. IUTAM Symp.
    Impact Dynamics*, Eds. ZHENG Zhemin, TAN Qingming, Peking Univ. Press,
    Beijing, 1994, pp. 224-233.
[9] J. Shioiri, K. Sakino, T. Santoh, Two Strain Rate Change Tests for
    Derivation of Constitutive Relationship of Materials at Very High Rates
    of Strain, *J. Physique* IV, Colloque C8, supplement au J. Physique III,
    Vol. 4, 1994, pp. C8-489-494
[10] R. Skalak, Longitudinal Impact of a Semi-Infinite Circular Elastic Bar,
    J. Appl. Mech., Vol. 24, *Trans. ASME*, Vol. 79, 1957, pp. 59-64.
[11] J. Shioiri, K. Satoh, An Ultrasonic Study of the Behaviour of Dislocations
    at Very High Rates of Strain, *Mechanical Properties at High Rates of
    Strain*, Ed. J. Harding, Conf. Ser. 70, 1984, Inst. Phys., London & Bristol
    pp. 89-96.
[12] J. Shioiri, H. Imaizumi, T. Muramatsu, Ultrasonic Study on Strain Rate
    Sensitivity in Aluminium at Strain Rates around 10000/s, *J. Physique* IV,
    Colloque C3, supplement au J. Physique III, Vol. 1, 1991, pp. C3-177-183.
[13] A. Kumar, R. G. Kumble, Viscous Drag on Dislocations at High Strain Rates in
    Copper, *J. Appl. Phys.*, Vol. 40, 1969, pp. 3475-3480.
[14] A. Hikata, R. A. Johnson, C. Elbaum, Interaction of Dislocations with Elec-
    trons and with Phonons, *Physical Review* B, Vol. 2, 1970, pp. 4856-4863.
[15] M. F. Ashby, H. J. Frost, The Kinetics of Inelastic Deformation above 0°K,
    Constitutive Equations in Plasticity, Ed. A. S. Argon, The MIT Press,
    Cambridge (Massachusetts) & London (England), 1975, pp. 117-147.
[16] Y. Aono, E. Kuramoto, K. Kitajima, Plastic Deformation of High-Purity Iron
    Single Crystals, *Rep. Res. Inst. Appl. Mech.*, Kyushu Univ., Vol. XXIX,
    No. 92, pp. 127-193.

Plastic Shearing at Very High Strain Rates, A Review

J.R.Klepaczko

Laboratory of Physics and Mechanics of Materials, ISGMP, Metz University,
Ile du Saulcy, F-57045 METZ CEDEX, France

Abstract

A review is presented on recent progress in shear testing of materials at high and very high strain rates. Some experimental techniques are discussed which allow for materials testing in shear up to 10^6 1/s. More detailed informations are provided on experimental technique based on the Modified Double Shear specimen loaded by direct impact, [14]. The role of adiabatic heating at different rates of shearing are discussed, including transition from pure isothermal to pure adiabatic deformation, [18]. Finally, a new development is discussed in determination of the Critical Impact Velocity in shear, [23],[24]. A comparison is discussed between recent experimental findings and a simple analytic estimation.

Keywords: Dynamic plasticity, Rate sensitivity, Adiabatic instability, Critical impact velocity in shear.

1. Introduction, General Remarks on Fast Shearing

Shear is the fundamental mode of plastic deformation in materials. Testing of materials in shear over a wide range of strain rate and temperature can provide a fundamental knowledge in contruction and improvement of constutive relations used nowadays in large numerical codes. The advances in experimental techniques made it possible to determine plastic properties of materials in wide range of strain rates, from 10^{-4} 1/s to 10^6 1/s, that is nine decimal orders. However, when the nominal strain rate, that is the mean strain rate over the gage length, is high enough or time of plastic deformation is relatively short (order of hundred of microseconds and less) adiabatic instability and strain localisation can occur and dominate the process of plastic deformation. At still higher strain rates in shear (impact velocities ~ 100 m/s and higher) plastic waves in a deforming material can completely change the mechanics of deformation.

On the other hand, it is well known that experiments with strain rates higher than 10^3 1/s constitute a formidable task concerning technical difficulties, and only a limited number of reliable experimental techniques provide reliable results. Each setup configuration, that is specimen geometry and type of loading, offers some advantages and defficiences in comparison to another one. Sometime the roles may be completely inverted. It is then very important to test the same materials at high strain rates with different experimental techniques and compare the results.

Most studies of both, constitutive modeling and adiabatic instabilities in shear are based so far on results from Split Hopkinson Torsion Bar technique (torsional Kolsky apparatus),[1],[2]. This experimental technique is quite reliable within a narrow range of strain rates, typically from $3*10^2$ 1/s to $2*10^3$ 1/s. A thin-walled tubular specimens of short length, from 2 to 5 mm, and wall thickness from 0.5 to 1.0 mm, are loaded by the incident shear wave. The incident, reflected and transmitted torsional waves are analysed in a similar way as in the Split Hopkinson Pressure Bar, [3]. The SHTB technique has been modified later to perform incremental/decremental strain rate tests (jump tests), [4].

The thin-walled tubular specimens were used much earlier to test materials at different strain rates with torsion machines and also at impact shearing, typically up to 10^2 1/s, for example [5,6,7].

The SHPB technique (original Kolsky apparatus) can also be used in shear by loading of specimens of special geometries. One of such techniques is so called "hat" specimen, for example [8],[9]. Because of a non-uniform deformation of the radial field within the shear zone in the "hat" specimen, determination of the shear stress vs. shear strain is very difficult and needs an application of an FE technique. This technique is relatively well suited for a study of controlled initiation of Adiabatic Shear Bands, [9].

A unique experimental technique is the pressure-shear plate configuration described in [10]. This technique is an attractive configuration for studying dynamic plastic flow at shear strain rates from 10^5 1/s to 10^6 1/s. In this experiment, the strain rate of the order 10^5 1/s is achieved by sandwiching a very thin specimen between two hard elastic plates. The specimen is attached to the flyer and launched with velocity of a few hundred meters per second. The flyer impacts in a vacuum at a small angle the stationary anvil. Because of the inclined impact the specimen is deformed in shear with relatively high component of pressure. The elastic wave profiles, shear and normal components, which are transmitted by the specimen and the anvil plate are recorded on the free external surface of the anvil. The wave analysis permits to find shear stress vs. shear strain characteristics at strain rates in excess of 10^5 1/s. Characterisation of marerials in shear at strain rates below 10^5 1/s must be complemented by a different experimental techniques, for example SHTB. Those two techniques are quite different, including way of loading and specimen geometries.

Because the effects of wave propagation in a specimen , every change of the specimen geometry introduces changes in the specimen respose to fast or impact loading. The best solution is to use *one* specimen geometry for the widest possible spectrum of strain rates, for example from 10^{-4} 1/s to 10^5 or even 10^6 1/s.

An effective specimen geometry which can be used in studies of dynamic plasticity at low, medium and high strain rates is the double-notch specimen. Such a specimen was first introduced in [11] to study dynamic plasticity of single crystals. Later the Double Shear specimen was applied, with the loading scheme consistig of the incident Hopkinson bar and transmitter Hopkinson tube, [12], to study the temperature and strain rate dependence of the yield stress of mild steel. A very small gage length of 0.84 mm was used in the latter study [12]. Because of a small gage length of the original DS specimen, application of the conventional mass velocities in the incident bar, from ~1.2 m/s to ~11 m/s provided the nominal strain rates in shear up to ~10^4 1/s. Determination of higher strains than a few percent from the DS specimen geometry, by use of the method of the net displacement between the Hopkinson bar and Hopkinson tube,[12], leads to large errors due to a non-uniform shear and severe plastic deformation of the specimen supports.

In order to study dynamic plasticity within wide range of strain rates and at large strains the concept of the double shear test was completely modified to load specimens of the same geometry at very different velocities, from quasi-static to impact.

2. Direct- Load Modified Shear Test

This relatively new experimental technique was briefly described in [13] and a complete outline is given in [14]. It combines some advantages in comparison to the original DS technique, [11],[12]. The scheme of the impact loading is shown in Fig.1. The fundamental change in comparison to the bar-tube configuration has been introduced by elimination of the incident tube and application of the direct impact. On the other hand, specimen geometry is different, the shearing zone has been enlarged to 2.0 mm and the external parts enforced, so that to eliminate plastification of the support regions. The Modified Double Shear specimen is shown in Fig.2. Since the direct impact has been applied the risetime of the incident wave has been reduced to ~2 μs. The flat-ended projectiles of different lengths made of maraging steel and of diameter $d_p = 10$ mm are launched from an air gun with desired velocity V_0 ; $1.0 < V_0 < 200$ m/s. Impact velocity is measured by three optic chaines : source of light, photodiode and input/output optic fibers. □Signals from the photodiodes are recorded by two time counters. The setup with three light axes makes it possible to determine acceleration/deceleration of a projectile just before impact, so the exact value of V_0 at the impact face of the MDS specimen can be found.

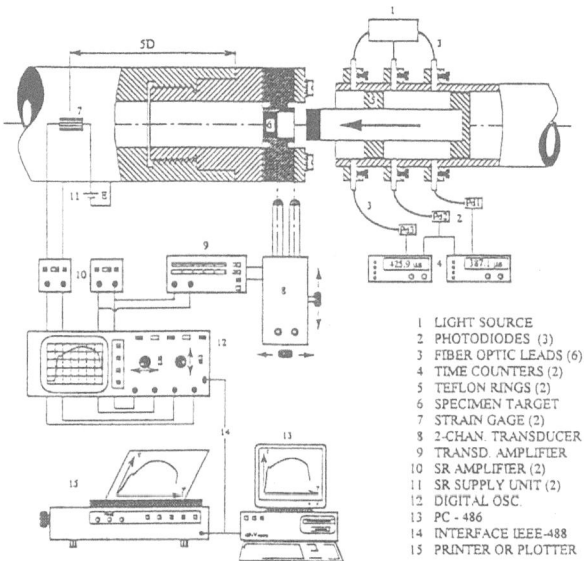

1 LIGHT SOURCE
2 PHOTODIODES (3)
3 FIBER OPTIC LEADS (6)
4 TIME COUNTERS (2)
5 TEFLON RINGS (2)
6 SPECIMEN TARGET
7 STRAIN GAGE (2)
8 2-CHAN. TRANSDUCER
9 TRANSD. AMPLIFIER
10 SR AMPLIFIER (2)
11 SR SUPPLY UNIT (2)
12 DIGITAL OSC.
13 PC - 486
14 INTERFACE IEEE-488
15 PRINTER OR PLOTTER

Figure 1: Configuration of experimental setup for direct impact on the MDS specimen, [14].

Axial displacement $\delta_A(t)$ of the central part of the specimen is measured as a function of time by an optical transducer, acting as a non-contact dislpacementy gage. Since the double channel transducer is used the second channel controls displacement $\delta_F(t)$ of the impact face of the MDS specimen. Measurements of V_0 and $\delta_F(t)$ makes it possible to determine the coefficient of restitution for each test. A black and white target is cemented on the side of the MDS specimen and at the same time the impact end of the projectile is black, in this way the non-contact displacement transducer reacts also to the movement of the impact face.

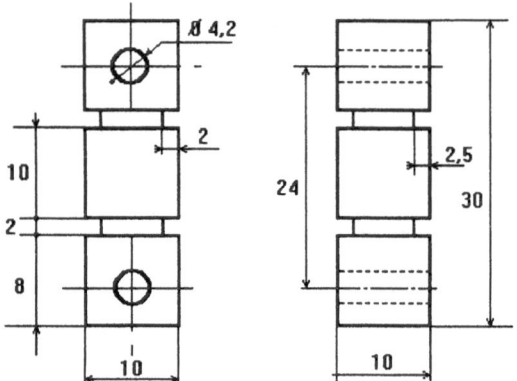

Figure 2: Geometry of the Modified Double Shear specimen, [14]

The axial force transmitted by the specimen symmetric supports into the Hopkinson tube can be determined as a function of time, F(t), from the transmitted longitudinal elastic wave $\varepsilon_t(t)$. The transmitted wave $\varepsilon_t(t)$ is measured by strain gages 7, DC supply units 11 and amplifier 10. All electric signals, that is voltages of displacements $\delta_A(T)$ and $\delta_F(t)$ and transmitted wave $\varepsilon_t(t)$ are recorded by the digital oscilloscope 12 and stored later in the PC hard disk for further analyses. After an analysis of recorded signals and elimination of time, a force-net displacement curve $F(\delta)$ can be constructed for each test and $\tau(\Gamma)$ and also $\dot{\Gamma}(\Gamma)$ characteristics determined, where τ is the shear stress, Γ strain, and $\dot{\Gamma} = d\Gamma/dt$ is the strain rate. The complete theory of the test is given in [14].

The experimental technique based on the direct impact on the MDS specimen has appeared to be quite effective and flexible in materials testing in shear at high strain rates, 10^3 1/s $< \dot{\Gamma} <$ 10^5 1/s. In addition, a special device has been constructed which permits for loading of the MDS specimen at low and medium strain rates, 10^{-4} 1/s $< \dot{\Gamma} < 5*10^2$ 1/s. A fast, closed-loop hydraulic testing machine is used together with this device, [13]. Thus, the experimental technique based on the MDS specimen assures a wide spectrum of strain rates, typically from 10^{-4} 1/s to 10^5 1/s .

Several alloys, mostly variety of steels, were tested by this method, almost all of them show a very high rate sensitivity above 10^3 1/s.

In materials testing at very different strain rates, additional effects like transition from isothermal to adiabatic deformation, adiabatic instability and localisation , and finally deformation wave trapping complicate interpretation of the experimental results. Some of those features are adressed in the next part of this review.

3. Adiabatic Heating, Instability and Localisation

It is well known that during plastic deformation of materials a large part of plastic work is converted into heat, [15]. When deformation is slow all heat generated is evacuated into sourroundings by heat diffusion or by direct emission. However, when the process of deformation is short enough, there is no time for evacuation and almost all plastic work is converted locally into volume heating. In certain range of strain rates the transition occurs between entirely isothermal and entirely adiabatic mode of plastic deformation. Some preliminary study on this subject were reported in [16] and [17].

Because this transition depends, in the first place, on geometry of the deformed body and also on the intensity of the heat extraction from the heated zones, a numerical analysis is very helpfull in determination in which region of strain rates the transition occurs. When the thickness of deformed layer is ~2.0 mm, like in the MDS specimen, the transition occurs at the following critical strain rates: copper ~85 1/s, aluminum ~68 1/s and a mild steel ~48 1/s [18].The finite differences technique has been applied, and the calculations were carried out at T_0 = 300 K, the initial temperature. The results support quantitatively the physical intuition that the critical strain rate increases in proportion to the thermal conductivity, which is the lowest for steel and the highest for copper. Since those values are based on the maximum temperature gradients, a complete transition into the entirely adiabatic conditions occurs at strain rates at least 5 1/s higher. Variations in the specimen geometry change also the transition region, [18].

Thermal softening during adiabatic heating leads directly to instability and localization, for example early works, [17,19,20]. In the case of pure shear the condition for stablity is reduced to maximum shear stress $d\tau = 0$, where τ is the shear stress, [19]. This condition can be rewritten into the form of zero tangent modulus

$$(\frac{dT}{d\Gamma})_A = 0 \text{ or } \theta_A(\Gamma_c) = 0 \tag{1}$$

where $d\tau/d\Gamma = 0$ is determined at $\Gamma=\Gamma_c$, the instability strain. A schematic stress-strain curves showing different stages of deformation during fast (adiabatic) shearing is shown in Fig.3.

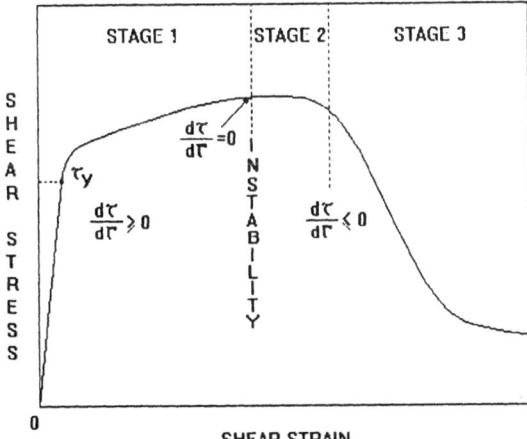

Figure 3:   Schematic stress-strain curve showing different stages of fast (adiabatic) shearing, τ_y is the yield stress and $d\tau/d\Gamma = 0$ is the instability point.

All stages shown in Fig.3 were identified by a high-speed photography during deformation of a tubular specimen made of HY-100 steel loaded in SHTB, [21]. During the first stage of deformation, that is in between the yield point (τ_y, Γ_y) and the instability point (τ_c, Γ_c) where $d\tau/d\Gamma = 0$, strain distribution over the gage length is homogeneous. During stage 2 shown in Fig.3 the strain distribution becomes to be inhomogeneous, a localization begins. Finally, during stage 3 a catastrophic shear localization occurs and flow stress drops rapidly, the Adiabatic Shear Band is well formed. However, during this stage the high-speed photography revealed some circumferential nonuniformities in formation of the ASB, [21]. The latest part of the stage 3 shows, typically this is observed for more ductile metals, a slowdown of the rate of the flow stress. Such behavior is caused by cooling of a very thin layer of material situated very closely to the ASB, [17].

The instability point (τ_c, Γ_c) can be found for different constitutive relations using eq.(1), for example [22]. Since in many theoretical studies the constitutive relation of the following structure is used

$$\tau = f_1(T)f_2(\Gamma)f_3(\dot{\Gamma}) \tag{2}$$

where f_1 is related to temperature T, f_2 is related to strain hardeneing and f_3 to strain rate, a more general analysis of the condition (1) have been discussed in [13]. Here, only the final result is given. Application of the condition for stability (1), and the form of eq.(2) for constitutive modeling, leads to the following result when the adiabatic process of deformation is assumed together with a constant strain rate, [13],

$$f_2(\Gamma_c) = \left[-\frac{(\partial f_2/\partial \Gamma)}{(\partial f_1/\partial T)} \frac{\rho(T)C_p(T)}{f_3(\dot{\Gamma})(1-\xi)} \right]^{1/2} \tag{3}$$

where the derivative $(\delta f_2/\delta \Gamma)$ is related to the tangent modulus (strain hardeneing) at constant strain rate, $(\delta f_1/\delta T)$ is related to the temperature sensitivity of flow stress at constant strain rate (thermal softening), ρ is the mass density, C_v is the specific heat and ξ is the coefficient of stored energy, [15]. Inversion of $f_2(\Gamma_c)$ permits to find in the explicit form the instability strain Γ_c. The expression in the square brackets $f_2(\Gamma_c) = [*]^{1/2}$ has real and imaginary parts. Existence of the real $f_2(\Gamma_c)$ is possible only if the argument in the expression $[*]^{1/2}$ is negative. Since ρ, C_v and $(1-\xi)$ must be always positive, and the expression which accounts for strain rate sensitivity $f_3(\Gamma)$ is also assumed positive, the only term which may be negative is $(\delta f_2/\delta \Gamma)/(\delta f_1/\delta T)$. The most common case is the thermal softening where $(\delta f_1/\delta T)$ is negative, of course if $(\delta f_2/\delta \Gamma)$ is at the same time positive. The role of a positive rate sensitivity is quite interesting, that is if $f_3(\Gamma)$ is an increasing function of strain rate $\dot{\Gamma}$. The positive rate sensitivity, that is an increase of the flow stress when the strain rate is increased, has a negative effect on the onset of adiabatic instability, that is Γ_c is reduced when strain rate is increased. The positive rate sensitivity increases production of plastic work converted into heat, this process accelerates formation of the instability, so values of Γ_c is a diminishing function of the nominal strain rate. Several industrial steels tested so far in shear at strain rates from 10^{-3} to $\sim 10^5$ 1/s with the MDS method,[14], showed a substantial evolution of the stress-strain curve. Schematic changes of the $\tau(\Gamma)$ curves observed for many industrial steels at increasing rates are shown in Fig.4.

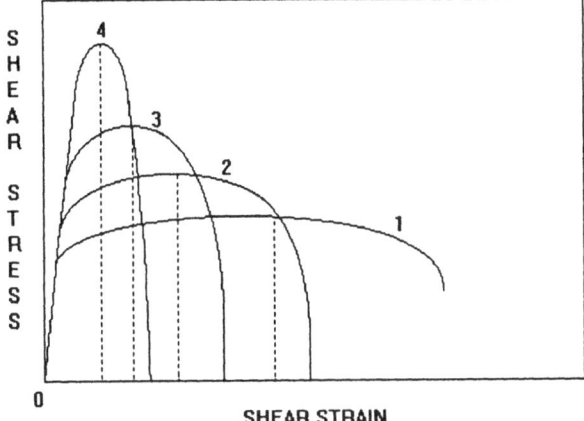

Figure 4: Schematic evolution of stress-strain curves of industrial steels at strain rates higher than $5*10^2$ 1/s; (1) ~$5*10^2$ 1/s ; (2) ~$5*10^3$ 1/s ; (3) ~$1*10^4$ 1/s ; (4) ~$5*10^4$ 1/s.

In this schematic figure the mean proportions are conserved to show the effect of the very high rate sensitivity on the yield stress and instability stress above ~10^3 1/s. On the contrary to the positive rate sensitivity of the crtical stress τ_c, the critical strains of instability Γ_c always show a tendency to diminish when the strain rate is above ~10^3 1/s. The same happens with the fracture strains. At very high strain rates this feature is even more abrupt, this occurs because of a superposition of ASB formation and plastic waves in shear.

4. Critical Impact Velocity in Plastic Shearing

It has been shown recently, [23] and [24], that during shear deformation imposed by a high-velocity, plastic waves excited in a deformed material can completely change the mechanics of plastic field. As a rule, an intense plastic deformation will appear near the impact end of a specimen. In the case of the MDS specimen with the gage length 2.0 mm the nominal strain rate when the plastic waves start to dominate is $5*10^4$ 1/s, that is the velocity of shearing is ~ 100 m/s.

Since formulation of the rate independent theory of elasto-plastic waves in solids by Karman, Taylor and Rakhmatulin, in late forties and early fifties, for a review see [25], it is known that the longitudinal plastic deformation can be localised in thin bars by a high velocity impact,[26]. This deformation trapping by longitudinal plastic waves is called the Critical Impact Velocity in tension,[26] and [27].

It has been shown recently that the Critical Impact Velocity in shear can be determined experimentally using the MDS specimen and the direct impact loading, [23] and [24]. An existence of the CIV in shear was predicted by FE numerical method in [28]. A more detailed analytic study was published elsewhere,[24], here only a brief discussion is offered.

It is clear that the CIV in shear is closely related with adiabatic heating and thermal softening. The CIV in shear is caused by an instantaneous instability and strain localisation superimposed on plastic wave propagation in shear. When the rate-independent wave propagation theory is applied to analyse CIV in shear the following formula is derived, [24]

$$V_c = \int_0^{\Gamma_c} C_2(\Gamma) d\Gamma \qquad (4)$$

where C_2 is the wave speed in shear and Γ_c is given by eq.(3). The wave speeds in shear are defined in the rate-inependent wave propagation theory as follows:
in the elastic range

$$C_2 = (\frac{\mu}{\rho})^{1/2} \qquad (5)$$

and in the plastic range

$$C_2(\Gamma) = (\frac{\theta(\Gamma)}{\rho})^{1/2} \qquad \text{where} \qquad \theta(\Gamma) = (d\tau / d\Gamma) \qquad (6)$$

If constitutive relation in the form of eq.(2) is used then the final formula for CIV in shear is obtained,[24]

$$V_c = \int_0^{\Gamma_c} \left\{ \frac{f_1 f_2 f_3}{\rho} \left[\frac{1}{f_2} \left(\frac{\partial f_2}{\partial \Gamma} \right) + f_2 f_3 \left(\frac{\partial f_1}{\partial T} \right) \frac{1 - \xi}{\rho} \right] \right\}^{\frac{1}{2}} d\Gamma \qquad (7)$$

the upper limit of integration is defined by eq.(3) in the inversed form $\Gamma_c = f_2^{-1}(*)$. Preliminary numerical calculations of the CIV for 1018 steel (French Standard XC18) have confirmed the usefulness of the analytic procedure presented above. The value of CIV obtained using formulas (4) and (7) with a simplified constutitve relation

$$\tau = BT^{-v} (\Gamma_0 + \Gamma)^n \dot{\Gamma}^m \qquad (8)$$

was $V_{cr} = 98.0$ m/s and $\dot{\Gamma}_c = 4.9*10^4$ 1/s for 2.0 mm gage length, the value very close to the one determined by the MDS direct impact technique, $V_{cr} \cong 90$ m/s. The experimental technique based on the MDS specimen and direct impact has been applied so far for determination of the CIV for three steels. Besides XC18 mild steel, the CIV determined for VAR4340 steel (52 HRC) was $\cong 140$ m/s and in the case of hot-rolled C-Cr steel $\cong 60$ m/s, [24].

5 Discussion and Conclusions

It has been shown in this review that materials testing in shear within wide range of strain rates, from quasi-static to impact and up to 10^5 1/s, is possible using the MDS specimen loaded at moderate rates or by a direct impact. Since duration of experiments occurs at very different time span, additional effects like transition from isothermal to adiabatic deformation, adiabatic instability and localisation, and finally CIV in shear change the deformation process. In each range of strain rates different processes may dominate. Indeed, the thermal coupling associated with heat production due to plastic deformation and heat conduction play an important role in dynamic plasticity during shearing. Characteristic values of the CIV in shear can be proposed as a new material constant.

Acknowledgement

This review was supported, and some results in the references cited here, in part by CNRS-DRET Program N° 972 "Impact on Materials".

References

[1] J.Duffy, J.D.Campbell and R.H.Hawley, On the Use of a Torsional Split Hopkinson Bar to Study Rate Effects in 1100-0 Aluminum, *J. Appl. Mech.*, Vol. 38, 1971, p.89.

[2] J.L.Lewis and J.D.Campbell, The Development and Use of a Torsional Hopkinson Bar Apparatus, *Exp. Mech.*, Vol.12, 1972, p.520.

[3] H.Kolsky, An Investigation of the Mechanical Properties of Materials at Very High Rates of Loading, *Proc. Phys. Soc., London*, Vol. 62-B, 1949, p. 676.

[4] R.A.Frantz and J.Duffy, The Dynamic Stress-Strain Behavior in Torsion of 1100-0 Aluminum Subjected to a Sharp Increase in Srain Rate, *J. Appl. Mech.*, Vol.39, 1972, p.939.

[5] J.R.Klepaczko, Rate Sensitivity and Strain Rate History Effects in Technically Pure Aluminum, *Ph. D. Thesis, Institute of Fund. Techn. Res., Polish Acd. Sci.*, Warsaw, 1965.

[6] J.R.Klepaczko, A New Device for Dynamic Torsional Tests, *Theoretical and Appl. Mech.*, Vol.5, 1967, p.198.

[7] J.R.Klepaczko, The Strain Rate Behavior of Iron in Pure Shear, *Int. J. Solids and Struct.*, Vol.5, 1969, p.533.

[8] K.H.Hartman, H.D.Kuntze and L.W.Meyer, High Strain Rate Deformation of Steel, *Shock Waves and High Strain rate Phenomena in Metals*, Plenum Press, N.Y., 1981, p.325.

[9] J.H.Beatty, L.W.Meyer, M.A.Meyers and S.Nemat-Nasser, Formation of Controlled Adiabatic Shear Bands in AISI 4340 High Strength Steel, *Shock Waves and High Strain Rate Phenomena in Metals*, Plenum Press, N.Y., 1992, p.645.

[10] R.J.Clifton and R.W.Klopp, Pressure-Shear Plate Impact Testing, *Metals Handbook*, Vol.8, ASM, Metals Park, 1985, p.230.

[11] W.G.Fergusson, F.E.Hauser and J.E.Dorn, Dislocation Damping in Zinc Single Crystals, *Brit. J. Appl. Phys.*, Vol.18, 1967, p.114.

[12] J.D.Campbell and W.G.Fergusson, The Temperature and Strain-Rate dependence of the Shear Strength of Mild Steel, *Phil. Mag.*, Vol.21, 1970, p.63.

[13] J.R.Klepaczko, Adiabatic Shear Bands, Review of Experimental Techniques and Results *Anniversary Volume LMA, Mechanics, Numerical Modelling and Dynamics of Materials*, CNRS Marseille, 1991, p.335.

[14] J.R.Klepaczko, An Experimental Technique for Shear Testing at High and Very High

Strain Rates, the Case of Mild Steel, *Int. J. Impact Engng.*, Vol.15, 1994, p.25.

[15] G.I.Taylor and H.Quinney, The Latent Energy Remaining in a Metal After Cold Working, *Proc. Roy. Soc., London,* Vol.143A, 1934, p.307.

[16] M.Kaminski, Coupling of Strain and Temperature fields in the Problem of Torsion of a Thin-Walled Tube, *Engng. Trans.,* Vol.24, 1976, p.185; (in Polish).

[17] J.Litonski, Numerical Analysis of Plastic Torsion Process With Account of a Heat Generated During Deformation, *IFTR Reports, N° 33,* Warsaw, 1985.

[18] O.Oussouaddi and J.R.Klepaczko, An Analysis of Transition Between Isothermal and Adiabatic Deformation For the Case of Torsion of a Tube, *Proc. 3rd Int. Conf. on Mech. and Phys. Behaviour of Materials Under Dyn. Loading,* Les éditions de physique, Les Ulis, 1991, p.C3-237.

[19] H.C.Rogers, Adiabatic Plastic Deformation, *Ann. Rev. Mat. Sci.,* Vol.9, 1979, p. 238.

[20] C.Zener and J.H.Hollomon, Effect of Strain Rate Upon Plastic Flow of Steel, *J. Appl. Phys.,* Vol.15, 1944, p.22.

[21] A.Marchand and J.Duffy, An Experimental Study of the Deformation Process of Adiabatic Shear Band Formation in a Structural Steel, *J. Mech. Phys. Solids,* Vol.36, 1988, p.251.

[22] R.Dormeval, The Adiabatic Shear Phenomenon, *Materials at High Strain Rates,* Elsevier Appl. Sci., London, 1987, p.47.

[23] J.R.Klepaczko, Plastic Shearing at High and Very High Strain Rates, *Proc. Conf. on Mech. and Phys. Behaviour of Materials Under Dyn. Loadings,* Les éditions de physique, Les Ulis, 1994, p.C8-35.

[24] J.R.Klepaczko, On the Critical Impact Velocity in Plastic Shearing, *EXPLOMET'95, Proc. Int. Conf. On Metallurgical and Materials Applications of Shock Waves and High Strain Rate Phenomena,* Elsevier Science, Amsterdam, 1995; (in print).

[25] N.Cristescu, *Dynamic Plasticity,* North-Holland Publ.Co., Amsterdam, 1967.

[26] Th. Kàrmàn and P.Duwez, The Propagation of Plastic Deformation in Solids, *J. of Appl. Phys.,* Vol.21, 1950, p.987.

[27] J.R.Klepaczko, Genaralized Conditions for Stability in Tension Tests, *Int. J. Mech. Sci.,* Vol.10, 1968, p.297.

[28] F.H..Wu and L.B.Freund, Deformation Trapping Due to Thermoplastic Instability in One Dimensional Wave Propagation, *J. Mech. Phys. Solids,* Vol.32, 1948, p. 119.

A Strain-Rate and Temperature Dependent Constitutive Model with Unified Sets of Parameters for Various Metallic Materials

Koji Mimura and Shinji Tanimura

Department of Mechanical Systems Engineering, University of Osaka Prefecture
1-1 Gakuen-cho, Sakai, Osaka 593, JAPAN

Summary

A simple but practical constitutive model describing strain rate and temperature dependencies of flow stress by unified sets of material parameters has been developed. In the present model, a hyperbolic relationship between quasistatic yield stresses and gradients of normalized strain rate sensitivity is assumed for the thermally activated mechanism, while a strain-rate power-law equation is employed for the viscous drag mechanism. Temperature dependence of flow stress and strain rate history effect are also taken into account in a simple form. Based on the proposed constitutive model, stress-strain relations of several metallic materials at high strain rates are simulated, and a representation of effect of strain rate, strain rate history and temperature on mechanical behavior is discussed.

Keywords: Rate-sensitive constitutive model, strain rate sensitivity, strain rate history effect, temperature dependence of flow stress.

1. Introduction

Most metallic materials, to a greater or lesser extent, show a significant dependence of flow stress on applied strain rates and temperatures. Such strain rate and temperature dependence of flow stress should be modeled properly if problems associated with dynamic loading under various temperatures are to be precisely analyzed and if structures subjected to dynamic load are to be safely designed. A number of attempts have been made to develop strain rate- or temperature-dependent constitutive equations. One of the basic constitutive models for elasto/viscoplastic body at high rates of strain had been given by Malvern [1], and then it was extended to three dimensional stress state by Perzyna [2]. In order to take advantage of these models, however, we have to find the concrete form of constitutive equations which represent the relation between flow stress and inelastic strain rates either theoretically or phenomenologically. Many excellent reviews have been made concerning the practical forms of constitutive equations; a review based on the thermal activation of dislocation motion has been given by Campbell [3]; an inclusive review by Perzyna [4] and a comprehensive review by Duffy [5] emphasizing the strain rate history effects; Klepaczko [6] has made a review on recent development in constitutive modeling for metallic materials; Harding [7] has given an inclusive review containing the mechanical behavior of composite and non-metallic materials. In this paper, a phenomenological constitutive equation which is very simple, but can describe mechanical behavior of various metals and alloys at high rates of strain by a unified set of material constants, is presented.

2. Constitutive Model

In order to describe the wide range of strain rate sensitivity, the following relation between the flow stress and the inelastic strain rate [8] is assumed.

$$\sigma = \sigma_s(\varepsilon^p, \ T) + C(\varepsilon^p, T)\ln\left(\frac{\dot{\varepsilon}^p}{\dot{\varepsilon}_s}\right) + B\left(\frac{\dot{\varepsilon}^p}{\dot{\varepsilon}_u}\right)^m \tag{1}$$

where, σ_s is the quasistatic yield stress as a function of inelastic strain ε^p and absolute temperature T, C is the strain rate sensitivity gradient (the tangential modulus of the excess stress $\sigma - \sigma_s$ and logarithm of inelastic strain rate $\ln(\dot{\varepsilon}^p)$ diagram), B and m are material constants, $\dot{\varepsilon}_s$ is the reference quasistatic strain rate and $\dot{\varepsilon}_u$ is the unit strain rate (=1/sec). Obviously, the second term on the right hand side of eq.(1) represents the effect of thermal activation of dislocation motion on the flow stress, while the third term represents the effect of viscous-drag on the flow stress.

Now, the interest is focused on the representation of the function C in eq.(1). In the ordinary identification process, the values of the function C are determined for individual materials, and this process requires a great number of experimental data. If we could find a simple and general relation which gives the value of C at an arbitrary strain and temperature for the wide range of materials by a single set of material constants, the time and effort may be gratefully saved.

In actual fact, there is a very interesting relation between the quasistatic yield strength and the strain rate sensitivity gradient. Now, let us consider the normalized strain rate sensitivity gradient $\tilde{C}$ as follows.

$$\tilde{C} = \frac{C}{\sigma_s} = \frac{(\sigma - \sigma_s)}{\sigma_s \ln(\dot{\varepsilon}^p/\dot{\varepsilon}_s)} \tag{2}$$

Figures 1 (a),(b) and (c) show relations between the normalized strain rate sensitivity gradients $\tilde{C}$ and the quasistatic yield strengths σ_s for many bcc, fcc and hcp metals at various strain states. Here, it should be noted that experimentally derived data in these figures are obtained by tensile or compression tests, not by torsional nor shear tests. Obviously, hyperbolic relations between $\tilde{C}$ and σ_s are observed in the figures. Namely, the strain rate sensitivities of metallic materials decrease with increasing quasistatic yield strength. Furthermore, it is found that for bcc metals (steels), these curves move toward the origin of the figure as deformation progresses, while for fcc (Al, Cu and their alloys) and hcp (α-titanium) metals, they move toward the upper-left corner as inelastic strain increases. These facts indicate that the strain rate sensitivity of bcc metals, in general, decreases with increasing strain, on the contrary, those of fcc and hcp metals increase with increasing strain. In spite of these differences between bcc, fcc and hcp metals, the important fact is that there is a possibility to represent the strain rate sensitivities of metallic materials whose lattice structures are of the same type by a single set of material constants, because the experimental points at the same strain state are on the same hyperbolic line. Taking account of these facts, we assume the following relation to represent the dependence of normalized strain rate sensitivity gradient on quasistatic yield strength.

$$\tilde{C} = \frac{(\sigma - \sigma_s)}{\sigma_s \ln(\dot{\varepsilon}^p/\dot{\varepsilon}_s)} = K(\varepsilon^p, \ T)\left(\frac{1}{\sigma} - \frac{1}{\sigma_{cr}}\right) \tag{3}$$

where σ_{cr} is the critical value of the quasistatic yield stress at which materials are no longer intensified by the strain rate effect, and $K(\varepsilon^p, T)$ is the coefficient that represents the dependence of the normalized strain rate sensitivity on the inelastic strain and the temperature defined as:

$$K(\varepsilon^P, T) = \left(\alpha \varepsilon^P + \beta\right)\left(\frac{T_R}{T}\right) \tag{4}$$

where α and β are material constants, T_R is a reference temperature. Substituting eqs.(3) and (4) into eq.(1), then we can obtain the following relation of the form:

$$\sigma = \sigma_s(\varepsilon^P, T) + (\alpha \varepsilon^P + \beta)\left(\frac{T_R}{T}\right)(1 - \sigma_s / \sigma_{cr})\ln\left(\frac{\dot{\varepsilon}^P}{\dot{\varepsilon}_s}\right) + B\left(\frac{\dot{\varepsilon}^P}{\dot{\varepsilon}_u}\right)^m \tag{5}$$

For further modification of the constitutive equation, we consider the strain rate history effect as well as the strain rate effect here. Referring to the formulation given by Campbell [6], we rewrite eq.(5) as follows:

$$\sigma = \sigma_s(\varepsilon^P, T) + B\left(\frac{\dot{\varepsilon}^P}{\dot{\varepsilon}_u}\right)^m$$
$$+ (1 - \sigma_s / \sigma_{cr})\left(\frac{T_R}{T}\right)\left[(\alpha \varepsilon^P + \beta)\ln\left(\frac{\dot{\varepsilon}_1^P}{\dot{\varepsilon}_s}\right) + \left\{\alpha(\varepsilon^P - \varepsilon_c^P) + \beta\right\}\ln\left(\frac{\dot{\varepsilon}_2^P}{\dot{\varepsilon}_1^P}\right)\right] \tag{6}$$

where ε_c^P is the strain at which the strain rate jump occurs, $\dot{\varepsilon}_1^P$ is the initial inelastic strain rate and $\dot{\varepsilon}_2^P$ is the subsequent inelastic strain rate. Of course, such a simple relation may express the results of strain rate jump tests insufficiently, however, the experimental tendencies such as positive strain rate history effects for fcc metals and negative ones for bcc metals are well described.

3. Identification of Material Constants

The value of σ_{cr} in eq.(6) may be discussed in conjunction with that of the ideal critical shear stress τ_c in metallic materials. According to the dislocation theory, the values of τ_c in pure metals are comparable to $1/20 \sim 1/30$ of the rigidity G. In other words, the value of σ_{cr} is within the range from $\sqrt{3}G/20$ to $\sqrt{3}G/30$ if von Mises yield criterion is assumed. However, the exact determination of

Table 1 Material constants used in the model

| Lattice structure | BCC | FCC | | HCP |
		Al and alloy	Cu and alloy	
σ_{cr}	4.0 GPa	1.5 GPa	1.5GPa	2.5 GPa
α	- 45.6	5.50	8.77	17.5
β	18.4	1.19	5.87	13.16

σ_{cr} is still difficult because actual metals are not always pure, and the effects of inclusions on strain rate sensitivity has clarified insufficiently. In this research, therefore, the values of σ_{cr} are determined by means of the least square method. Obtained values for bcc, fcc and hcp metals are shown in Table 1. It is found that identified values of σ_{cr} are of order of $1/50 \sim 1/30$ of $\sqrt{3}G$.

Once σ_{cr} was identified, the determination of α and β in eq.(6) is not difficult. Figures 2 (a), (b) and (c) indicate the relations between the coefficient K in eq.(3) and inelastic strain ε^P for bcc, fcc and hcp metals. In the figures, the linear dependence of K on ε^P is clearly shown. By using the least square

72

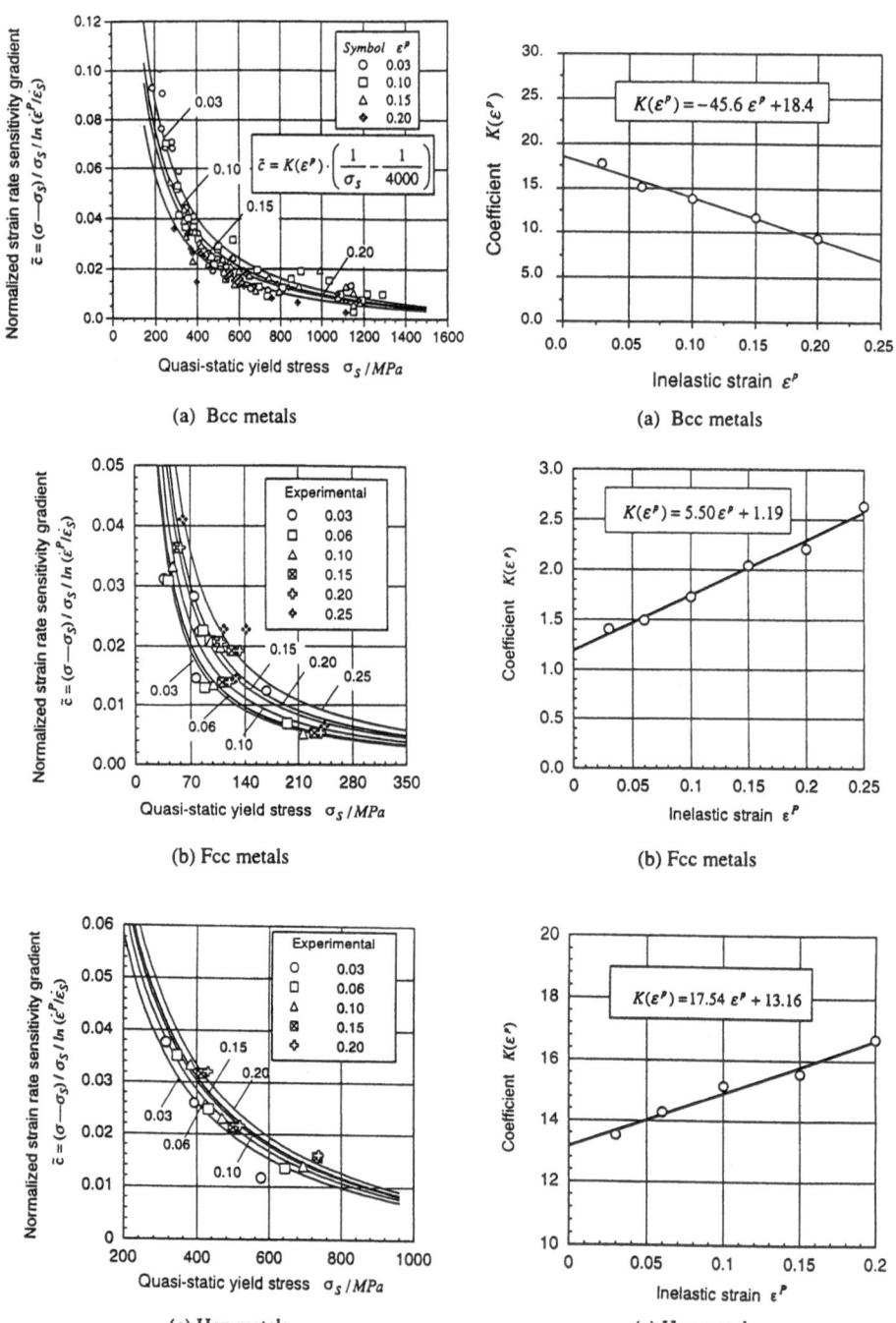

(a) Bcc metals

(a) Bcc metals

(b) Fcc metals

(b) Fcc metals

(c) Hcp metals.

(c) Hcp metals

Figure 1 Hyperbolic relationship between the normalized strain rate sensitivity gradients and the quasistatic yield stress. (a) bcc metals, (b)fcc metals and (c) hcp metals.

Figure 2 Linear dependence of the coefficient K on inelastic strain.
(a) bcc metals, (b) fcc metals and (c) hcp metals.

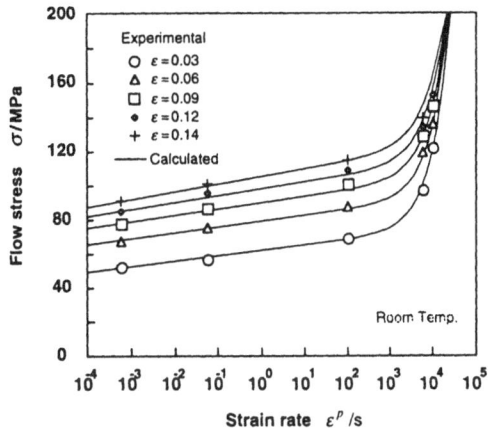

Figure 3 Comparison of the simulated and experimentally derived $\sigma - \ln(\dot{\varepsilon}^p)$ diagram for commercially pure aluminum (99.98w%).

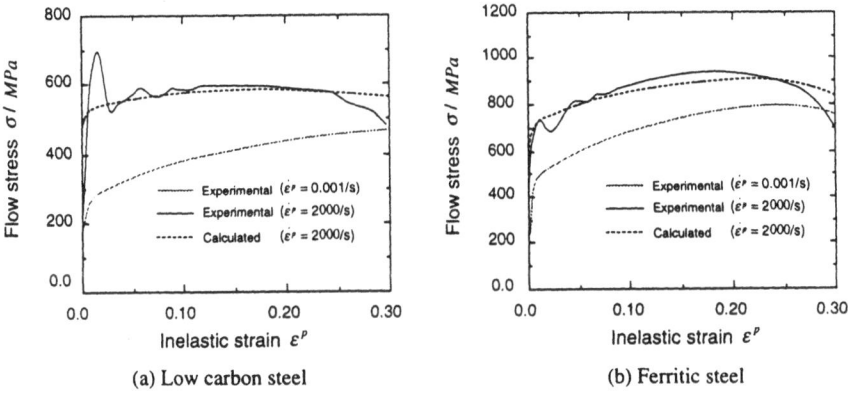

(a) Low carbon steel

(b) Ferritic steel

Figure 4 Stress-strain relations of bcc metals at high rates of strain.
(a) Low carbon steel (2000/s) and (b) ferritic steel (2000/s)

method, the specified sets of α and β for bcc, fcc and hcp metals are obtained, and are shown in Table 1. As you can see, there are two sets of α and β values for fcc metals; one is for aluminum and its alloys, and the other is for copper and its alloys. Here, it should be noted that the difference of the sign on α is closely related to the strain rate history effect as well as the dependence of strain rate sensitivity on the inelastic strain. If the value of α is negative, a calculation of strain rate jump based on eq.(6) depicts the negative strain rate history effect, and vice versa.

4. Results and Discussion

In order to validate the applicability of the proposed constitutive equations, the relation between flow stresses and inelastic strain rates is simulated for commercially pure aluminum (99.98wt%). Results are shown in Fig. 3. In the calculation, $B = 4.74 \times 10^{-3}$ and $m = 1$ in eq.(6) are assumed so that the calculated curves at considerably high strain rates above 5×10^3/s just fit to experimental data. As you can see, the simulated results are in good agreement with the experimental ones over a wide range of strain rates. For further verification of constitutive equations, stress-strain relations for bcc, fcc and

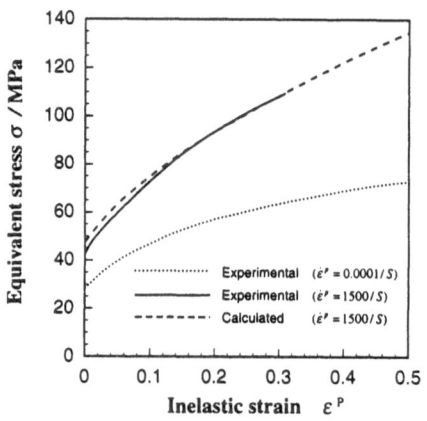

(a) Commercially pure aluminum

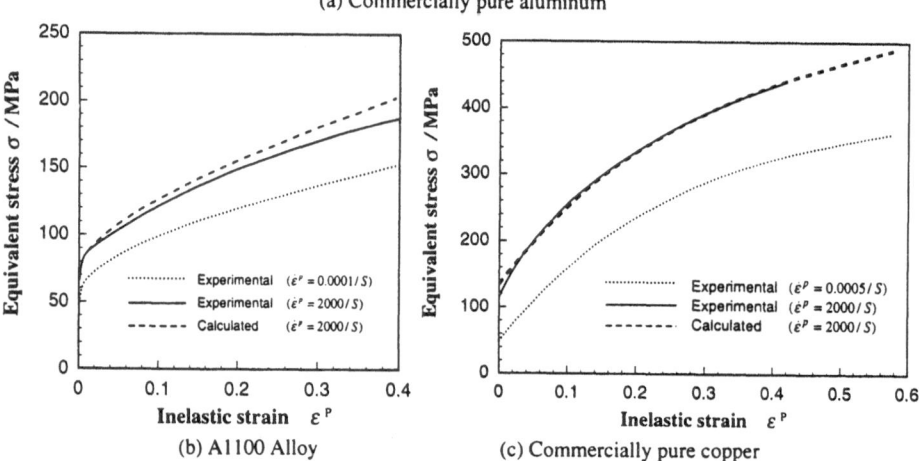

(b) A1100 Alloy

(c) Commercially pure copper

Figure 5 Stress-strain relations of fcc metals at high rates of strain.
(a) Commercially pure aluminum, (b) A1100 alloy and (c) commercially pure copper

hcp metals at constant high strain rates are calculated from the quasistatic curves. Comparison with experimental results are shown in Figs. 4(a),(b) (for steels), Figs. 5(a),(b),(c) (for aluminum, aluminum alloy, copper, respectively) and Fig. 6 (for titanium).

As shown in the figures, calculations are relatively in good agreement with experimental results. In many cases, the errors between calculations and experiments are within 10% or below. By comparing Fig. 4(a) with (b), it is found that the higher quasistatic yield stress leads to the less strain rate dependence. Furthermore, it is found that calculations well express the difference in the inelastic strain dependence of strain rate sensitivities between bcc, fcc and hcp metals. Namely, in the case of bcc metals (Figs. 4(a) and (b)), the differences in the flow stress between high strain rates and the quasistatic strain rate decrease with increasing strain; in the case of fcc metals (Figs. 5(a),(b) and (c)), by contrast, these differences increase with increasing strain; in the case of hcp metal (Fig. 6), the difference is almost constant. Simulated results well express these experimental tendencies.

In order to verify the multiple effects of temperature, dynamic stress-strain relations of A2017

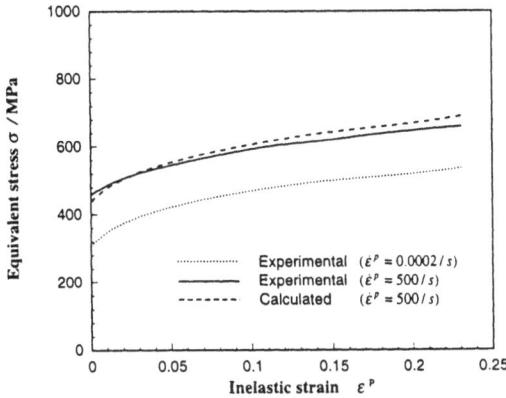

Figure 6 Stress-strain relations of α-titanium (hcp) at high rates of strain.

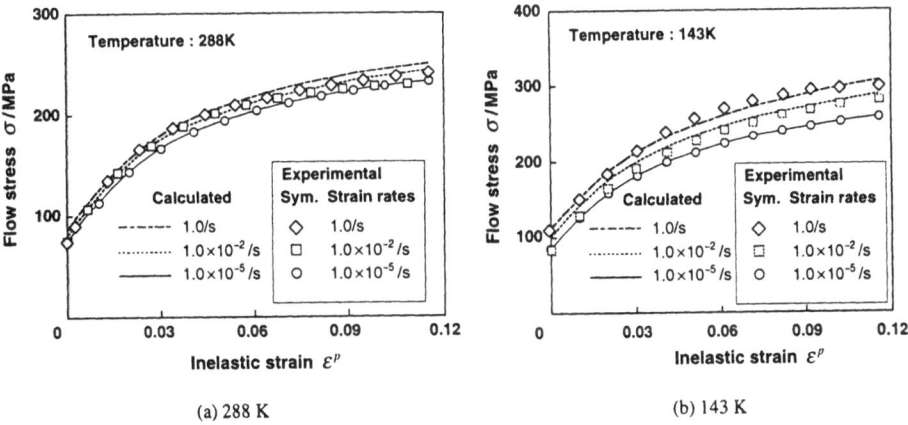

(a) 288 K   (b) 143 K

Figure 7 Temperature dependence of A2017 aluminum alloy under dynamic strain rates,
(a) at 288K and (b) at 143K .

aluminum alloy at room and lower temperatures are simulated. Results are shown in Fig. 7(a) for 288K and (b) for 143K. Obviously, significant multiplication of strain rate sensitivity occurs at lower temperatures. Notwithstanding the very simple formulation for temperature dependence in the constitutive equation, calculated results show fairly good agreement with experimental results.

Finally, the representation of strain rate history effects is investigated. Figures 8(a) and (b) show the typical results of strain rate jump tests obtained by Campbell and Eleiche. Their original data was obtained in shear mode, therefore, the translation from shear stress to equivalent stress and shear stain to equivalent strain was performed. Here, it should be noted that the material constants shown in Section 3 are derived from tension/compression tests, and in the case of shear or torsional mode, values of α and β must be slightly modified. In the calculations, $\alpha/\sqrt{3}$ and $\beta/\sqrt{3}$ instead of α and β are employed. In this point, further investigations may be necessary. Although the errors between calculations and experiments are not small, the calculated results successfully describe the experimental tendencies of positive (fcc; copper) and negative (bcc; mild steel) strain rate history effects.

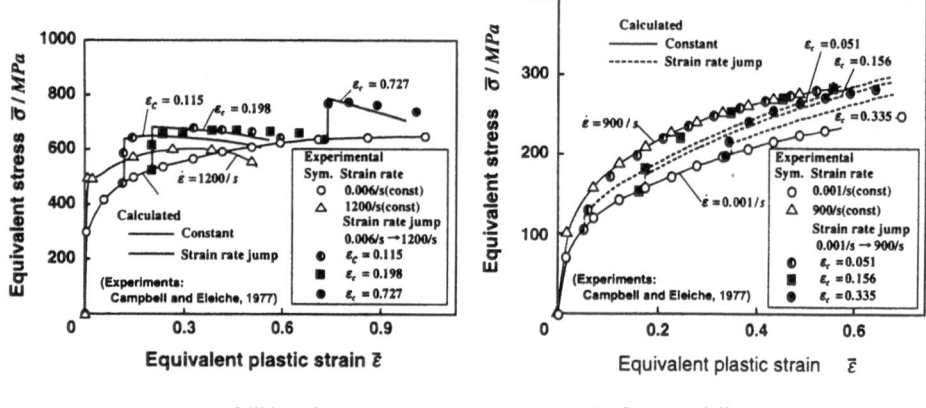

(a) Mild steel (b) Commercially pure copper.

Figure 8 Simulations of strain rate jump on (a) mild steel, and (b) commercially pure copper.

5. Conclusion

A simple but practical rate and temperature dependent constitutive model has been presented. In this model, dynamic behaviors of metals of which lattice structures are of the same type are described by using the unified set of material constants. In order to validate the applicability of the proposed constitutive model, stress-strain relations of several metals at high rates of strain were simulated and were compared with the experimental ones. In many cases, the errors between simulations and experiments were within 10%, and the simulations well express the experimental tendencies. For further verification of the constitutive equation, its temperature dependence and applicability to strain rate history effect were also investigated.

References

[1] Malvern, L.E., The propergation of longitudial wave of plastic deformation in a bar of materials exhibiting a strain rate effect, J. Appl. Mech., 18, 1951, pp203-208

[2] Perzyna, P., The constitutive equation for rate sensitive plastic materials, Quart. Appl. Math., 20, 1963, pp321-332

[3] Campbell, J.D., Dynamic plasticity of Metals, Conference Series No.46, Springer-Verlag, 1970, pp1-15

[4] Perzyna, P., The constitutive equations describing thermo-mechanical behavior of materials at high rates of strain, Inst. Phys. Conf. Ser. No.21, Bristorl, 1974, pp138-153

[5] Klepaczko, J.P., The newest development on physically based constitutive modeling in dynamic plasticity, Mechnical Behaviour of Materials-IV (Proc. of 6th Int. Conf.), Pergamon Press, 1991, pp235-244

[6] Duffy, J., The Dynamic Plastic Deformation of Metals, Report No. AFWAL-TR-82-4024

[7] Harding, J., Materials at High Strain Rates, Ed. by Blazynski, T.Z., Elsevier, 1987, pp133-186

[8] Tanimura, S., Constitutive models for dynamic behavior of materials, Proc. of Int. Sym. on Impact Engng, Vol. I, 1992, pp17-26

An energetic characterization of the constitutive equations for plastic strings on plastic foundations

I. Suliciu

Institute of Mathematics, P.O. BOX 1–764, RO–70700 Bucharest, Romania.

Summary

A continuous initial or initial–boundary value problem for an elastic viscoplastic string on an elastic viscoplastic foundation is formulated and an energy identity/inequality is deduced. Bounds in energy of the solutions are derived and the stress deviation of the viscoplastic model from the plastic one is estimated. As a farther limiting case similar results are derived for the rigid perfect plastic case. It is also shown that the energy identity/inequality is applicable to the discontinuous problem of impulsive loading of a rigid perfect plastic model when the horizontal motion of the string is neglected and that the total energy is preserved in this case.

Keywords: Plastic/viscoplastic strings on plastic/viscoplastic foundations, energy estimates, energetic upper bounds, impulsive loadings

1. Introduction

Since the results of this paper are closely related to those in [1] we use here the same notations. We analyze how the rate independent model used there, which is a rigid perfect plastic one, can be obtained as a limiting case of an elastic viscoplastic model when Maxwell's viscosity coefficients and the dynamic elastic moduli of the rate dependent model tend to infinity. The main tool we use is an energy identity/inequality, which derivation is based on the second law of thermodynamics. Among other things, we show that in the case of the rigid perfect plastic constitutive equation the energy identity is a balance law, i.e. it holds for smooth solutions as well as across all discontinuities compatible with the model. Thus, we have one of the main result of this paper, for an impulsive loading problem. In terms of the energetic measure of Section 5, which may be experimentally determined this result reads: *the sum of the total elongation of the string and the area between the final shape of the string and its initial undeformed position can not exceed the square of the impulsive load intensity* . Another important result is the energetic characterization of the rigid perfect plastic solution presented in [1] in the case when the horizontal motion of the string is neglected. *The total energy of the initial data is preserved all the time* and that solution appears as an energetic upper bound for the solution of the problem when the horizontal motion is not neglected.

2. Description of the problem.

We consider a viscoplastic string of finite or infinite length on a viscoplastic foundation of the same finite or infinite length. The string is continuously or impulsively loaded, as in [1] see also [2] for detailed justifications of the relation between plastic string on plastic foundations and different engineering structures subjected to mass impact, pressure pulses or contact explosions approximated as ideal impulsive loadings (we here deal with). We treat first, a continuous initial–boundary value problem and then extend the results concerning the free energy functions to the discontinuous problems as impulsive loadings of [1] are.

The system of governing equations is formed from the balance law of momentum:

$$\varrho\frac{\partial^2 \tilde{x}}{\partial \tilde{t}^2} - \frac{\partial}{\partial \tilde{X}}\left(\tilde{T}\frac{\partial \tilde{x}}{\partial \tilde{X}}\right) = 0, \quad \varrho\frac{\partial^2 \tilde{y}}{\partial \tilde{t}^2} - \frac{\partial}{\partial \tilde{X}}\left(\tilde{T}\frac{\partial \tilde{y}}{\partial \tilde{X}}\right) = \varrho(q - p). \tag{1}$$

Conjugated to the stress $\tilde{T}$ we use as in [1] the strain measure

$$2\tilde{\varepsilon}(\tilde{x},\tilde{t}) = \left(\frac{\partial \tilde{x}}{\partial \tilde{X}}\right)^2 + \left(\frac{\partial \tilde{y}}{\partial \tilde{X}}\right)^2 - 1. \tag{2}$$

For a discussion on the use of other stress and strain measures see [1] and also §5 below. The viscoplastic constitutive equations for the string and for the foundation are taken of Sokolovskii type [6], i.e.,

$$\frac{\partial \tilde{T}}{\partial \tilde{t}} - \bar{E}\frac{\partial \tilde{\varepsilon}}{\partial \tilde{t}} = -\bar{k}\bar{g}(\tilde{T}), \quad \frac{\partial q}{\partial \tilde{t}} + \bar{E}_q\frac{\partial \tilde{y}}{\partial \tilde{t}} = -\bar{k}_q\bar{g}_q(q); \quad \text{with}$$

$$\bar{g}(\tilde{T}) = 0 \text{ for } \tilde{T} \in [0,\bar{N}], \text{ and } \bar{g}(\tilde{T}) > 0 \text{ for } \tilde{T} > \bar{N}, \tag{3}$$

$$\bar{g}_q(q) = 0 \text{ for } q \in [0,q_o], \text{ and } \bar{g}_q(q) > 0 \text{ for } q > q_o.$$

The properties of $\bar{g}(\tilde{T})$ and $\bar{g}_q(q)$ together with the constitutive equations $(3)_{1,2}$ express the fact that the string can only elongate and change its shape but it can not be compressed while the foundation can only be compressed but not elongated.

The pressure $p(\tilde{X},\tilde{t})$ in $(1)_2$ we take of the following pulse form

$$p(\tilde{X},\tilde{t}) = \begin{cases} p_o, & \text{for } \tilde{x} \in (-X_o, X_o), \ \tilde{t} \in (0,t_o), \\ 0, & \text{otherwise}, \end{cases} \quad \text{or} \quad p(\tilde{X},\tilde{t}) \equiv 0. \tag{4}$$

In what follows in the this section we consider the zero pressure case. The case $(4)_1$ will be treated in details in §5.

The quantities ϱ, $\bar{E}$, $\bar{E}_q$, $\bar{k}$, $\bar{k}_q$, p_o, X_o, t_o, $\bar{N}$ and q_o are positive constants with the following meanings: $\bar{E}$ and $\bar{E}_q$ are Young's moduli of the string and of the foundation, respectively; $\bar{N}$ and q_o are their yield limits, while $\bar{k}$ and $\bar{k}_q$ are Maxwell's viscosity coefficients of the string and of the foundation. p_o, X_o, t_o have obvious meanings. For the independent variables $\tilde{X}$, $\tilde{t}$ we may have: $\tilde{X} \in (-L,L)$ where $0 < X_o \leq L < \infty$ or $L = \infty$ and $\tilde{t} \geq 0$.

First we shall consider the following continuous but otherwise arbitrary initial data

$$\tilde{x}(\tilde{X},0) = \tilde{x}_o(\tilde{X}), \quad \frac{d\tilde{x}_o(\tilde{X})}{d\tilde{X}} - 1 \geq 0, \quad \tilde{y}(\tilde{X},0) = \tilde{y}_o(\tilde{X}), \quad \frac{\partial \tilde{x}}{\partial \tilde{t}}(\tilde{X},0) = \tilde{v}_1^o(\tilde{X}),$$

$$\frac{\partial \tilde{y}}{\partial \tilde{t}}(\tilde{X},0) = \tilde{v}_2^o(\tilde{X}), \quad \tilde{T}(\tilde{X},0) = \tilde{T}_o(\tilde{X}) \geq 0, \quad \tilde{\varepsilon} = \tilde{\varepsilon}_o(\tilde{X}) \geq 0, \quad q(\tilde{X},0) = q_o(\tilde{X}) \geq 0. \tag{5}$$

In the case when $0 < X_o \leq L < \infty$ we consider the following boundary conditions,

$$\tilde{x}(-L,\tilde{t}) = 0, \quad \tilde{x}(L,\tilde{t}) = 0, \quad \tilde{y}(-L,\tilde{t}) = 0, \quad \tilde{y}(L,\tilde{t}) = 0, \tag{6}$$

and require to the above initial data to be such that we do not introduce shock waves in the string at $\tilde{t} = 0+$. The case when such shock waves are present will be discussed in §4–5.

We introduce the following notations:

$$C_o = \sqrt{\bar{N}/\varrho}, \ \ C_E = \sqrt{\bar{E}/\varrho}, \ \ \varepsilon_Y = \bar{N}/\bar{E}, \ \ a = C_o^2/(q_o X_o), \ \ I_o = a\bar{I}_o/C_0,$$

$$E = \bar{E}/(a^2\bar{N}) = 1/(a^2\varepsilon_Y), \ \ E_q = (X_o/C_o)^2 \bar{E}_q, \ \ k = X_o\bar{k}/C_o, \ \ k_q = X_o\bar{k}_q/C_o, \tag{7}$$

We also introduce (as in [1]) the dimensionless variables and functions

$$X = \bar{X}/X_o, \ \ t = C_o\bar{t}/X_o, \ \ x = a\bar{x}/X_o, \ \ y = a\bar{y}/X_o, \ \ S = \bar{T}/(\varrho C_o^2), \ \ \varepsilon = a^2\bar{\varepsilon},$$

$$Q = q/q_o, \ \ g(S) = \bar{g}(\varrho C_t^2 S)/(\varrho C_t^2), \ \ g_q(Q) = \bar{g}_q(q_o Q)/q_o, \tag{8}$$

in order to write the initial value problem (1)-(6) (when $L = \infty$) or the initial-boundary value problem (1)-(7) (when $0 < X_o \leq L < \infty$) under the dimensionless form:

$$\frac{\partial u_1}{\partial t} - \frac{\partial v_1}{\partial X} = 0, \ \ \frac{\partial v_1}{\partial t} - \frac{\partial}{\partial X}(Su_1) = 0, \ \ \frac{\partial u_2}{\partial t} - \frac{\partial v_2}{\partial X} = 0, \ \ \frac{\partial v_2}{\partial t} - \frac{\partial}{\partial X}(Su_2) = Q,$$

$$\frac{\partial S}{\partial t} - E\left(u_1\frac{\partial u_1}{\partial t} + u_2\frac{\partial u_2}{\partial t}\right) = -kg(S), \ \ \frac{\partial Q}{\partial t} = -E_q v_2 - k_q g_q(Q) \tag{9}$$

$$u_1(X,0) = u_1^o(X) \geq a, \ \ u_2(X,0) = u_2^o(X), \ \ v_1(X,0) = v_1^o(X), \ \ v_2(X,0) = v_2^o(X),$$

$$S(X,0) = S_o(X) \geq 0, \ \ \varepsilon = \varepsilon_o(X) \geq 0, \ \ Q(X,0) = Q_o(X) \geq 0;$$

where

$$u_1 = \frac{\partial x}{\partial X}, \ \ u_2 = \frac{\partial y}{\partial X}, \ \ v_1 = \frac{\partial x}{\partial t}, \ \ v_2 = \frac{\partial y}{\partial t}. \tag{10}$$

and the functions $g(S)$ and $g_q(Q)$ are derived from $(3)_{3,4}$ in an obvious way. For the case of finite string the boundary conditions are

$$v_1(\pm\ell, t) = 0, \ \ v_2(\pm\ell, t) = 0, \ \ , 1 \leq \ell = L/X_O < \infty. \tag{11}$$

3. Energy identity/inequality

The second law of thermodynamics for isothermal processes is equivalent here (see also [3] and the references given there) with the existence of a free energy function $\psi(\varepsilon, S, y, Q)$ verifying

$$\frac{\partial}{\partial t}\psi(\varepsilon, S, y, Q) \leq S\frac{\partial\varepsilon}{\partial t} - Q\frac{\partial y}{\partial t} \tag{12}$$

for the pairs $(\varepsilon(t) = (u_1^2 + u_2^2 - a^2)/2, S(t))$ $(y(t), Q(y))$ verifying the constitutive equations $(9)_{5,6}$. One can show that the function $\psi(\varepsilon, S, y, Q)$ must verify

$$\frac{\partial\psi}{\partial\varepsilon} + E\frac{\partial\psi}{\partial S} = S, \ \ \frac{\partial\psi}{\partial y} - E_q\frac{\partial\psi}{\partial Q} = -Q,$$

$$k\frac{\partial\psi}{\partial S}g(S) + k_q\frac{\partial\psi}{\partial Q}g_q(Q) \geq 0, \tag{13}$$

since the rates $\partial\varepsilon/\partial t$ and $\partial y/\partial t = v_2$ can be given arbitrarily.

There are many functions ψ verifying (13) (see [3]); a convenient one for what follows is

$$\psi(\varepsilon, S, y, Q) = \frac{S^2}{2E} + \frac{Q^2}{2E_q} + \varphi(S - E\varepsilon) + \varphi_q(Q + E_q y) \tag{14}$$

where $\varphi(\tau)$ and $\varphi_q(r)$ are determined as in (3.16) of [3].

We multiply equations $(9)_2$ by v_1 and $(9)_4$ by v_2 and sum them up to obtain, by taking also into account $(9)_{1,3}$, (10) and (13), the following energy identity/inequality

$$\frac{\partial e^{\cdot}}{\partial t} - \frac{\partial}{\partial X}\left(S(u_1 v_1 + u_2 v_2)\right) = -kg(S)\frac{\partial \psi}{\partial S} - k_q g_q(Q)\frac{\partial \psi}{\partial Q} \leq 0. \tag{15}$$

where $e^{\cdot}$ is the density of the total energy e of the string and they are defined as

$$e^{\cdot} = \frac{v_1^2 + v_2^2}{2} + \psi(\varepsilon, S, y, Q), \quad e(t) = \int_{-\ell}^{\ell} e^{\cdot}(X, t)dX. \tag{16}$$

Then for the string with fixed ends (11) and the initial data $(9)_{7-12}$ we have from (15)

$$0 \leq e(t) \leq e(0). \tag{17}$$

We introduce the overstress $\Delta_1(S)$ and over foundation reaction force $\Delta_2(Q)$ as

$$\Delta_1(S) = \begin{cases} S - S_Y, & \text{if } S > S_Y, \\ 0, & \text{if } 0 \leq S \leq S_Y, \end{cases} \quad \text{and} \quad \Delta_2(Q) = \begin{cases} Q - Q_Y, & \text{if } Q > Q_Y, \\ 0, & \text{if } 0 \leq Q \leq Q_Y, \end{cases} \tag{18}$$

respectively. For more complicated overstress functions see [3].

On the other hand one can directly write using (14) (see also [3])

$$\frac{\partial \psi}{\partial S} = \phi_1(S - S_R(\varepsilon)), \quad \frac{\partial \psi}{\partial Q} = \phi_2(Q - Q_R(y)), \quad \phi_1 \geq \frac{1}{E}, \quad \phi_2 \geq \frac{1}{E_q},$$

$$S_R(\varepsilon) = \begin{cases} E^{\circ}\varepsilon, & 0 \leq \varepsilon \leq \sigma_Y/E^{\circ} \\ S_Y = 1, & \varepsilon > \sigma_Y/E^{\circ} \end{cases}, \quad Q_R(y) = \begin{cases} -E_q y, & 0 \leq -y \leq y_Y = 1/E_q^{\circ} \\ Q_Y = 1, & -y > y_Y \end{cases} \tag{19}$$

where $\phi_1(\varepsilon, S)$ and $\phi_2(y, Q)$ are discontinuous functions but lower bounded as above and $0 < E^{\circ} < E$, $0 < E_q^{\circ} < E_q$.

Now, the right hand side of (15) can be written as

$$kg(S)\frac{\partial \psi}{\partial S} + k_q g_q(Q)\frac{\partial \psi}{\partial Q} \geq kg(S)\phi|S - S_R(\varepsilon)| + k_q g_q(Q)\phi|Q - Q_R(y)|$$

$$\geq \frac{k}{E}\Delta_1(S)g(S) + \frac{k_q}{E_q}\Delta_2(Q)g_q(Q) \geq 0, \tag{20}$$

by the use of (18) and the form of $g(S)$, $g_q(Q)$. We return back to the energy identity (15) which we integrate with respect to X on $(-\ell, \ell)$ and with respect to t on a time interval $(0, t)$ and take into account (20) to get

$$\int_0^t \int_{-\ell}^{\ell} \Delta_1(S(X, s))g(S(X, s))dX ds \leq \frac{E(e(0) - e(t))}{k},$$

$$\int_0^t \int_{-\ell}^{\ell} \Delta_2(Q(X, s))g_q(Q(X, s))dX ds \leq \frac{E_q(e(0) - e(t))}{k_q}. \tag{21}$$

Formulas (21) show that, when k and k_q become very large the overstress functions $\Delta_1(S)$ and the over foundation reaction force $\Delta_2(Q)$ become very small and the viscoplastic models tend to the rate independent elastic perfect plastic models. The way the elastic perfect plastic model is approached by the viscoplastic one depends on the way g depends on S and g_q on Q. For the most examples considered in the literature [3,6]

$$g(S) = \Delta_1^n(S), \quad \text{we also take} \quad g_q(Q) = \Delta_2^n(Q). \tag{22}$$

There are also two extreme examples, that of [4,5]. Some more details are presented in [3,6]. When g and g_q depend on Δ_i, $i = 1$, 2 as in (22) then the solution of viscoplastic problem (9), approaches in L^{1+n} sense on the domain $(-\ell, \ell) \times (0, t)$, the solution of the rate-independent elastic perfect plastic problem when k and k_q tend to infinity. If in addition we take $E \to \infty$ and $E_q \to \infty$ we obtain the rigid perfect plastic model of [1]. Now, from (14) we obtain

$$\bar{\psi}_\infty(\varepsilon, y = \lim_{E, E_q \to \infty} \psi(\varepsilon, S, y, Q) = S_Y|\varepsilon| + Q_Y|y| \tag{23}$$

and remark that it is a free energy for both rigid viscoplastic and rigid perfect plastic models.

4. The structure of the shock waves

Across a shock wave $(x = \zeta(t), t)$ propagating with the speed $C = d\zeta/dt$, the balance laws and the compatibility conditions (9) and the energy identity (15) whenever it is satisfied by the discontinuous solution, lead to the following jump conditions:

$$C[\![v]\!] + [\![\tau S\sqrt{a^2 + 2\varepsilon}]\!] = 0, \quad [\![v]\!] + C[\![\tau\sqrt{a^2 + 2\varepsilon}]\!] = 0,$$
$$C[\![e']\!] + [\![\tau \cdot v\, S\sqrt{a^2 + 2\varepsilon}]\!] = 0, \quad \text{where} \quad \tau = (u_1, u_2)/\sqrt{a^2 + 2\varepsilon}, \quad v = (v_1, v_2). \tag{24}$$

For the way these jump conditions are obtained see for instance [7–9]. From $(24)_{1-2}$ we get

$$\left(C^2 - S^+\right)\sqrt{a^2 + 2\varepsilon^+}\tau^+ = \left(C^2 - S^-\right)\sqrt{a^2 + 2\varepsilon^-}\tau^-. \tag{25}$$

As τ^+, τ^- are unit vectors (25) implies

$$\left(C^2 - S^+\right)\sqrt{a^2 + 2\varepsilon^+} = \left(C^2 - S^-\right)\sqrt{a^2 + 2\varepsilon^-} = B. \tag{26}$$

Now, the structure of the shock waves is as follows (see [7–9]). *Case I*, when $B \neq 0$; then $\tau^+ = \tau^-$ and ε, S, v may jump and such a discontinuity is called a longitudinal shock wave and it propagates with the speed $C^2 = [S^+\sqrt{a^2 + 2\varepsilon^+} - S^-\sqrt{a^2 + 2\varepsilon^-}]/[\sqrt{a^2 + 2\varepsilon^+} - \sqrt{a^2 + 2\varepsilon^-}] = C_\ell^2$. *Case II*, when $B = 0$; then from (26) $S^+ = S^-$ and the string may have a jump in shape at $x = \zeta(t)$ from τ^- to τ^+ which propagates with the transversal wave speed $C^2 = S = C_t^2$. This last case has two subcases. *Case IIa*, if $S \neq 1$ then $\varepsilon^+ = \varepsilon^-$ (see also the jump conditions below imposed by constitutive equations) and τ may jump from τ^- to τ^+. In this case the wave is called transversal shock wave since it affects the shape of the string only. *Case IIb*, if $S = S_Y = 1$ then we may have jumps for both strain ε and shape τ, i.e. $\varepsilon^+ \neq \varepsilon^-$ and $\tau^- \neq \tau^+$. This is called a mixed longitudinal and transversal shock wave and the wave is mixed only when the stress equals the yield stress of this stress measure.

There are also two limiting cases when $C = 0$ or $C = \infty$

$$C = 0 \Rightarrow [\![v]\!] = 0, \ \tau^+ = \tau^-, \ [\![S\sqrt{a^2 + 2\varepsilon}]\!] = 0; \quad C = \infty \Rightarrow [\![v]\!] = [\![\tau]\!] = 0, \ [\![\varepsilon]\!] = 0, \tag{27}$$

and in the last case S may jump. We also have from the constitutive equations $(9)_{5,6}$

$$S^+ - E\varepsilon^+ = S^- - E\varepsilon^-, \quad Q^+ + E_q y = Q^- + E_q y, \quad \text{for} \ C \neq 0. \tag{28}$$

Based on the above relations one can prove that across a transversal shock wave and across mixed longitudinal–transversal shock waves the energy jump condition $(24)_3$ is a consequence of the jump conditions $(24)_{1,2}$.

Unfortunately in the case $C = C_\ell$ in (24) the energy jump condition $(24)_3$ is not always a consequence of the jump conditions $(24)_{1,2}$. In other words the energy identity (15) is a consequence of the governing system of equations $(9)_{1-6}$ for smooth solutions but it may not be a consequence of the same equations for discontinuous solutions without additional conditions.

5. Impulsive loading problem

We assume as in [1] that initially the string is a straight line at rest along OX axis and it is uniformly prestressed at yield stress $\tilde{T} = \tilde{N}$. For the pulse pressure $(4)_1$ we introduce the total impulse $I(X)$ (see [1]) defined as

$$I(X) = \int_{0-}^{\infty} p(X,t)dt = \int_{0-}^{t_o} p(X,t)dt = \begin{cases} p_o t_o & \text{if } X \in (-X_o, X_o), \\ 0, & \text{otherwise} \end{cases} \tag{29}$$

and require p_o be such that

$$\lim_{t_o \to 0} p_o t_o = \bar{I}_o = \frac{C_o}{a} I_o = const. \tag{30}$$

The initial value problem (9) (where of course, in eq. $(9)_4$ instead of Q we must put $Q - p/q_o$), when $p(X,t)$ satisfies (30) may be called an impulsive loading initial value problem while the initial-boundary value problem (9)+(4) when p(X,t) satisfies (30) may be called an impulsive loading initial-boundary value problem.

Following the same procedure as in [1] (see also the way the eqs. (13) of [1] were obtained) an impulsive loading problem can be transformed in a discontinuous initial or initial-boundary value problem which in dimensionless form is similar to (9), i.e.,

$$\frac{\partial u_1}{\partial t} - \frac{\partial v_1}{\partial X} = 0, \quad \frac{\partial v_1}{\partial t} - \frac{\partial}{\partial X}(Su_1) = 0,$$

$$\frac{\partial u_2}{\partial t} - \frac{\partial v_2}{\partial X} = 0, \quad \frac{\partial v_2}{\partial t} - \frac{\partial}{\partial X}(Su_2) = Q,$$

$$\frac{\partial S}{\partial t} - E\left(u_1 \frac{\partial u_1}{\partial t} + u_2 \frac{\partial u_2}{\partial t}\right) = -kg(S), \quad \frac{\partial Q}{\partial t} = -E_q v_2 - k_q g_q(Q) \tag{31}$$

$$u_1(X,0) = u_1^o(X) = \sqrt{a^2 + 2\varepsilon_Y/a^2}, \quad u_2(X,0) = 0, \quad v_1(X,0) = 0,$$

$$v_2(X,0) = v_2^o(X) = -I(X) = -\begin{cases} I_o, & X \in (-X_o, X_o), \\ 0, & \text{otherwise,} \end{cases}$$

$$S(X,0) = S_o(X) = S_Y = 1, \quad \varepsilon = \varepsilon_o(X) = \frac{1}{a^2}\varepsilon_Y, \quad Q(X,0) = 0.$$

For the case $1 \le \ell < \infty$ we have to add the boundary conditions (11).

We recall, see also [1] and [7-9] the references given there, that the shock wave structure is easier examined by using different conjugated stress–strain measures as $\sigma - \ni = (1/a)\sqrt{u_1^2 + u_2^2} - 1$ which are related to our $S - \varepsilon$ by

$$\sigma = S\sqrt{a^2 + 2\varepsilon}/a \quad \ni = \sqrt{a^2 + 2\varepsilon}/a - 1. \tag{32}$$

In terms of σ, $\ni$ the longitudinal and transversal shock wave speeds are $C_\ell^2 = [\![\sigma]\!]/[\![\ni]\!]$, $C_t^2 = \sigma/(1+\ni)$, while jump condition $(28)_1$ reads

$$\sigma - (\ni^3 + 2\ni^2)/(2\varepsilon_Y) = \sigma_o - (\ni_o^3 + 2\ni_o^2)/(2\varepsilon_Y).$$

where $(\sigma, \ni) = (\sigma^+, \ni^+)$ and $(\sigma_o, \ni_o) = (\sigma^-, \ni^-)$. Thus

$$\sigma = f(\ni) = (\ni^3 + 2\ni^2)/(2\varepsilon_Y), \quad \text{has } f''(\ni) = (3\ni + 2)/\varepsilon_Y > 0 \quad \text{for } \ni \ge 0, \tag{33}$$

which means that at any intersection of shock waves which leads to an increase in strain, i.e., to a loading process, a shock wave will be generated and any such intersection which leads to unloading will generate rarefaction waves (see [8–9]).

The above property and those proved in §4 allows us to conjecture that the energy identity (15) with ψ defined by (14) holds for the elastic-perfect plastic impulsive problem (31) when the constitutive equations $(31)_{5,6}$ are substituted by

$$\dot{S} = \begin{cases} 0 & \text{if } S = 1 \text{ and } \dot{\varepsilon} > 0, \\ E\dot{\varepsilon} & \text{if } \begin{cases} S = 1 \ \& \ \dot{\varepsilon} \leq 0, \\ S < 1 \ \& \ \varepsilon \geq 0, \end{cases} \end{cases} \text{ or } , \quad \dot{Q} = \begin{cases} 0 & \text{if } Q = 1 \text{ and } v_2 < 0, \\ -E_q v_2 & \text{if } \begin{cases} Q = 1 \ \& \ v_2 \geq 0, \\ Q < 1 \ \& \ y \leq 0; \end{cases} \end{cases} \text{ or } \qquad (34)$$

where $\dot{S} = \partial S/\partial t$ etc. Then for the string with fixed ends (11) and the initial data $(31)_{7-12}$ we have from energy inequality (15), with the use of notations (7)

$$0 \leq e(t) \leq e(0) = I_o^2 + \varepsilon_Y \ell/a^2 . \qquad (35)$$

We discuss now the rigid perfect plastic impulsive loading problem of [1] but without neglecting the horizontal motion of the string. The energy identity/inequality (15) with $\psi(\varepsilon, S, y, Q) = \psi_\infty(\varepsilon, y)$ given by (23) is used. In this case the energy jump condition $(24)_3$ is a consequence of $(24)_{1,2}$ as the longitudinal shock wave propagates with infinite speed (see $(27)_2$) and along the transversal shock wave $(24)_3$ is verified. As for this model $\varepsilon_Y = 0$, we have from (15), (23) and taking into account the initial-boundary data $(31)_{7-12}$ - (11) that the total energy $\bar{e}(t)$ for the rigid perfect plastic material becomes

$$\bar{e}(t) = \int_{-\ell}^{\ell} \left[\frac{1}{2}(v_1^2(X,t) + v_2^2(X,t)) + S_Y|\varepsilon| + Q_Y|y(X,t)| \right] dX \leq I_o^2. \qquad (36)$$

For large t the motion of the string stops, i.e., $v_1(X,t) = 0$ and $v_2(X,t) = 0$ and with the notations

$$e_e(I_o) = S_Y \int_{-\ell}^{\ell} \varepsilon(I_o, X, \infty) dX, \quad e_s(I_o) = Q_Y \int_{-\ell}^{\ell} |y(I_o, X, \infty)| dX, \quad e(I_o, \infty) = \bar{e}(\infty) \qquad (37)$$

(36) reads

$$e(I_o, \infty) = e_e(I_o) + e_s(I_o) < I_o^2 . \qquad (38)$$

When $1 \leq \ell < \infty$ and the strain $\varepsilon(I_o, X, \infty)$ is not too large one can write

$$e_e(I_o) = S_Y a^2 \int_{-\ell}^{\ell} (ds - dX) = 2a^2 S_y(\mathcal{L} - \ell), \quad \text{for } 0 \leq \varepsilon \ll \frac{a}{2} \qquad (39)$$

where $2\mathcal{L}$ is the total length of the deformed string. Thus $e_e(I_o)$ may be interpreted as an energetic measure of the total elongation of the string while the energy $e_s(I_o)$ as a measure of the area comprised between the final shape of the string and its initial position. Then the inequality (38) says that the sum of the above two energetic measures can not exceed the energetic measure of the impulsive load intensity.

As previously mentioned, in [1] the impulsive loading problem for an infinite rigid-perfect plastic string was treated. The longitudinal part $u_1^2 - a^2$ of the strain $\varepsilon(X, I_o, t)$ was neglected and the final shape of the string determined, i.e., the function $y(X, I_o) = y(X, \infty)$ was explicitly constructed. We reproduce that solution below. Since the form of $y(X, I_o)$ depends on the position of I_o in $(0,1)$ or $(1,3/2)$ or $(3/2, \infty)$ we have the following three cases:

The case $I_o \in (0,1]$:

$$y(X, I_o) = \begin{cases} -\frac{I_o^2}{2}, & \text{for } X \in [0, 1 - I_o), \\ \frac{1}{2}(X^2 - 1) + (I_o - 1)(X - 1), & \text{for } X \in [1 - I_o, 1 - 2I_o/3), \\ -\frac{I_o^2}{6} + \frac{X-1}{8}(X - 1 + 4I_o), & \text{for } X \in [1 - 2I_o/3, 1), \\ \frac{3}{8}(1 - X^2) + \frac{3+2I_o}{4}(X - 1) - \frac{I_o^2}{6}, & \text{for } X \in [1, 1 + 2I_o/3), \\ 0, & \text{for } X \geq 1 + 2I_o/3. \end{cases}$$

The case $I_o \in (1, 3/2]$:

$$y(X, I_o) = \begin{cases} \frac{1}{2}(X - 1)^2 + I_o(X - 1), & \text{for } X \in [0, 1 - 2I_o/3), \\ \frac{1}{8}(X - 1)^2 + \frac{I_o}{2}(X - 1) - \frac{I_o}{6}, & \text{for } X \in [1 - 2I_o/3, 1), \\ -\frac{3}{8}(X - 1)^2 + \frac{I_o}{2}(X - 1) - \frac{I_o^2}{6}, & \text{for } X \in [1, 1 + 2I_o/3), \\ 0, & \text{for } X \geq 1 + 2I_o/3. \end{cases}$$

The case $I_o > 3/2$:

$$y(X, I_o) = \begin{cases} \frac{1}{8}(X - 1)^2 + \frac{I_o}{2}(X - 1) - \frac{3}{8} - \frac{1+2I_o}{2} \ln\left(\frac{\sqrt{1+2I_o}}{2}\right), & \text{for } X \in [0, 1), \\ -\frac{3}{8}(X - 1)^2 + \frac{I_o}{2}(X - 1) - \frac{3}{8} - \frac{1+2I_o}{2} \ln\left(\frac{\sqrt{1+2I_o}}{2}\right), & \text{for } X \in [1, 2), \\ -\frac{X^2}{4} + \frac{1+2I_o}{4} - \frac{1+2I_o}{2} \ln\left(\frac{\sqrt{1+2I_o}}{X}\right), & \text{for } X \in [2, \sqrt{1 + 2I_o}), \\ 0, & \text{for } X \geq \sqrt{1 + 2I_o}. \end{cases}$$

When we compute and sum up $e_e(I_o)$ and $e_s(I_o)$ to get by (38) the total energy $e(I_o, \infty)$ we arrive at a rather surprising result that

$$e(I_o, \infty) = e_e(I_o) + e_s(I_o) = I_o^2, \tag{40}$$

which means that *the rigid perfect plastic model in an impulsive loading problem preserves the total energy when the horizontal motion of the string is neglected (i.e. when $x(X, t) = X$ for all times $t \geq 0$)*. This result has certain connection with the conclusion of Section 4 that for the transversal shock waves the energy jump conditions is a consequence of the balance of momentum jump conditions $(24)_1$ and kinematical jump conditions $(24)_2$.

Acknowledgements: The comments of Prof. T. Wierzbicki from MIT-USA on an earlier draft of this work as well as the late discussions with Prof. M. Mihăilescu-Suliciu from Institute of Mathematics-Romania are greatly appreciated.

References

[1] M. Mihăilescu-Suliciu, I. Suliciu, T. Wierzbicki and M. S. Hoo Fatt, Transient response of an impulsively loaded plastic string on a plastic foundation, Q. Appl. Math., to appear

[2] T. Wierzbicki and M. S. Hoo Fatt, Int. J. Impact Engng. Vol. 13, 1993, pp. 215–241.

[3]. I. Suliciu and M. Şabac, J. Math. Anal. Appl. **131**, 1988, pp. 354–372.

[4] S.Kuriyama and K. Kawata, J. of Appl. Phys. **44**, 1973, pp. 3445–3454.

[5] I. Suliciu, Mech. Res. Comm. **1**, 1974,pp. 101–105.

[6] N. Cristescu and I. Suliciu, "Viscoplasticity", Nijhoff, The Hague, 1982.

[7] N. Cristescu, Dynamic Plasticity, North-Holland, Amsterdam, 1967.

[8] M. Mihăilescu and I. Suliciu, J. Math. Analysis Appl. **52** pp. 10-24, 1975.

[9] M. Mihăilescu and I. Suliciu, Rev Roum. Pures et Appl. **20**, pp. 551-559, 1975.

Loss of localization at high crack speeds

K.B. Broberg

Department of Mathematical Physics, University College Dublin
Belfield, Dublin 4, Ireland

Summary

At the IUTAM Symposium on "High Velocity Deformation of Solids" in Tokyo 1977, I argued that localization of the process region at a crack edge to a size determined by an intrinsic length parameter of the material was lost at a very high crack speeds [1]. A few years later several experimental results appeared, giving indirect support for such a loss by showing lack of a unique relation between stress intensity factor or surface roughness and crack velocity [2],[3],[4],[5]. Some of the experiments also showed a remarkable preference for constant crack velocity in the high velocity region. These observations were later parallelled by numerical simulations by Johnson [6],[7]. The constant velocity reached did not seem to be a material property. Here it is argued that the preference for a constant velocity is in accordance with a perception of the process region as a continuum at very high crack velocities, a consequence of the lost significance of an intrinsic length parameter.

Keywords: Dynamic crack growth, process region, strain localization, energy flux, material length parameter.

1   Introduction

At the IUTAM Symposium on "High Velocity Deformation of Solids" in Tokyo, August 24-27, 1977, I argued that the size of the process region at fast crack growth would not be "given by intrinsic dimensions of the material" as at slow crack growth, but be "governed by the strength of the surrounding stress-strain field" [1]. It seemed for a while that complete anarchy prevailed, contrary to the current belief in the 1970s that, under conditions of small scale yielding, the energy dissipation at the crack edge was uniquely related to the crack velocity. A few years later several experimental results appeared, showing an apparent lack of a unique relation, cf. [2],[3],[4],[5],[8]. In some of these experiments the outcome was that

1. the crack accelerates to a constant velocity

2. this constant velocity is different for different experimental conditions, although the material is the same

3. the crack face roughness and the stress intensity factor increase during the constant velocity phase.

In some cases the stress intensity factors, which were measured by caustics, increased considerably during the constant velocity phase. However, doubts have been raised about the accuracy of such measurements, see e.g. [9],[10],[11],[12],[13],[14],[15],[16]. Fortunately, measurements of the stress intensity factors are not crucial: already the measurements of the crack face roughness, which could increase considerably during the constant velocity phase, seem to provide conclusive evidence of increasing process region size and thereby also of increasing energy dissipation.

In experiments by Arakawa and Takahashi [5] the setup was such that the crack accelerated to a maximum velocity and then decelerated. Generally the crack face roughness reached a maximum after the maximum velocity had been reached, very clearly documented in some cases. This indicated that the process region, and thereby the energy dissipation per unit of crack advance, continued to grow during the initial stages of deceleration.

The apparent preference for crack growth towards a constant velocity, except at clearly decreasing stress intensity factors, was well known already from the experiments by, for instance, Schardin [17],[18], but it was anticipated that this constant velocity, which hardly exceeded 70 % of the Rayleigh wave velocity, was a material property, the maximum crack velocity for the material. However, the experiments by Ravi-Chandar [2],[3] resulted in different constant velocities, depending on the acceleration history.

Washabaugh and Knauss [19], who merged two PMMA plates together to form one large plate with a weak joint, were able to obtain speeds above 90 % of the Rayleigh wave velocity for crack propagation along the joint. Thereby a length parameter, the joint thickness, was introduced, and although this parameter was not intrinsic to the material, but to the body geometry, it served the purpose of limiting the process region height to a constant value.

While the theory presented at the 1977 symposium got experimental support, the outcome was not anarchy but an astonishing preference for a constant crack velocity, obtained under increasing energy dissipation. What can explain this preference?

2 The loss of localization at high crack velocities

The motivation for the loss of localization at high crack velocity [1] was essentially based on a cell perception of the material. The material is divided into cells, each one containing one dominant kernel of micro-separation. A cell is either in a *cohesive state*, which implies that an extension requires a load increase, or in a *decohesive state*, in which continued extension occurs under decreasing load, see Fig. 1. More precisely, under load controlled conditions a cell would be stable in the cohesive state, but unstable in the decohesive state. The conditions *in situ* are, of course, between load and grip control. A micro-separation may be opened in a cell before it has reached the decohesive state. The cell size or, equivalently, the distance between micro-separation kernels, usually particles, provides the intrinsic dimension of the material. The *process region* can be defined as the assembly of cells ahead of the crack which have reached the decohesive state.

A cell can be rearded as the smallest material unit needed for understanding the fracture characteristics of the material, although, of course, the cells vary in size and composition, so that a number of cells would be needed to accurately represent the material from the fracture point of view.

At low crack speeds the process region is localized to essentially only one layer of cells, see Fig. 2, because the unloading associated with their decohesion prevents offside cells to reach the decohesive state. The energy dissipation in the process region per unit of crack growth is then determined: it equals the dissipation needed to bring one cell into complete decohesion, coalescence with the main crack, divided by the cell length (the size in the crack direction). At small scale yielding, the size of the plastic region outside the process region is determined by the process region size, which implies that *the total energy dissipation per unit of locally steady crack edge advance at small scale yielding is controlled by the intrinsic length parameter of the material at each given velocity in the low velocity range.* The restriction to "each given velocity" has to be made because of rate sensitivity of materials, inertia effects and velocity dependence of the stress distribution in the vicinity of the dissipative region.

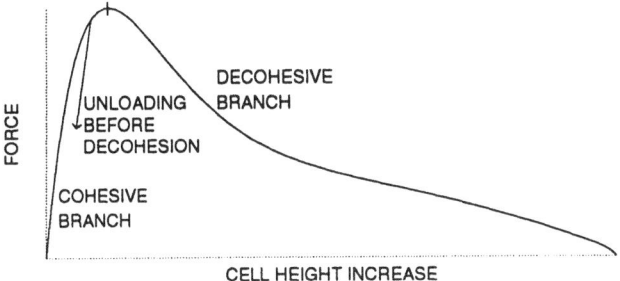

Fig. 1: Force on a material cell containing one kernel of micro-separation. The ascending part represents stable cohesive response. The descending part represents decohesion, which would be unstable under load controlled conditions. Cells inside the process region belong to the decohesive branch, cells outside to the cohesive branch. Unloading of cells who have reached the decohesive state follows the decohesive branch, and other cells are unloaded along the line indicated.

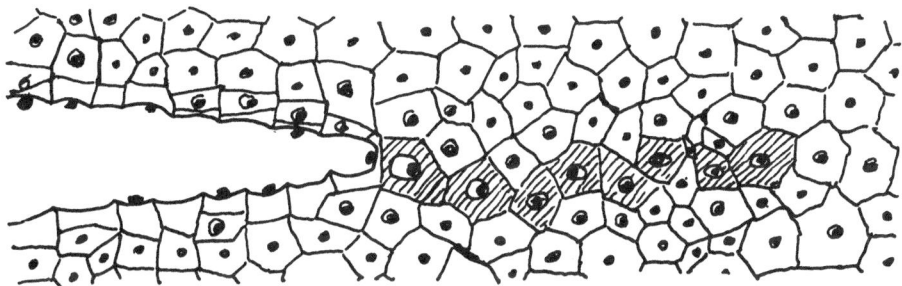

Fig. 2: At slow crack growth the process region is localized to essentially one layer of cells. Thus, for each given velocity, the energy dissipation in the process region, and thereby, for small scale yielding also the total energy dissipation, is controlled by the cell height, an intrinsic length parameter of the material.

At high velocities unloading from cells in the layer straight ahead of the crack is communicated to offside cells via stress waves and, if the crack speed is high enough, this communication may arrive to late to prevent neighbouring offside cells to reach the decohesive state. The localization of the process region to one cell layer is lost. At very high crack speeds it may consist of several layers, see Fig. 3.

The energy dissipation at small scale yielding is no longer controlled by an intrinsic length parameter at very fast crack growth, but then, how is it controlled? The dissipation is obviously no longer a material property, but "governed by the strength of the surrounding stress-strain field", as previously anticipated [1]. Obviously some mechanism maintains constant crack velocity, except at clearly decreasing stress intensity factor.

Numerical simulations of the experiments reported by Ravi-Chandar [2] and by Ravi-Chandar and Knauss [3] were performed by Johnson [6],[7],[20], using a cell model of the material. In agreement with the experimental results they showed that

1. the crack accelerates to a constant velocity

2. this constant velocity is different for different loading histories, although the material parameters are the same

3. the stress intensity factor as well as the process region height increases during the constant velocity phase.

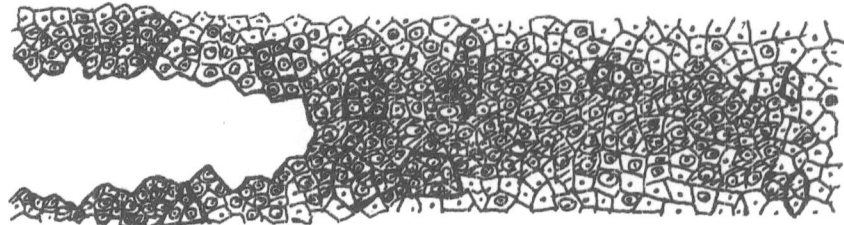

Fig. 3: At very fast crack growth the process region may be very large and consist of several cell layers. The significance of the individual cell size is then lost. The energy dissipation in the process region is no longer a material property.

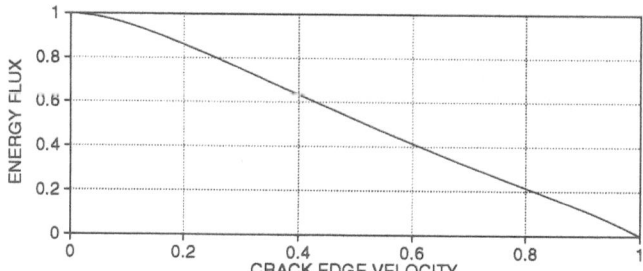

Fig. 4: Energy flux into the crack edge of a symmetrically expanding mode I crack at small scale yielding as a function of the crack velocity. The energy flux is normalized with respect to its value at slow crack expansion and the velocity is normalized with respect to the Rayleigh wave velocity. Poisson's ratio equals 1/4, if plane strain prevails, or 1/3, in the plane stress approximation for a thin plate. Note that the non-normalized flux increases in proportion to the crack length at constant velocity.

Johnson [7] also simulated crack propagation allowing only one row of cells straight ahead of the crack to reach the decohesive state. This corresponds to the experiments by Washabaugh and Knauss [19]. The result was in full accordance with what was expected and later confirmed by these experments, although the numerical capacity did not allow the simulations to continue towards the neighbourhood of the Rayleigh speed. However, no constant terminal velocity was obtained – the crack accelerated during the whole simulation.

3   Why constant crack speed?

The energy flux to the dissipative region at the crack edge [21],[23] is velocity dependent, see Fig. 4. Study now an expanding crack, in a large plate, subjected to a constant remote mode I load and accelerating from a static state. The increasing crack length works towards increasing energy flux into the crack edge, whereas the increasing crack velocity works in the opposite direction, due to the descending flux *vs* velocity curve, Fig. 4. In the absence of a unique relation between energy dissipation and velocity, the crack has different options to accommodate these two influences. Two extreme options can be singled out. One corresponds to the minimum energy dissipation, which is given by the requirement of at least one layer of cells in the process region for crack growth. In this option the crack would limit the energy flux into its edge by increasing its velocity as much as is needed for the minimum energy dissipation, which might be somewhat, though not strongly, velocity dependent. The crack would then accelerate towards the Rayleigh wave velocity, provided that branching does not occur, In the other extreme option, which corresponds to a terminal constant velocity, the crack would meet the increase of energy flux into its edge by increasing the energy dissipation as much as is needed to keep its velocity constant.

The theory advanced in [1] indicates that the first extreme option is the only possible choice at low crack velocities when the energy dissipation at each given velocity is determined by an intrinsic length parameter of the material, thus a material property. On the other hand it does not follow from the theory that any of the extreme options would be necessarily chosen at high crack velocities. Any result between the two extremes would be in accordance with the theory, but reality, as it appears, does not tolerate such anarchic behaviour. It chooses the second extreme for sufficiently high crack velocities. Why?

One simple answer would be: "Why not?". Another answer is found by creating an idealized model, like most common continuum models in crack mechanics. Such models use constitutive equations for continua, elastic or plastic, extrapolated to infinite strains, and the process region is assumed to be point-sized. A finite size process region cannot be regarded as a continuum at low crack velocities, but at high crack velocities it can. This is, of course, a consequence of the lost significance of the intrinsic length parameter, the cell height [1]. Concepts like stress and strain could then be applied to the process region, and constitutive equations, limited to highly dynamic states, could, in principle, be established.

Regard now the problem of a crack, symmetrically expanding from zero length with given constant velocity in the high velocity range. Remote mode I loading is assumed. The crack faces are traction free and the material, which is regarded as a continuum, is described by elastic-plastic constitutive equations outside the process region. Inside the process region the continuum equations, which in principle could be established for sufficiently high crack velocities are assumed to hold, regardless of the size of the process region, which of course, is very small during an initial stage. Assume also that no rate effects, like visco-plasticity, are involved. Then an obvious solution consists of complete self-similarity. Such a solution would ideally be smooth with stress magnitudes decreasing towards zero as the crack edge is approached. Crack opening would thus occur automatically: no crack growth criterion would be involved. The solution is therefore unique. Other solutions exist, of course, under other conditions, specifying different kinds of crack growth criteria.

Compare this idealized model with a real situation, a crack symmetrically expanding under remote mode I loading from a finite length. During the initial stages of the acceleration phase the process region size does not change very much: it consists of essentially one layer of cells. During later stages it grows until it comprises so many cells that it starts behaving like a continuum. The conditions for the real case then approach those in the idealized case, and constant crack edge velocity should eventually be approached.

Note now that the idealized problem may well have a unique solution for a range of velocities in the high velocity region, but it is not possible to find a preference for one particular velocity, except that the velocity must be sufficiently high for the continuum description to apply. What, then, determines the constant velocity in a particular case?

The answer has to do with the way in which the self-similar state is generated. This situation is not unique in crack mechanics: a similar one is known for steady state crack propagation. Thus the idealized problem of steady state unidirectional mode II crack motion possesses a unique solution for each given sub-Rayleigh velocity, at infinitesimally small scale yielding. In a corresponding real case, such as earthquake slip, the crack starts expanding, but is converted after a while, for instance due to an obstacle on one side, to unidirectional motion, eventually approaching steady state. The constant velocity reached and the crack length (the two are inter-related, see [23] depend on conditions during the initial transient phase, for instance the distance to the obstacle from the crack origin [24].

So far the discussion has concerned a symmetrically expanding crack under remote loading, but it is reaonable to assume that the main conclusions should hold also for other body and loading geometries, such as crack face loading, as in the experiments by Ravi-Chandar [2].

For the continuing discussion, it might be helpful to make a distinction between the absolute (linear) size of the process region and the relative (linear) size, which is the absolute size divided by the crack length. At self-similar motion the relative size of the process region is constant.

If the initial acceleration is very high, the absolute size of the process region increases very rapidly. Then the transition towards constant velocity occurs already when the crack length is relatively small. Thus, the relative size of the process region, when constant velocity is approached, is very large, compared to a case when constant velocity is reached after a lower initial acceleration, or, equivalently, at a lower applied load. Thus it could be expected that the constant velocity obtained increases with the applied load. This is certainly usually the case, judging from experimental results [2],[3],[4] and numerical simulations [6],[7],[20].

One series of simulations by Johnson [6] resulted in the unexpected outcome that the constant terminal velocity decreased somewhat when the remote load was increased above a certain limit, whereas it had been increasing at lower loads. This might be due to what was called "barrier effect" in [1], but would nowadays be called "shielding".

A requirement for crack growth is that a certain part of the energy flux from the stress-strain field penetrates the process region and reaches the cells immediately ahead of the crack edge. At high velocities there might be a tendency of trapping the energy flux somewhere between the process region periphery and the crack edge. In a real case (as opposed to the idealized continuum model) this implies excessive coalescences between micro-separations inside the process region. Such coalescences may be related to a strain criterion: for a void growth mechanism it is generally anticipated that a cell collapses at a strain of 20-30 %. The simulations by Johnson [6] showed a smaller process region at the load for which the highest constant velocity was obtained than at the larger load, unable to produce an equally high or higher velocity. They also showed a tendency towards offside coalescences at the higher load.

If shielding really plays the role indicated by theoretical speculations and numerical investigations, then there seems to be a highest possible sub-Rayleigh mode I crack velocity in each given material, appearing at some load magnitude for a favourable body and loading geometry.

References

[1] K.B. Broberg, On the behaviour of the process region at a fast running crack tip. In *High Velocity Deformation of Solids*, edited by K. Kawata and J. Shioiri, Springer-Verlag, Berlin Heidelberg, 1979, pp. 182-194.

[2] K. Ravi-Chandar, An experimental investigation into the mechanics of dynamic fracture. *Ph.D. Thesis*, California Institute of Technology, Pasdena, California, 1982.

[3] K. Ravi-Chandar, W.G. Knauss, An experimental investigation into dynamic fracture, in four parts. *Int. J. Fract.*, I Crack initiation and arrest: Vol. 25, 1984, pp. 247-262, II Microstructural aspects, Vol. 26, 1984, pp. 65-80, III On steady-state crack propagation and crack branching, Vol. 26, 1984, pp. 141-154, IV On the interaction of stress waves with propagating cracks, Vol. 26, 1984, pp. 189-200.

[4] J.F. Kalthoff, On some current problems in experimental fracture mechanics. In *Workshop on Dynamic Fracture*, California Institute of Technology, Pasadena, California, 1983, pp. 11-35.

[5] K. Arakawa, K. Takahashi, Relationships between fracture parameters and fracture surface roughness of brittle polymers. *Int. J. Fract.*, Vol. 48, 1991, pp. 103-114.

[6] E. Johnson, Process region changes for rapidly propagating cracks. *Int. J. Fract.*, Vol. 55, 1992, pp. 47-63.

[7] E. Johnson, Process region influence on energy release rate and crack tip velocity during rapid crack propagation. *Int. J. Fract.*, Vol. 61, 1993, pp. 183-187.

[8] K. Takahashi, K. Arakawa, Dependence of crack acceleration on the dynamic stress-intensity factor in polymers. *Experimental Mechanics*, Vol. 27, 1987, pp. 195-200.

[9] L.B. Freund, Some theoretical results on the dependence of the dynamic stress intensity factor on crack tip speed. In *Workshop on Dynamic Fracture*, California Institute of Technology, Pasadena, California, 1983, pp. 129-136.

[10] J.W. Dally, W.L. Fourney, G.R. Irwin, On the uniqueness of the stress intensity factor – crack velocity relationship. *Int. J. Fract.*, Vol. 27, 1985, pp. 159-168.

[11] W. Yang, L.B. Freund, Transverse shear effects for through-cracks in an elastic plate. *Int. J. Solids and Structures*, Vol. 21, 1985, pp. 977-994.

[12] A.J. Rosakis, K. Ravi-Chandar, On crack-tip stress state: an experimental evaluation of three-dimensional effects. *Int. J. Solids and Structures*, Vol. 22, 1986, pp. 121-134.

[13] C.C. Ma, L.B. Freund, The extent of the stress intensity factor field during crack growth under dynamic loading conditions. *J. Appl. Mech.*, Vol. 53, 1986, pp. 303-310.

[14] L.B. Freund, A.J. Rosakis, The structure of the near-tip field during transient elastodynamic crack growth. *J. Mech. Phys. Solids*, Vol. 40, 1992, pp. 699-719.

[15] C. Liu, A.J. Rosakis, L.B. Freund, The interpretation of optical caustics in the presence of dynamic non-uniform crack-tip motion histories: A study based on a higher order transient crack-tip expansion. *Int. J. Solids Struct.*, Vol. 30, 1993, pp. 875-897.

[16] S. Aoki, Y. Nonoyama, K. Amaya, Boundary element study on the optical method of caustic for measuring fast crack propagation toughness K_{ID}. *Int. J. Fract.*, Vol. 71, 1995, pp. 379-390.

[17] H. Schardin, Ergebnisse der kinematographischen Untersuchung des Glasbruchvorganges. *Glastechnische Berichte*, Vol. 23, 1950, pp. 1-10; 67-79; 325-336.

[18] H. Schardin, Velocity effects in fracture. In *Fracture*, edited by B.L. Averbach, D.K. Felbeck, G.T. Hahn and D.A. Thomas, John Wiley & Sons, New York, 1959, pp. 297-329.

[19] P.D. Washabaugh, W.G. Knauss, A reconciliation of dynamic crack velocity and Rayleigh wave speed in isotropic brittle solids. *Int. J. Fract.*, Vol. 65, 1994, pp.97-114.

[20] E. Johnson, Process region influence on crack branching. *Int. J. Fract.*, Vol. 57, 1992, pp. R27-R29.

[21] K.B. Broberg, On the speed of a brittle crack. *J. Appl. Mech.*, Vol. 31, 1964, pp. 546-547.

[22] K.B. Broberg, Discussion of fracture from the energy point of view. *Recent Progress in Applied Mechanics*, edited by B. Broberg, J. Hult and F. Niordson. Almqvist and Wiksell, Stockholm, 1967, pp. 125-151.

[23] K.B. Broberg, On transient sliding motion. *Geophys. J. R. astr. Soc.*, Vol. 52, 1978, pp. 307-432.

[24] E. Johnson, Initiation of unidirectional slip. *Geophys. J. Int.*, Vol. 101, 1990, pp. 125-132.

Crack Growth Equations

Andrzej Neimitz

Dept. of Mechanical Eng. Kielce University of Technology, Kielce, POLAND

Summary

In the paper the equations of stable and unstable crack growth are reconsidered in the framework of thermodynamics. The new approach is proposed that is tested for the fast crack growth. The results obtained reflect qualitatively the main features of this process.

Keywords: Crack growth equations, fast crack growth, Dugdale model

1 Introduction

An ultimate purpose of each engineer - designer is to find a real response of a structural member to external loading. They are mostly concerned how the geometry changes under given loading conditions and at which moment of the loading history the structure may fail due to excessive irreversible deformations or fracture. In many cases the answers for the above questions can be obtained utilizing tools established within the frame of mechanics of deformable materials i.e. continuum mechanics. The set of equilibrium equations supplemented by relations between strain and displacement components and constitutive equations can in principle be solved for given boundary conditions. The constitutive equations are understood here as a kinematic response of material to stresses $\sigma_{ij}(x_i)$ at a particular point x_i within it. Thus, we again have the relation between the measure of loading (but now internal) and kinematic response (strains or strain rates) but now confined to an arbitrary point within material and independent of element geometry. Thus, the constitutive equation characterizes the properties of materials. In this paper we will concentrate on fracture mechanics which is very often classified as a separate interdisciplinary field of science. However, its backbone has been located within the frame of the mechanics of solids. The main goal of fracture mechanics is to predict the failure of structural or machine member under applied loading due to the evolution of the crack located within material. Thus, again the response like relation should be formulated. Response of the kinematical quantity, now the crack length or crack length (area) growth rate to external loading. This relation will be called here the crack evolution equation. It is neither a constitutive equation as defined above nor response relation between external forces and geometry change defined e.g. in theory of elasticity. It is expected that crack growth evolution equation defines material constants or parameters characterizing fracture toughness or resistance to the crack growth that are geometry independent (similarity to constitutive equations). These constants (parameters) are defined independently of classical constitutive equations and enrich our knowledge about material itself. In fracture mechanics one limits one's attention to a crack itself. Crack surfaces are in fact the part of the element surfaces (external or internal) but because of stress and/or strain concentration in front of its edge they require a special attention. In the present paper the nature of the crack evolution equations is discussed. A new approach is shown that has been applied to the fast crack growth analysis.

2 Local and Global Approach to Fracture Mechanics.

The process of fracture with a crack present within material is in most cases the local process that takes place just ahead of the crack tip. Thus, the natural tendency that has been observed in fracture mechanics is to find one parameter (or more) which could characterize the process itself being at the same time uniquely connected to external loading and geometry of the specimen. On the other hand the critical value of this parameter should be a material constant or parameter preferably independent of specimen geometry. For selected group of materials - brittle and quasi brittle this approach has been successful to predict the crack growth initiation moment and to characterize initial stage of the crack growth. However, fracture analysis started from the global approach proposed by Griffith in 1921 [1]. Griffith analyzed brittle material: glass and his famous hypothesis was that just before the onset of crack growth the crack was in equilibrium and in such a state the total potential energy of a material had a minimum value. Utilizing the principle of energy conservation he introduced the length of the crack as an internal state variable of the system under consideration. Griffith approach was later extended to quasi brittle materials by Irwin and Orowan. At that time the analysis was purely within mechanics of deformable materials and the notion of internal variables was not mentioned. The classical Griffith - Irwin - Orowan criterion of the crack growth initiation can be written in the form

$$G(applied\ loading,\ a,\ geometry) = 2\gamma(Griffith) = G_{IC}(Irwin,\ Orowan) \tag{1}$$

where $G = -\partial P/\partial a$ and P is potential energy $P = W - L$, W is the work done by external forces, L is the strain energy stored in the body, γ is a specific surface energy of a material, G_{IC} is a specific energy dissipated when crack moves over distance da and it also includes irreversible plastic dissipation. G is a global quantity and it depends on the overall geometry of the element. In contrast the quantities γ or G_{ic} (i = I, II, III for three elementary modes of loading) are considered to be material constants if certain geometrical requirements are satisfied: the plain strain exists. It is due to Irwin that local approach supplemented the global one in fracture mechanics. Utilizing Williams and Sneddon solutions of the boundary value problems where crack entered as a part of the specimen boundary he noticed the universal character of the stress field in front of the crack and introduced the concept of the stress intensity factor (SIF) K that was an amplitude of the singular stresses. He also derived a relation between G and K (e.g. $G_I = K_I^2/E$ for plane strain and Mode I). The Griffith and Irwin approach was followed by Rice [2] and Hutchinson [3] and Rice and Rosengren [4] to extend analysis onto nonlinear materials. The J integral was introduced (an equivalent definition to G) which being a global quantity serves as well as an amplitude of the singular HRR field. For crack initiation criteria both local and global approaches are equivalent. The local and global quantities are related to each others and their critical values may be considered (under certain restrictions) as material constants. It must be emphasized here that in deriving relations between local and global quantities it was always assumed that global quantity (G or J) is related to the local one (K or J) if the former represents also an energy flux directly and only to the crack tip. Computing the energy flux into the crack tip the first law of thermodynamics is usually utilized along with the field theory and crack length cannot be introduced as an internal variable. It enters solution as a parameter characterizing the strength of singularity in front of the crack tip. However, independently of the fact whether crack length can be or not an internal state variable it always enters fracture criteria and its change is the response to the external loading.

3 The Global Approach to Fracture Mechanics - Thermodynamics of Fracture Process

The fracture criterion or crack growth evolution equation are those physical postulates that distinguish fracture mechanics from others branches of mechanics of solids. The fracture process itself is a very complex one involving variety of physical phenomena at the microscale. They include several different mechanisms of a new surface creation, plastic dissipation, voids and microcrack nucleation and growth, heat flow, kinetic energy generation and possibly others. These processes are usually not confined to the crack tip only. Thus, an approach based on the continuum mechanics only may not be sufficient to understand the fracture correctly and to formulate the proper evolution equation. The thermodynamics may serve as an additional source of understanding.

"Classical" approach based on continuum mechanics only led to a series of fracture mechanics criteria and evolution equations that were simply an extension of the original Griffith idea of the energy conservation expressed either in local or in global quantities. Without going into extended analysis of the experimental data there are enough experimental evidences to say that the stable crack initiation criteria predict sufficiently accurate the local instability moment provided the geometry of the specimen satisfies the plane strain condition. The more brittle material is the more precise prediction can be expected. The equations proposed for stable crack growth reflect the main features of the real process. They are based on a R-curve concept that in fact represents the sequence of the equilibrium states. These equations are based on global quantities mostly and are strongly geometry dependent (the CTOA being the local quantity when computed is related to the global quantities). They have no predictive power - except for the first stage of the crack growth that does not exceed 6% of unbroken ligament. The onset of unstable crack growth could be defined according to mentioned criteria if the R - curves were properly defined. Finally, the equations for unstable crack growth do not reflect either quantitive or qualitative features of the fast crack growth. Here the big gap between theory and experiment is observed. Thus, we can see that the further from the equilibrium the poorer agreement between theory and experiment.

Rice was one of the first who reconsidered the process of the crack growth in terms of thermodynamics. His main result [5] was the equation relating the crack driving force with internal entropy production rate in the form

$$\dot{S}_i = (G - 2\gamma)\dot{a} \geq 0 \tag{2}$$

In his analysis Rice assumed that the fracture process is isothermal and totally reversible including creation of a new surface. Thus his free Helmholtz energy consisted of reversible elastic strain energy and reversible surface energy $2\gamma a$ (the unit thickness was assumed). He introduced three internal state variables: temperature T, work conjugate displacement Δ and crack length (writing about crack length a as an internal state parameter Rice writes these words within quotation marks). In derivation of (2) all processes were taken as reversible ones thus, one could not see a physical reasons for creation of internal entropy. Eq (2) should be considered in the form proposed as a lower limit for all crack propagation cases. Its form is well known in thermodynamics of irreversible process taking place very close to the equilibrium. $G - 2\gamma$ can be interpreted as a thermodynamical (or generalized) force, $\dot{a}$ as a power conjugate flow (or rate, flux, current). If such interpretation is adopted, during the equilibrium all quantities are equal to zero $\dot{S}_i$, $\dot{a}$ and $G - 2\gamma$. In this way Griffith criterion was obtained. However the question arises: is the definition of thermodynamic force $G - 2\gamma$ correct? Let us reconsider Rice's derivation with fewer restrictions imposed on reversibility of various processes taking place within externally loaded material containing crack. Let us introduce the set of two internal variables: external force, conjugate displacement Δ, temperature T and internal parameter - length of the crack a.

The state functions can be defined as a functions of internal variables and parameter: internal energy $U(\Delta, T, a)$, Helmholtz free energy $\Psi(\Delta, T, a)$ and entropy $S(\Delta, T, a)$:

The first law of thermodynamics can be written as follows:

$$\dot{U} = F\dot{\Delta} + \dot{Q} \tag{3}$$

where F is external force and Q is a heat supply. The second law of thermodynamics is adopted in the form:

$$\dot{S} = \dot{S}_i + \dot{S}_r = \dot{S}_i + \frac{\dot{Q}}{T} \geq 0 \tag{4}$$

where $\dot{S}_i$ is internal entropy production rate, $\dot{S}_i \geq 0$ and equality sign applies to reversible processes. Free energy is $\Psi = U - TS$. Combining (3) and (4) one obtains

$$\dot{U} - T\dot{S} = F\dot{\Delta} - T\dot{S} + \dot{Q} = F\dot{\Delta} - T\dot{S}_i \tag{5}$$

where $(-T\dot{S})$ was added to the left and right hand sides of equation. Relation (5) can be rewritten in the form

$$\left[\frac{\partial U}{\partial \Delta}|_{a,T} - T\frac{\partial S}{\partial \Delta}|_{a,T} \right] \dot{\Delta} + \left[\frac{\partial U}{\partial T}|_{a,\Delta} - T\frac{\partial S}{\partial T}|_{a,\Delta} \right] \dot{T} + \left[\frac{\partial U}{\partial a}|_{T,\Delta} - T\frac{\partial S}{\partial a}|_{T,\Delta} \right] \dot{a} = F\dot{\Delta} - T\dot{S}_i \tag{6}$$

in (6) the term: $[\partial U/\partial \Delta - T \, \partial S/\partial \Delta]\dot{\Delta}$ is a product of quasi - conservative force F_q and displacement rate $\dot{\Delta}$. It represents reversible power. The second term is always equal to zero since $[\partial U/\partial T - T \, \partial S/\partial T]$ is always equal to zero. The third term represents reversible power of internal forces due to internal parameter $[\partial U/\partial a|_{T,\Delta} - T \, \partial S/\partial a|_{T,\Delta}]\dot{a} = L_q|_{T,\Delta}\dot{a}$. The first law of thermodynamics can now be given in the form:

$$(F - F_q|_{a,T})\dot{\Delta} - L_q|_{\Delta,T}\dot{a} = T\dot{S}_i \tag{7}$$

If $\dot{\Delta} = 0$ we have:

$$L_q\dot{a} = -T\dot{S}_i \tag{8}$$

Thus internal forces can be a source of dissipation provided that:

$$L_q\dot{a} = -L_d\dot{a} \tag{9}$$

where L_d is dissipative internal force. Relation (9) must hold because internal power due to internal forces must be equal to zero since it does not enter the first fundamental law [6]. We can now write

$$L_d\dot{a} = T\dot{S}_i \quad ; \quad \dot{a}, S_i \geq 0 \tag{10}$$

and $$L_d = -\left[\frac{\partial U}{\partial a} - T\frac{\partial S}{\partial a} \right] = -\frac{\partial \Psi}{\partial a}|_{\Delta,T} \tag{11}$$

From (8) and (10) it follows that if test is performed at constant displacement (fixed grips) dissipation is driven by internal forces only equal to $\partial \Psi/\partial a|_{\Delta,T}$. For stationary case and the moment of crack initiation one can write

$$L_d = -L_q = -\frac{\partial \Psi}{\partial a}|_{\Delta,T} = -\frac{\partial U}{\partial a}|_{\Delta,T} = G \tag{12}$$

where G is so called the energy release rate. It can also be shown that for the case of constant external load (dead load case) at the onset of the crack growth

$$L_d = \frac{\partial U}{\partial a} = G \tag{13}$$

Usually, for stable crack growth neither external load nor external displacement is constant. It is so for real materials when other forms of dissipation then creation of a new crack surfaces are observed. For this case relation (7) can be considered as a crack growth equation. The left hand side of (7) is an internal and external power driving a crack. The right hand side of (7) represents a dissipation but now understood as:

$$T\dot{S}_i = L_d \dot{a} + F_d \dot{\Delta} \tag{14}$$

where $F_d = F - F_q$ can be called external dissipative force in contrast to L_d an internal dissipative force. The term $F_d\dot{a}$ can describe dissipation due to plastic deformation, voids nucleation and growth or other forms of dissipation. If equation (7) is divided by $\dot{a}$ it can be rewritten as

$$C = \frac{(F - F_q)d\Delta}{da} - L_q = \frac{T\dot{S}_i}{\dot{a}} = D \tag{15}$$

where $C = (F - F_q)d\Delta/da - L_q$ is the driving energy rate due to external and internal forces. $D = TS_i/\dot{a} = L_d + F_d d\Delta/da$ is dissipation rate. The symbols C and D were adopted from Turner's model [7]. In fact Eq (7) strongly supports Turner's model of the "two step elementary crack increment" consisting of the damage accumulation in front of the crack tip at constant crack length followed by the crack jump at constant displacement. Eq (7) is a more general form of a Rice's equation (2) derived without making assumption on reversibility of the fracture process. Here 2γ is replaced by $L_q = -L_d \equiv 2\gamma$.

The general structure of the crack growth equation that will be used in the present paper is as follows:

$$\zeta\dot{a} = D \tag{16}$$

where ζ will be called the crack driving force; $\zeta\dot{a} = (F - F_q)\dot{\Delta} - L_d\dot{a}$ and the first term on the *rhs* is understood as a part of the work rate due to external forces that is not reversibly stored within the material. The second term represents the release of the internal power associated with the material separation. ζ should be computed for a given boundary value problem. D is called dissipation function and should be computed independently of ζ from the micro - macro mechanics approach taking into account evolution of the dominant dissipation process.

4 Local Approach to Fracture Mechanics

Fracture of material containing macrocrack is a local process limited to the closest neighbourhood of the crack tip. In the second paragraph the relation of the local to global approach was shortly discussed. It can be shown that both relation between SIF-K and ERR-G as well as the definition of the J integral follow from the first law of thermodynamics and are proportional to the energy flux $\dot{\Gamma}$ to the crack tip.

$$\dot{\Gamma} = \dot{a}G = \lim_{C \to 0} \int_C \left[\rho \left(e + \frac{1}{2}\dot{u}_i\dot{u}_i \right) \dot{a}_j + \sigma_{ij}\dot{u}_i - h_j \right] n_j ds \tag{17}$$

where C is arbitrary contour around the crack tip, e is internal energy per unit mass. For steady - state above results can be written in the form

$$G = \lim_{C \to 0} \int_C \rho \left[\left(e + \frac{1}{2}\dot{a}_m\dot{a}_k u_{i,m}u_{i,k} \right) l_j^v - \sigma_{ij}l_k^v u_{i,k} - h_j/\dot{a} \right] n_j ds \tag{18}$$

where l_j^v is a unit vector in direction of crack propagation. For quasi static conditions without heat exchange $G = J$ and is called the energy release rate or crack driving force and is utilized to write the crack evolution equation based on a R-curve concept. It is required that integral over contour C is path independent in order to perform the limit procedure. The path independence is satisfied only for steady state and for elastic materials. For inelastic materials according to the deformation theory of plasticity the above integral is path dependent for growing cracks and relation between the local and global approach is broken. For this case the another integral is introduced and G is a sum of the far field contour integral J_{ff} analogous to the integral (17) (except the contour) and the area integral I [8].

$$T^{\bullet} = G = J_{ff} + I \tag{19}$$

The integral I contains terms characterizing the reversible terms of the stored energy and irreversible ones $I = I_{ir} + I_{re}$. Thus

$$G = J_{ff} + I_{ir} + I_{re} \tag{20}$$

Now if the crack evolution equation is written in the form (the R curve concept):

$$G = 2\gamma \tag{21}$$

where 2γ is here a material property characterizing the energy rate associated with a new crack surface formation, and Eq (20) is used one obtains

$$\left(J_{ff} - I_{re}\right) - 2\gamma = I_{ir} \tag{22}$$

The above relation when multiplied by $\dot{a}$ is similar to the Eq (7) with the similar physical interpretation of particular terms within it. Thus it is suggested that equation of crack growth should be used rather in the form of Eq (7) than in the forms stemming from the R-curve concepts. Moreover we extend this form to the case of the fast crack growth although it may be a very strong assumption. The fast crack growth is a strongly nonlinear process being far away from the equilibrium and the state functions e.g. free energy are not well defined in this case.

5 The Fast Crack Growth

In order to test the proposed·crack evolution equation the fast crack growth has been analyzed. The aim was to reanalyze the process of the fast crack growth in the frame of the model proposed and to reduce the gap between the theory and experiment observed so far. It is very important, in particular after a several experimental studies on fast crack growth that were published recently [9] [10] [11]. From these papers the following conclusions can be drawn.
- The best known argument against the existing theories based on the linear theory of elasticity and assumption that the effective fracture energy (2γ) is not dependent on the crack tip speed is observation that the limit crack velocity is much lower than the Rayleigh or shear wave speeds. Such a velocity limit is predicted by the theory. Usually, we observe speeds less than half of those values.
- Another important discrepancy between the theory and experiment is that experiments show the lack of the unique dependence between the crack tip speed and DSIF. Such a dependence is predicted by the theory
- According to the theory the energy flux to the crack tip is not a function of the crack tip acceleration. In the experiment the acceleration/deceleration are observed and the change of the crack tip speed does not follow the SIF - velocity relations. Quite recently a results by Freund and Rosakis [12] introduced the acceleration to the equations for the stress distribution in front of a crack tip.

The first discrepancy listed has already been resolved by Slepyan (1991, 1993), [13] [14] who introduced the principle of the maximum energy dissipated. He defined the excess of the energy flux M as:

$$M = \left(G - 2\gamma\right)\dot{a} \tag{23}$$

and from the condition $\partial M/\partial \dot{a} = 0$ computed the maximum crack tip speed in agreement with experiment. Slepyan's equation (23) is in fact Rice's equation (2) or equation (7) but for linear elastic materials. Slepyan principle defines the atractor state to which the growing crack tends and it is in fact the principle of maximum entropy production rate.

In the present analysis we adopt the crack growth equation in the form (7). We will analyze the fast crack growth without restrictions imposed on the crack tip speed. It will be controlled by the crack evolution equation and boundary conditions that may change in time. We allow for plasticity in front of the crack tip but the small scale yielding is assumed. The specimen is assumed to be large enough in order to avoid interaction between the crack tip with waves reflected from the specimen boundaries. The closed form solution to define some quantities entering Eq (7) has been obtained but for the simplest loading: Mode III. Plasticity in front of the crack tip has been modelled according to Dugdale - Panasyuk strip yield zone (s.y.z) model. The left hand side of the Eq (7) has been interpreted as an energy flux into the s.y.z. The length of this zone may change in time. The energy entering s.y.z is totally dissipated there. The energy flux divided by the crack tip speed is called the generalized crack driving force (in fact this force drags both crack tip and plastic zone). Dissipation function on the right hand side of Eq (7) when divided by crack tip speed is called the crack growth resistance force. According to the earlier discussion we assume that the functional form and quantities entering dissipation function should be computed from the separate analysis taking into account microstructural processes in front of the crack tip accompanying the crack growth. In the present analysis we assume that plastic deformation and the new surface creation are the dominant dissipation processes. We will understand the specific surface energy according to the Irwin - Orowan interpretation rather than the Griffith one. Thus it is reasonable to assume that the specific surface energy g is a function of the crack tip speed: $g = g(\beta_T)$ where $\beta_T = \dot{a}_T/c_T$, c_T is the speed of transverse wave, $\dot{a}_T$ is speed of the crack tip. We expand this quantity around the steady state and compute the rate of it to obtain:

$$D = \frac{1}{B}\beta_t c_T \Gamma_{SS}(\beta_T) + \frac{1}{B}\frac{dM}{dt}\left(\beta_T - \beta_t^{ss}\right) + \frac{1}{B}M\dot{\beta}_T + \dots \tag{24}$$

where B is the specimen thickness, $\Gamma_{SS} = dg(\beta_T)/da$, $M \equiv dg(\beta_T)/d\beta_T$ and is here interpreted as a product of ξ - the specific energy of plastic zone formation per unit mass and m the mass of the plastic zone. We will call M as an equivalent mass of the plastic zone in front of the crack tip. In the computations we assume that ξ is known and constant while $m = \alpha \rho r_p^2 B$, where ρ is material density within the plastic zone and r_p is the length of the plastic zone, α is constant and was assumed here as equal to 1. The length of the plastic zone can be computed for the crack model assumed and is equal to [15]:

$$r_p = \frac{\pi}{8}\frac{K_{III}^2}{\tau_f^2}\left(1 - \beta_T\right) \tag{25}$$

where K_{III} is instantaneous static stress intensity factor, τ_f is the yield stress in pure shear. The $\Gamma_{SS}(\beta_T)$ is unknown at the time being and in computer simulation performed to trace the crack growth trajectories various forms of this function were assumed. To compute the energy flux into plastic zone $\zeta\dot{a}$ the formula (17) was used but with contour of integration surrounding closely the plastic zone [15]. This contour has finite, changing size since the length of the strip yield zone changes in time. The stress and displacement components were computed according to procedure introduced by Achenbach and Neimitz [15], Neimitz [16]. The energy flux to the plastic zone is equal:

$$\dot{a}\zeta_{III} = \beta_T c_T \frac{K_{III}^2}{2\mu} \left[\frac{1-\beta_T}{1+\beta_L}\right]^{\frac{1}{2}} \left\{ 2 - \left[\frac{1+\beta_T}{1+\beta_L}\right]^{\frac{1}{2}} + \frac{2}{3}\frac{(\beta_L-\beta_T)}{\beta_T} + \right.$$

$$\left. - \left[\frac{\beta_L-\beta_T}{\beta_T}\right]^{\frac{1}{2}} \left[\frac{1+\beta_L}{\beta_T}\right]^{\frac{1}{2}} \mathrm{arctg}\left[\frac{\beta_L-\beta_T}{1+\beta_T}\right]^{\frac{1}{2}} \quad for \ \beta_T < \beta_L \right.$$

$$\left. + \left[\frac{\beta_T-\beta_L}{\beta_T}\right]^{\frac{1}{2}} \left[\frac{1+\beta_L}{\beta_T}\right]^{\frac{1}{2}} \ln\left[\frac{(1+\beta_T)^{\frac{1}{2}} + (\beta_T-\beta_L)^{\frac{1}{2}}}{(1+\beta_L)^{\frac{1}{2}}}\right] \quad for \ \beta_T > \beta_L \right.$$

$$(26)$$

where $\beta_L = \dot{a}_L/c_T$ is a speed of a tip of the strip yield zone.

If $\beta_L = \beta_T$ and K_{III} is constant we obtain the formula known for steady state crack growth. In Eq (26) one may observe that the energy flux into the strip yield zone does not contain the crack tip acceleration. It is due to the simplified model of the crack growth kinetics (see Fig. 1) where the "real" crack tips (leading and trailing) trajectories were approximated by a piece - wise linear functions. Thus the crack tip acceleration enters the crack growth equation only through function D. While performing computer simulation this assumption does not influence essentially the results obtained since the linear segments change their slopes along the real trajectory. However, substituting to (26)

$$\beta_L = \beta_T + \frac{dr_P}{dt}\frac{1}{c_T} \qquad \text{and} \qquad \frac{dr_P}{dt} = \frac{\pi}{8}\frac{2K_{III}}{dt}(1-\beta_T) - \frac{\pi}{8}\frac{K_{III}^2}{\tau_f^2}\dot{\beta}_T$$

one may estimate the influence of the crack tip acceleration on the crack growth equation. The left hand side of Eq (16) is not sensitive to the crack tip acceleration up to the level of $10^6 \div 10^7$ m/s^2. The right hand side of Eq (26) does not "feel" acceleration up to the level of 10^2 m/s^2. The computer simulation has been performed utilizing the model of the crack growth kinetics shown in the Figure 1.

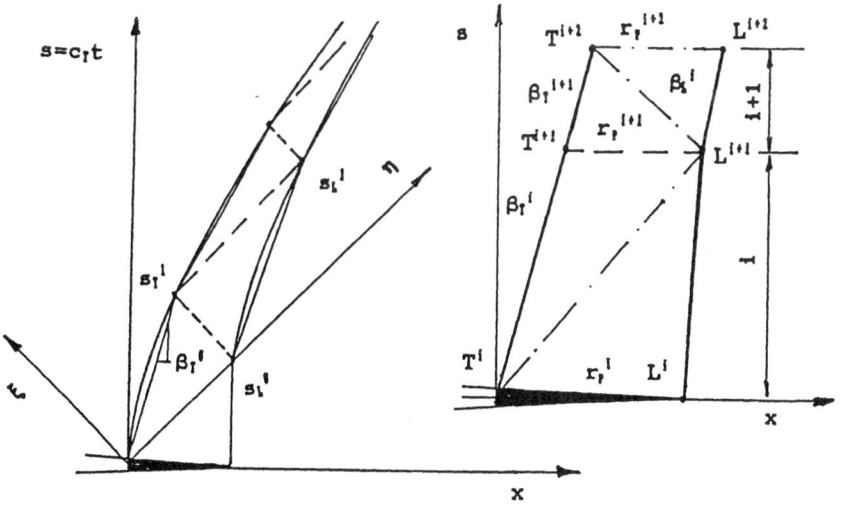

Figure 1: Schematic representation of the leading and trailing edges trajectories used in the computation

At the critical moment the trailing edge of the crack starts moving with the speed β_T^0 while the leading edge is in rest. The information that the trailing edge started its motion travels to the loading edge with a speed of sound in elastic material. At the moment s_L^0 the leading edge starts motion with the same speed as trailing edge and sends information about that to the trailing edge. The signal reaches the trailing edge at the moment s_T^1 which changes its speed according to the crack growth equation, sends the signal to the loading edge and the situation repeats itself. The algorithm of the computer simulation of the fast crack growth according to the presented model was given by Neimitz and Lis [17]. The main purpose of the computer simulation was to verify the proposed model of the crack growth and compare results with experimental data obtained by other authors. However, the comparison may be only qualitative since experiments were performed for Mode I not for Mode III that were analyzed in the paper. The computer simulation should show the main features of the fast crack growth phenomenon. They are: a) at the beginning of the fast crack growth an rapid crack tip acceleration is observed in a very short period of time up to 5-6 μsec, b) at the end of this first stage of propagation the acceleration decays and the mean crack tip speed remains almost constant, c) the crack tip speed is not a unique function of the stress intensity factor, d) some authors claim that the constant crack tip speed value depends on a loading conditions at the onset of the fast crack growth (e.g. Kalthoff) [18] other that it depends on a material only (Fineberg et al [9]), e) Fineberg et al observe sharp velocity oscillations around mean slightly increasing speed of the crack.

To perform simulation typical material constants were adopted K_{IIIC} E, v, σ_y. We changed the stress intensity factor rate (increasing, constant and decreasing) and the material function $\Gamma_{ss}(\beta_T)$. The later one was assumed in various functional forms as shown in the figure 2.

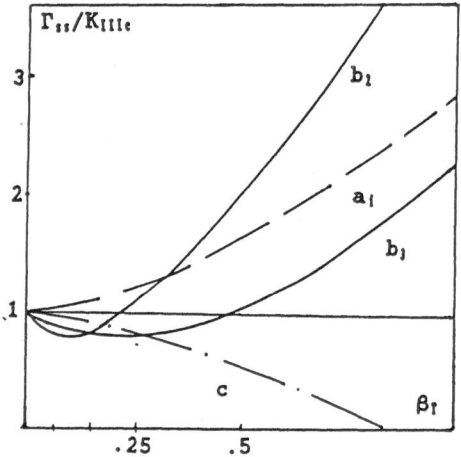

	Γ_{ss}/K_{IC}	β_T
b_1	0.9	0.05
b_2	0.9	0.15
b_3	0.9	0.25
b_4	0.7	0.05
b_5	0.7	0.15
b_6	0.7	0.25
b_7	0.5	0.05
b_8	0.5	0.15
b_9	0.5	0.25

Figure 2: An example material curves $\Gamma_{ss}(\beta_T)/K_{IIIC}$ used in computer simulations

In this paper results with curves a_i, b_i will be discussed only. b_i - type curves differ from each other by location of a minimum ($0.05\beta_T$, $0.15\beta_T$, $0.25\beta_T$) and the value of a function at the minimum (0.9, 0.7, 0.5). We assume that the function Γ_{ss} may first decrease with increasing β_T since there might be less time for evolution of plastic deformation and then increase due to voids, microcracks creation and branching.

We also tested the crack propagation at different loading conditions at the onset of the fast crack growth assuming $K_{init} = mK_{IIIC}$ ($m = 1.0, 1, 2, \ldots 2.0$). The case for which $m > 1$ can be obtained if we blunt slightly the crack tip.

The analysis of numerous results of the computer simulations suggests that the shape of the material functions $\Gamma_{SS}(\beta_T)$ adopted may be of b or a type. In both cases we observe a rapid crack acceleration in the first stage of a crack growth but for type - a curve it is possible only if $m >$ 1.0. Moreover, the crack growth in a material characterized by a curve is slightly more sensitive to the change of the stress intensity factor than for b curves. It is not in agreement with experimental observations as mentioned earlier. These behavior is not observed when type b curve is used in computer simulations. Typical plots of the crack tip speeds v.s. time are shown in the example figures. In the Fig. 3 two curves represent two extreme materials modeled by b_6 and b_2 curves. The external loading is chosen to satisfy condition of constant stress intensity factor during crack growth $K_{in} = K_{IIIC}$. The shape of the curves is similar. They differ only by the value of constant crack tip speed. One observes a rapid crack acceleration in the first stage of propagation and later a constant crack tip speed. In the Fig. 4a and 4b the influence of the stress intensity factor rate dK_{III}/dt on a shape of the β_T vs.t curves for a selected material (curve b5) is shown. One may observe that the constant crack tip speed is not much sensitive to the change of stress intensity factor. The curves are similar both for increasing, decreasing or constant K_{III}. It is so if the rates are not rapid. For rapid SIF increase one observes arrest of the crack. The length of the plastic zone rapidly increases and dissipation is mostly due to plastic deformation. This behavior is similar to the one observed in Mode II loading and during adiabatic shear bands creation. For rapid drop of the SIF the crack is arrested shortly after the onset of crack propagation. The loading conditions at the onset of the crack growth have the decisive effect on constant average crack tip speed. In the Fig. 5 the curves β_T vs.t are shown for two different values of m. This behavior is similar to that observed by Kalthoff. For large m's the oscillations of the crack tip speed may be observed Fig. 6. This behavior was observed by Fineberg et al [9] High m's represent situation where large amount of energy is stored within the specimen before the onset of the crack growth.

The results of computer simulation show that the crack growth equation proposed in this article and the approach to the modelling of the dissipation function provide a proper qualitative description of the fast crack growth. They encourage to undertake more efforts to investigate the more realistic Mode I case of loading.

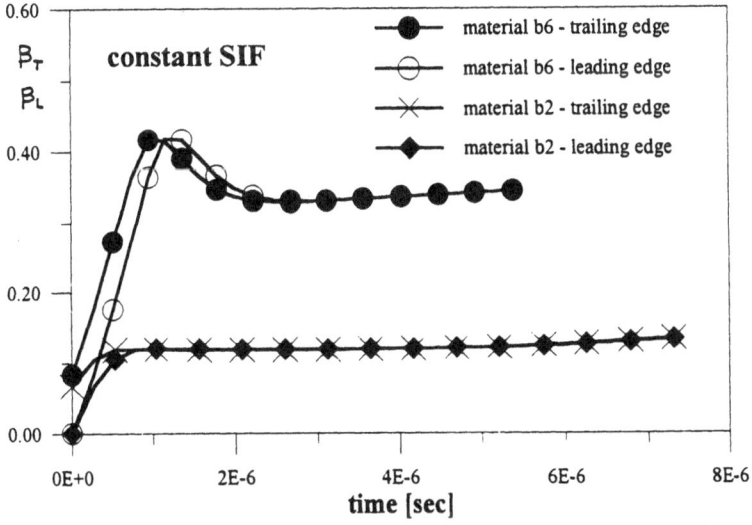

Figure 3: Influence of type of the material on the crack tip speed

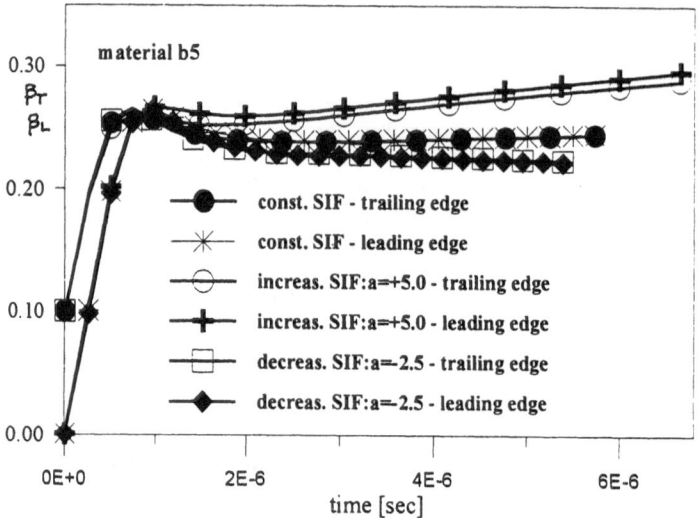

Figure 4a

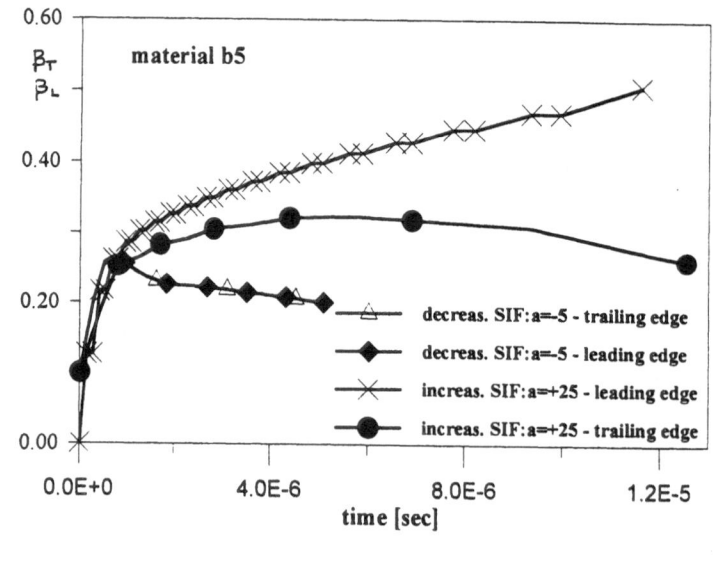

Figure 4b

Figure 4: Influence of the external loading on the crack tip speed

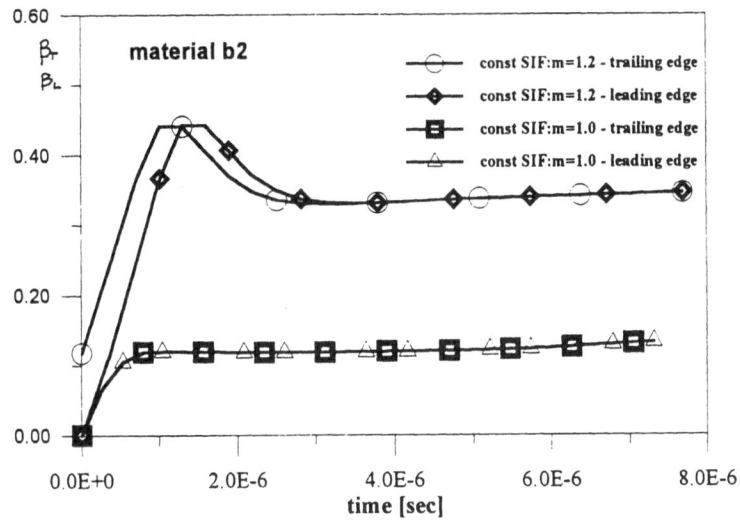

Figure 5: Influence of the loading at initiation on the crack tip speed

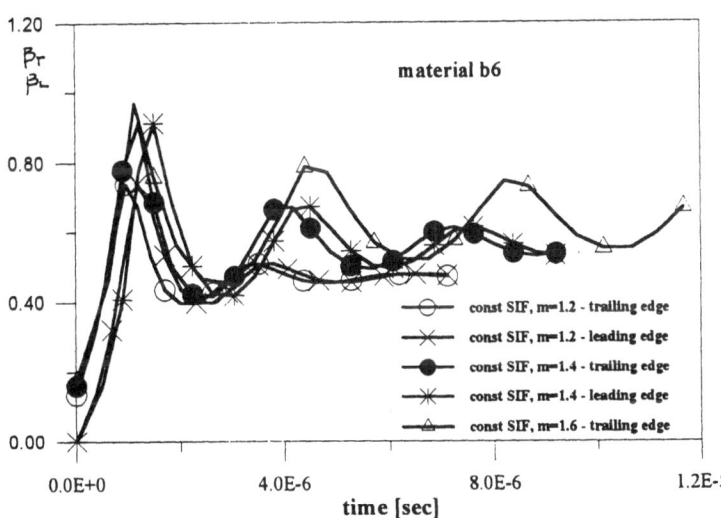

Figure 6: Oscillation of the crack tip speed

Acknowledgement
This work was performed under grant No. 234/T07/95/08 sponsored by Polish Research Committee (KBN).

References

[1] A. A. Griffith, The phenomenon of rapture and flow in solids, *Philosophical Transactions of the Royal Society*, London, A221, 1920, pp. 163-198.

[2] J. R. Rice, A Path Independent Integral and the Approximate Analysis of Strain Concentrations by Notches and Cracks, *J. Appl. Mech.*, 1968, pp. 379-386.

[3] J. W. Hutchinson, Singular Behaviour at the End of a Tensile Crack in a Hardening Material, *J. Mech. Phys. Solids*, Vol. 4, 1968, pp. 13-31.

[4] J. R. Rice, G. F. Rosengren, Plane Strain Deformation Near a Crack Tip in a Power Law Hardening Material, *J. Mech. Phys. Solids*, Vol. 4, 1968, pp. 1-12.

[5] J. R. Rice, Thermodynamics of the quasi-static growth of Griffith cracks, *J. Mech. Phys. Solids*, Vol. 26, 1978, pp. 61-78.

[6] H. Ziegler, *An Introduction to Termomechanics*, North Holland, 1977.

[7] C. E. Turner, A Re-assessment of Ductile Tearing Resistance, *Fracture Behaviour and Design of Materials and Structures. Proc. of ECF8*, Vol. 2, 1990, pp. 933-949, 951-968.

[8] S. N. Atluri, *Computational methods in the mechanics of fracture*, Vol. 2, 1968, North-Holland, pp. 122-165.

[9] J. Finberg, S. P. Gross, M. Marder, H. L. Swinney, Instability in the Propagation of Fast Cracks, *Physical Review B*, Vol. 45, 1992, pp. 5146-5154.

[10] K. Takahashi, K. Arakawa, Dependence of Crack Acceleration on the Dynamic Stress-Intensity Factor in Polymers, *Experimental Mechanics*, Vol. 27, 1987, pp. 195-200.

[11] K. Ravi-Chandar, W. G. Knauss, An Experimental Investigation into Dynamic Fracture: I. Crack Initiation and Arrest, *International Journal of Fracture*, Vol. 25, 1984, pp. 247-262.

[12] L. B. Freund, A. J. Rosakis, The Structure of the Near-tip Field During Transient Elastodynamic Crack Growth, *J. Mech. Phys. Solids*, Vol. 40, 1992, pp. 699-719.

[13] L. I. Slepyan, Crack dynamics in elastic-plastic body, *Mekhanika Tverdogo Tela* (English translation), Vol. 11, 1976, pp. 126-134.

[14] L. I. Slepyan, Principle of Maximum Energy Dissipation Rate in Crack Dynamics, *J. Mech. Phys. Solids*, Vol. 41, 1993, pp. 1019-1033.

[15] J. D. Achenbach, A. Neimitz, Fast Fracture and Arrest According to the Dugdale Model, *Engineering Fracture Mechanics*, Vol. 14, 1981, pp. 385-395.

[16] A. Neimitz, Analysis of the Crack Motion with Varying Velocity According to the Dugdale-Panasyuk Model, *Engineering Fracture Mechanics*, Vol. 39, 1991, pp. 329-338.

[17] A. Neimitz, Z. Lis, Simulation of Fast Dugdale Crack Motion with Varying Velocity, *Dynamic Failure of Materials*, Elsevier, 1991, pp. 362-377.

[18] J. F. Kalthoff, J. Beinert, S. Winkler, Measurements of Dynamic Stress Intensity Factors for Fast Running and Arresting Crack in Double-cantilever-beam Specimens, *Fast Fracture and Crack Arrest*, American Society for Testing and Materials, Philadelphia, 1977, pp. 161-176.

Dynamic Fracture Toughness and Crack Propagation in Brittle Material

Tadashi SHIOYA and Fenghua ZHOU

Department of Aeronautics and Astronautics, University of Tokyo
7-3-1 Hongo, Bunkyo-ku, Tokyo 113, JAPAN

Summary

Experiments of crack propagation in PMMA plates are performed from which the dynamic fracture toughness G_c of the material is obtained. The relationship between G_c and crack velocity v_0 is associated to the characteristic appearance on the fracture surface. Periodic patterns are observed on the fracture surface which suggest local oscillation of crack velocity. A model for analyzing crack propagation by global energy equilibrium concept is proposed, from which crack motion equation is deduced. Unstable crack propagation and local velocity oscillation is explained using this equation and the particular $G_c(v_0)$ relationship.

Key Word: Brittle fracture, dynamic fracture toughness, crack motion equation, propagation instability

1. Introduction

The problem of dynamic crack propagation is complicated by two factors: 1) the fracture toughness G_c of the material is crack velocity dependent, *i.e.*, a faster moving crack generally consumes more energy than a slower one; 2) the crack propagation changes the mechanical field around the crack tip and therefore changes the energy release rate at the tip. The first factor is of material nature. There have been a lot of experimental work and discussions on the subject (for example, [1] - [3]). The second factor is an elasto-dynamical problem and detailed analysis has been made (as a good reference, [4]). However, there are still many problems unsolved yet, such as: whether a unique $G_c \sim v_0$ relationship exists; whether and if yes, why unstable crack propagation happens. In the present paper, a series of experiments of crack propagation in brittle polymer (PMMA) plates are performed, the *dynamic fracture toughness* G_c of the material is evaluated as an increasing function of crack velocity v_0. The fracture surface of the specimen is examined after the tests and periodic patterns are observed in certain crack velocity range. The $G_c \sim v_0$ relationship is associated to the characteristic appearance of the crack surface, while the periodic patterns observed suggest the local unstable crack propagation. For the problem of crack propagation in a fixed sided strip, an analyzing model based on the global energy equilibrium concept is proposed

105

from which the motion equation of the crack system is derived. It is seen that the crack system has an equivalent mass corresponding to the crack acceleration. Using this equation, the crack propagation can be determined provided that $G_c(v)$ relationship is given. For a unique increasing relationship of $G_c(v)$, there is a solution of steady crack propagation, however, if $G_c(v)$ relationship is also disturbed by crack acceleration, unstable crack propagation will appear and the crack will run with local velocity oscillation.

2. Experiments

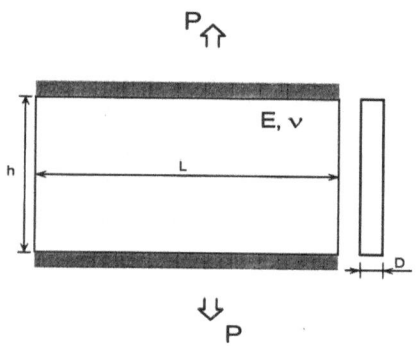

Experiments of crack propagating in the fixed-sided plate are performed. The PMMA plate of the thickness D (=3mm) is used as the testing material. Effective specimen area is rectangular shape with length L and width h, as Figure 1 shows. The sides of the specimen are chucked by a pair of heavy grips so that fixed boundary condition applies.

Fig. 1 Test configuration and specimen dimensions

Before crack initiation, the specimen is preloaded by certain boundary displacement. The total tensile load P is recorded. Therefore, the elastic energy stored in unit length of the tensioned plate is:

$$W_0 = \frac{\sigma^2 h}{2(1-v^2)E} = \frac{P^2 h}{2(1-v^2)EL^2 D^2} \qquad (1)$$

where E is Young's module and v is Poisson's ratio of the material.

When a small crack is initiated at the middle point of one specimen end, it propagates straight across the specimen. The history of crack propagation is recorded with the conductive lines drawn on the specimen surface, and crack velocity is calculated from this record. Typical records of crack propagation velocity along the specimen length are shown in Figure 2.

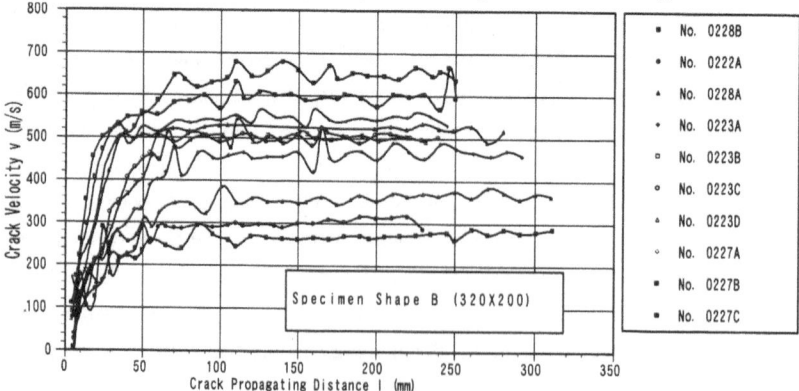

Fig. 2 The crack propagation velocity along the specimen (specimen type B)

From this figure, it is seen that after a short acceleration phase, the crack attains an (average) steady velocity v_0. For the specimen with the same geometric dimensions $(L \times h)$, different v_0 are obtained when the load magnitudes (P) is varied. Specimens with five sets of geometric dimensions are tested and similar results are obtained. In summary, the steady velocity v_0 are varied according to the variation of P, L and h. Because in the phase of the steady propagation, the *energy stored in unit length of specimen* W_0, as calculated by (1), is the same to the *energy consumed by the crack*, the *dynamic fracture toughness* G_c is measured. Each test data is plotted in $G_c \sim v_0$ coordinate system, as Figure 3 shows. It is seen that all the points are located nearing a monotonically increasing curve. As a comparison, the same tests data are also plotted in the stress $(=P/LD) \sim v_0$ coordinate system, as shown in Figure 4, the points scatter heavily. It is concluded therefore that the energy stored in the specimen governs the crack propagation velocity, and there exists an (average) unique relationship between G_c and v_0.

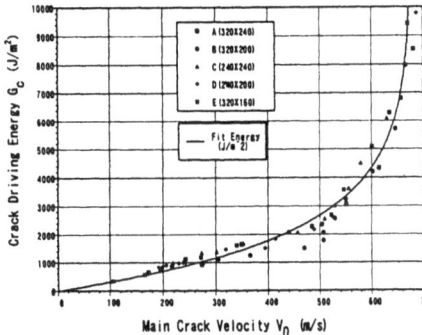
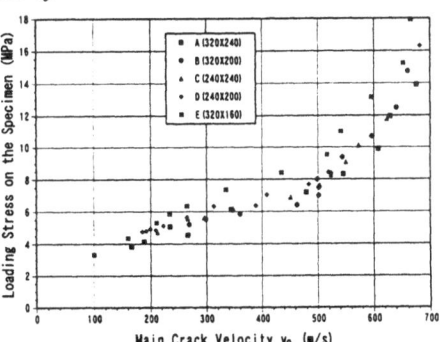

Fig. 3 $G_c \sim v_0$ relationship of PMMA Fig. 4 Test data plotted in $\sigma \sim v_0$ pair

Two phenomena are observed in Fig. 3: 1) G_c value approaches zero when crack velocity is very small, which means that the material is *very brittle* so that the static toughness is almost zero. This tendency is also demonstrated by the fact that a very small crack initiation ($1 \sim 2$ mm) will start the fracture process. 2) G_c approaches infinite at certain v_0 value. This tendency represents the limiting on crack velocity by the material nature, and is observed by many researchers [2], [3]. To emphasizing these two aspects, $G_c \sim v_0$ relationship of Fig. 3 is fitted by following empirically expression:

$$G_c(v) = G_0 \log\left(\frac{v_L}{v_L - v}\right) \qquad (2)$$

where v_L and G_c are the material constants, with the empirical values $v_L = 675$ m/s, $G_c = 2000$ J/m² for the PMMA material been tested.

The fracture surface of the specimen is examined microscopically after the test. Depending on the crack velocity, various characteristic fracture surface appears. When velocity increases from low to high in order, following microscopic fracture modes exist: 1) *mirror* ($v_0 = 100 \sim 200$ m/s), 2) *parabola patterns* from sparsely distributed to densely distributed ($v_0 = 200 \sim 480$ m/s), 3) *periodic patterns* with shifting parabola patterns and crazes ($v_0 = 480 \sim 600$ m/s), 4) *local (unsuccessful) branching* ($v_0 = 600 \sim 650$ m/s) and finally 5) *global branching* ($v_0 > \sim 650$ m/s). The microscopic characteristics is similar to that observed by other researchers (*e.g.*, [1], [2]). These mode changes with crack velocity are associated directly to the rapidly increasing $G_c \sim v_0$ relationship shown in Fig 3 [5].

The specific phenomenon is the periodic patterns appeared in the fractured surface at certain velocity range (v_0=480~600 m/s). It seems that the crack propagation is not stable in the microscopic scale, although in the macroscopic scale it is uniform. Such

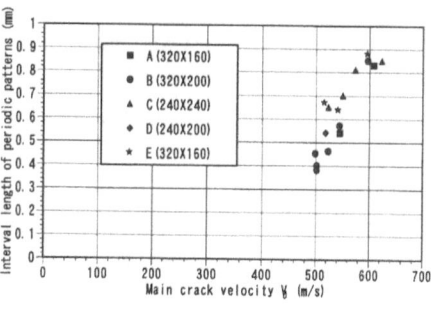

Fig. 5 $\lambda \sim v_0$ relationship

phenomenon is also reported recently by other researchers in detail [6, 7]. Figure 5 shows the relationship between the space interval length λ of the periodic patterns and the average propagation velocity v_0, which is similar to that given by Washabaugh and Knauss [7].

3. Analysis

3.1 Analytical model and energy terms

A model for analyzing crack propagation in the fixed-sided strip is proposed. First consider the plane problem shown in Figure 6: In the middle line of a strip loaded by constant boundary displacement δ_0, A crack is propagating uniformly with the velocity v_0; the material of the strip is elastic, with shearing module μ, Poisson ratio v and mass density ρ. A x-y coordinate system fixed at the crack tip is established.

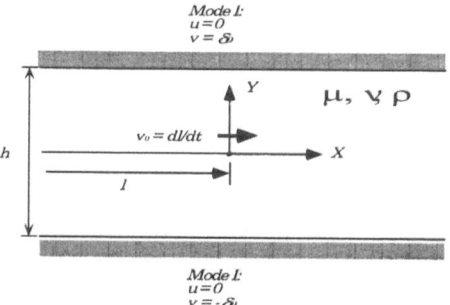

Fig. 6   Crack propagation analyzing model

Far ahead of the crack tip in the strip, the energy stored in the unit length of the strip is

$$W_0 = \frac{\kappa + 1}{\kappa - 1} \frac{\mu \delta_0^2}{h} \qquad (3)$$

where $\kappa = (3-v)/(1+v)$ for plane stress case and $\kappa = 3-4v$ for plane strain case.

The analytic problem of Fig. 6 is the same to that analyzed by Fujimoto et. al.[8] and Liu et. al.[9]. Because the crack is propagating in constant speed v_0, the differential with the time is changed into the differential with the x-coordinate: $\frac{\partial}{\partial t} = -v_0 \frac{\partial}{\partial x}$. Therefore the dynamic problem is changed into a static problem without time variable. Unlike [8] and [9] where complicated mathematical deductions are performed, the numerical finite differential method is used here to calculate the elastic displacement field (u, v) directly.

The global kinetic energy T and strain energy U of the system are evaluated by the following formulae:

$$T = \int_{-H/2}^{H/2} dy \int_{-\infty}^{+\infty} dx \left\{ \frac{1}{2} \rho \dot{u}_i \dot{u}_i \right\}$$

$$U = \int_{-H/2}^{H/2} dy \int_{-\infty}^{+\infty} dx \left\{ \frac{1}{2} \sigma_{ij} \varepsilon_{ij} \right\} \qquad (4)$$

and the total energy is: $E = T + U$

The strain energy U and the total energy E of the system defined above are not convergent, however, since only the changes of these energy terms with crack velocity are invoked, finite integration region is chosen. These energy terms are evaluated by their difference from the static value (when $v_0 = 0$). Figure 7 shows the change of the energy terms with the crack velocity, where crack velocity is non-dimensionalized by the shearing wave velocity $c_t = (\mu/\rho)^{1/2}$ and the energy terms by the $E_0 = W_0 h = (\kappa+1)\mu\delta_0^2/(\kappa-1)$

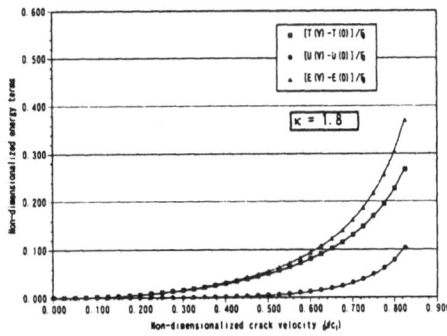

Fig. 7 Energy terms change with crack velocity

3.2 Motion equation of crack

Crack propagating with varying speed is considered now. It is assumed that the above results of system energy terms are still applicable, which implies that the crack velocity does not change radically. Therefore, the unsteady crack propagation is treated as the change between a series of *quasi-steady* states. The crack length l and the crack velocity v (hereafter without subscript $_0$) are the *state variables* (or the *generalized coordinates*) of the system, and the total energy of the system E is the function of (l, v), as expressed in Fig 7. Since during crack propagation, no work is done by the external force acting on the system, the energy necessary to create the new fracture surface is only supplied by the system energy storation, as:

$$dE + dW_d = 0 \qquad (5)$$

where dW_d is the energy consumption by unit length of the new fracture surface. $dW_d = G_c dl$, when G_c is the fracture toughness. The change of system energy dE is expressed as,

$$dE = \left(\frac{\partial E}{\partial v}\right)_l dv + \left(\frac{\partial E}{\partial l}\right)_v dl = \left(\frac{\partial E}{\partial v}\right)_l dv - W_0 dl$$

Then the energy equilibrium equation (5) yields:

$$\frac{1}{v}\left(\frac{\partial E}{\partial v}\right)_l \frac{dv}{dt} = W_0 - G_c(v) \qquad (6)$$

or $M(v)\dfrac{dv}{dt} = W_0 - G_c(v) \qquad (6')$

where $M(v) = \dfrac{1}{v}\dfrac{\partial E}{\partial v}$ is the *generalized mass* of the crack system.

From the numerical results, it is seen that when the crack velocity v is small, each energy term is approximately proportional to the square of v. If the energy factor $e(v)$ is

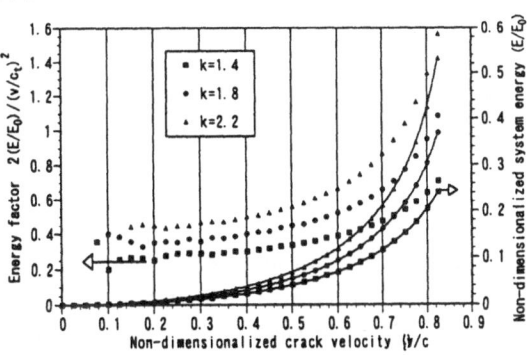

Fig. 8 The system mass E(v) and the mass factor e(v)

defined by: $E(v) = \dfrac{1}{2}v^2 e(v)$, then the generalized mass is expressed by: $M(v) = e(v) + \dfrac{1}{2}ve'(v)$.

The functions $E(v)$ and $e(v)$ are demonstrated in figure 8. It is seen that $e(v)$ is almost constant when $v/c_t<0.4$, so that $M(v)$ is also approximately constant at this velocity range. When $v/c_t<0.4$, the non-dimensionalized value of generalized mass, $m(v) = M(v)\dfrac{c_t^2}{E_0}$ is: $m{\sim}0.28$ for $\kappa{=}1.4$; $m{\sim}0.37$ for $\kappa{=}1.8$ and $m{\sim}0.45$ for $\kappa{=}2.2$.

4. Explanation of crack propagation instability phenomenon

Traditional explanation for the crack propagation instability phenomenon is the stress wave reflection from the plate surface [10], while recently Washabaugh and Knauss suggested that the phenomenon is inherent during the fracture process [7]. The work of Abraham et al [11] showed that crack instability is the immanent one even in the atomic scale. In the following, we try to explain the unstable phenomenon through the stability analysis on the above crack motion equation.

The motion equation of the crack (6') is similar to that of a mass point, with $M(v)$, W_0 and $G_c(v)$ be understood as the *generalized mass*, *generalized driving force* and *generalized resistance force*. The existence of a positive M implies that crack accelerates when it is *over-driven* ($W_0>G_c$), and decelerates when it is *under-driven* ($W_0<G_c$). The uniform propagation velocity v_0 is kept when $G_c(v_0)=W_0$. It is proved that v_0 is the stable solution since $G_c(v)$ is an increasing function.

However, the existence of a unique increasing $G_c{\sim}v$ relationship has been questioned[3, 12]. The real time measurement of the *dynamic stress intensity factor* K_{ID} by caustics showed that K_{ID} is influenced by crack acceleration [13, 14]. In the present work, the $G_c(v)$ relationship is obtained by the experimental points, which represents the macroscopic (average) crack propagation behavior. Nevertheless, the possibility of local velocity oscillation cannot be excluded (Fig. 2). Suppose that in the microscopic scale, the $G_c(v)$ relationship is (locally) acceleration dependent, and $G_c(v, dv/dt>0) < G_c(v, dv/dt<0)$, as [13, 14] suggested. In other words, at the same velocity, an accelerating crack consumes more energy than a decelerating one. It is seen that the solution of (6') is unstable, implying that local instability of crack velocity will appear. To demonstrate this effect, the $G_c(v)$ relationships is separated according to the sign of crack acceleration:

$$G_c(v) = G_0 \log\left(\frac{v_{L1}}{v_{L1}-v}\right) \quad \frac{dv}{dt} > 0$$

$$G_c(v) = G_0 \log\left(\frac{v_{L2}}{v_{L2}-v}\right) \quad \frac{dv}{dt} < 0 \tag{7}$$

where the material constants are chosen as $v_{L1}{=}700$ m/s, $v_{L2}{=}650$ m/s. The relationship is shown in figure 9, which qualitatively resembles the experimental results of

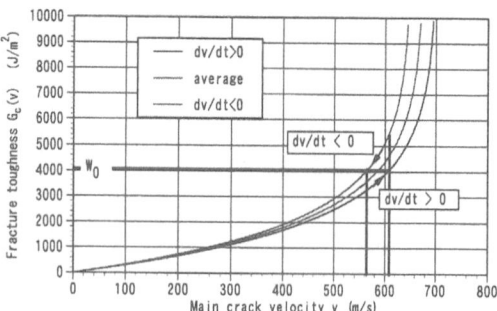

Fig. 9 $G_c(v)$ relationship influenced by acceleration

Takahashi et. al.[13, 14].

The motion equation (6') is rewritten as: $\dfrac{h}{c_i^2} m(v) \dfrac{dv}{dt} = 1 - \dfrac{G_c(v)}{W_0}$ (6'')

into which (7) is substituted. For the PMMA material, the material constants are chosen that $m(v)=0.4$, $c_i=1350$ m/s, $h=200$ mm. The crack velocity is then calculated by numerically integrating (6''). To simulate the unstable phenomenon, a small random perturbation value ($< \pm$ 5%) is added to the right side oh (6'') in each time step, propagating velocity of the crack is shown in figure 10, where it is seen that the crack propagation is not stable and crack velocity is oscillating.

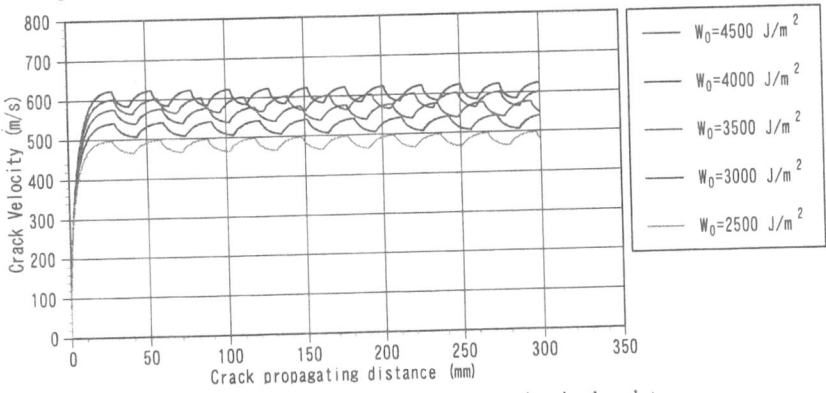

Fig. 10 Simulation of crack propagation in the plate

5. Conclusions and discussions

Following results are obtained in the present work:

1) Experiments of crack propagation in brittle PMMA plate were performed. The relationship between fracture toughness G_c and average crack velocity v is obtained (Fig. 3 and Eqn. 2);

2) Periodic patterns appeared in the fracture surface in certain velocity region (480~600 m/s) are observed which suggest local velocity instability. The interval of patterns (λ) is measured (Fig. 5);

3) A model of analyzing crack propagation in fixed-sided strip is proposed (Fig. 6) which derives the motion equation of the crack (Eqns 6, 6' or 6'');

4) If $G_c \sim v$ relationship is influenced by crack acceleration (Eqn. 7 and Fig. 9), as some researchers have observed [13, 14], the crack propagation will be unstable and local velocity oscillation will appear (Fig. 10).

Some discussions are made below:

1) The non-uniqueness of $G_c(v)$ relationship may come from the micro-mechanism of fracture in the crack tip zone. Arakawa and Takahashi performed detailed experimental observations and found that the surface toughness scale also depends on the crack accelerating state[20]

2) From figure 10, it is seen that the space interval of crack velocity oscillation is about 10 mm, which is one order larger than the observed periodic patterns (Fig. 5). The reason may be that the analytic solution obtained from 3.1 is not suitable in the case of unstable crack propagation,

112

because of radical crack velocity change. To obtain quantitative agreement, the analysis should be more localized near the crack tip for the *crack driving force* and the *equivalent mass*. The fundamental idea of the present paper is that when the inertia effect of material (the *generalized mass* term of Eqn. 6) is coupled with the microscopic process which happens in the crack tip (The non-uniqueness of $G_c \sim v$ relationship, Eqn. 7 and Fig. 9), unstable state of crack propagation appears (Fig. 10). Such coupling can be viewed as a kind of self-organization mechanism of crack propagation.

References

[1] J. Carlsson, L. Dahlberg and F. Nilsson, Experimental studies of unstable phase of crack propagation in metals and polymers, *Dynamic Crack Propagation*, Noordhoff Int. Pub., Leyden, pp165~181, 1973.

[2] W. G. Knauss and K. Ravi-Chandar, Some basic problems in stress wave dominated fracture, *International Journal of Fracture*, **27**, pp127-143, 1985

[3] J. W. Dally, W. L. Fourney and G. R. Irwin, On the uniqueness of the stress intensity factor-crack velocity relationship, *International Journal of Fracture*, **27**, pp159-168, 1985

[4] L. B. Freund, *Dynamic fracture mechanics*, Cambridge University Press, 1990

[5] T. Shioya, F. Zhou and R. Ishida, Micro-cracking process in dynamic brittle fracture, *DYMAT Journal*, **2**, No. 3, 1995

[6] J. Fineberg, S. P. Gross, M. Marden and H. L. Swinney, Instability in the propagation of fast cracks, *Physical Review B*, **45**, pp5146~5154, 1992

[7] P. D. Washabaugh and K. G. Knauss, Non-steady, periodic behavior in the dynamic fracture of PMMA, *International Journal of Fracture*, **59**, pp189~197, 1993

[8] K. Fujimoto and T. Shioya, Elastic analysis of dynamic crack propagation in fixed sided plates, *Proceedings of 20th Japan Congress Material Research*, pp49~58, 1985

[9] X. Liu and M. Marden, The energy of a steady-state crack in a strip, *Journal of Mechanics and Physics of solids* **39**, No. 7, pp947~961, 1991

[10]A. K. Green and P. L. Pratt, Measurement of the dynamic fracture toughness of PMMA by high-speed photography, *Engineering Fracture Mechanics*, **6**, pp71~80, 1974

[11]F. F. Abraham, D. Brodbeck, R. A. Rafey and W. E. Rudge, Instability dynamics of fracture: A computer simulation investigation, *Physical Review Letters* **73**, No. 2, pp272~275, 1994

[12]L. Dahlberg, F. Nilsson and B. Brickstad, Influence of specimen geometry om crack propagation and arrest toughness, *Crack Arrest Methodology and Applications, ASTM STP 711*, pp89~108, 1980

[13]K. Takahashi and K. Arakawa, Dependence of crack acceleration on the dynamic stress-intensity factor in polymers, *Experimental Mechanics* **27**, pp195~199, 1987

[14]K. Arakawa and K. Takahashi, Relationships between fracture parameters and fracture surface toughness of brittle polymers, *International Journal of Fracture*, **48**, pp103~114, 1991

On the Behavior of Crack Surface Ligaments

P. Nilsson, K.-G. Sundin and P. Ståhle

Department of Solid Mechanics, Luleå Institute of Technology, 971 87 Luleå, Sweden

Summary

Studies of cleavage fracture surfaces show that parts of the surface consist of plastically formed ridges, presumably traces of plastically torn ligaments. These ligaments are assumed to be formed between the areas fractured in cleavage. The tearing is studied both numerically and experimentally. An elastic visco-plastic material model is adopted for finite element calculations. The results show that relatively large amounts of energy are consumed during the tearing process. Further, the energy consumption is increasing rapidly with increasing tearing rate. A few preliminary experiments are made. The computed behavior is partly verified. The implications for slow stable crack tip speeds at dynamic fracture are discussed.

Keywords: Dynamic fracture, visco-plasticity, fracture processes, crack surface ligaments.

1 Introduction

In many situations of fast crack growth in structural steels, rate effects during plastic straining can not be ignored [1, 2]. Cases where the strain rate sensitivity is strong and, thus, the plastic straining has a limited effect on the stress distribution have been treated by Freund and Hutchinson [3]. They considered the energy rate balance necessary for steady state crack growth. Elastic energy release rate G was assumed to be equal to work rate, G_{vp}, due to viscous deformation in the plastic zone in addition to a crack tip driving force G_{tip} (see Fig. 1). Stable crack tip speeds were found to be possible only above 0.55 c_r where c_r is the Rayleigh wave speed. Below 0.55 c_r the speed will be unstable. The crack tip will immediately accelerate to a speed higher than 0.55 c_r or come to an arrest.

In structural steels, crack tip speeds higher than 0.55 c_r are seldom observed. Crack arrest often occurs at speeds as low as 0.05 - 0.1 c_r It is believed that the rate sensitivity of the fracture processes may provide an explanation.

The crack tip is assumed to propagate through coalescence of trans granular micro cracks. Due to granular mismatch at grain boundaries, unbroken parts will remain,

114

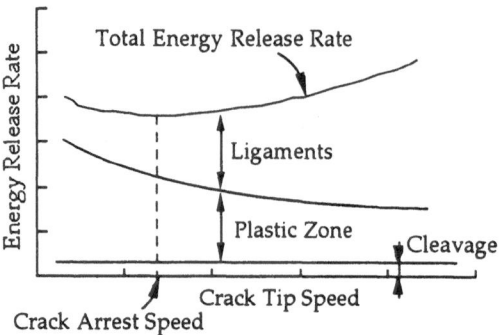

Fig. 1 Energy balance during fast crack growth.

connecting the upper and lower crack surfaces. During the increasing separation of the crack surfaces the unbroken parts will become ligaments, bridging the gap between the separating parts on opposite sides of the crack. Most of the energy consumed in the fracture process region is assumed to be due to the extension of ligaments.

In this paper small ligaments are studied as they deform plastically to the point where very little remains of its initial cross-sectional area. The energy consumption is calculated as a function of the ligament extension rate. At crack tip speeds of practical interest the energy release rate in the ligaments is found to be comparable to total energy release rates in structural steels. In the discussion part of this paper implications for stable crack growth are considered.

An attempt is made to verify the assumed material model. A few high loading rate tensile tests were performed on small test specimens. The preliminary results lend some confidence to the numerical results.

2 The Model

A single ligament in a large body is considered (see Fig. 2a). Coordinates $x = x_1$ and $y = y_1$ are used. Surfaces at $|x| \geq a$, $y = 0$ are assumed to be traction free. An elastic visco-plastic material model is used [4]. Strain rates are decomposed into an elastic part, $\dot{\varepsilon}_{ij}^e$, and a visco-plastic part, $\dot{\varepsilon}_{ij}^p$,

$$\dot{\varepsilon}_{ij} = \dot{\varepsilon}_{ij}^e + \dot{\varepsilon}_{ij}^p \ . \tag{1}$$

The elastic strain rates are given by Hooke's law using Young's modulus E and Poisson's ratio ν. The visco-plastic strain rates are defined as follows

$$\dot{\varepsilon}_{ij}^p = \dot{\gamma}_0 \left(\frac{\sigma}{\sigma_Y} - 1 \right)^n \frac{s_{ij}}{\sigma} \ , \tag{2}$$

where σ is von Mises yield stress $\dot{\gamma}_0$ is the strain rate sensitivity, σ_Y is the yield stress, n

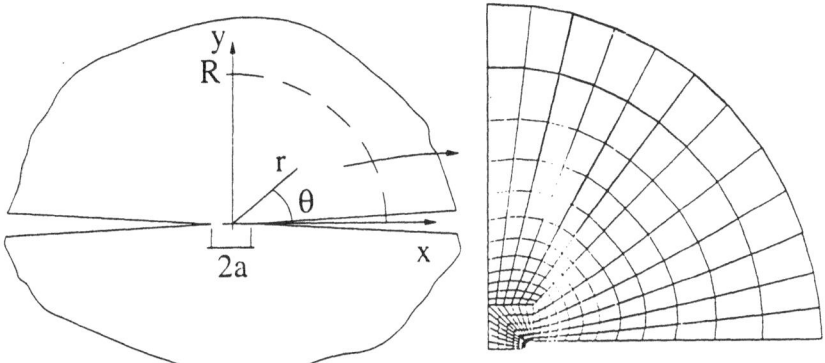

Fig. 2. a) The geometry and b) FE-mesh.

is the strain rate exponent and s_{ij} is the deviatoric stress [5]. For the study the following material parameters are used: $E/\sigma_Y = 954$, $\nu=0.3$ and $n=5$.

The structure is subjected to a remotely applied tension parallel with the y-axis.

The elastic solution is given by the potential function ψ given as follows

$$\Psi = \frac{i P(1 - v^2)}{\pi v E \sqrt{z^2 - a^2}} \quad , \tag{3}$$

where $z = x + iy$ [6]. The load P is defined by

$$P = \int_{-a}^{a} \sigma_y \big|_{y=0} \, dx \quad . \tag{4}$$

The stresses in the entire body is given by

$$\sigma_x + \sigma_y = 2\,\text{Re}(\Psi) \quad , \tag{5}$$
$$\sigma_y + i\tau_{xy} = \text{Re}(\Psi) - iy\,\Psi'$$

A polar coordinate system is attached to $x = y = 0$ as Fig. 2a shows. The solution (5) can be expanded for $r/a \to \infty$. The dominating term for the displacement v in the y-direction is obtained as follows

$$v = \frac{P(1 + v)}{\pi E}\left\{2(1 - v)\ln\left(\frac{2r}{y}\right) - \sin^2(\theta) - 1\right\} \tag{6}$$

The non-linear problem is analyzed numerically. For this reason a finite part bounded by $r \le R$ is studied. Due to symmetries across $x=0$ and across $y=0$ the study is further limited to the part $x \ge 0$ and $y \ge 0$. The elastic solution is employed for a boundary layer analysis, where (5) is applied as a boundary condition at $r = R$. A prescribed displacement rate $\dot{v}_0$ is used as a load parameter. Thus, the following boundary conditions are used

$$u = 0 \quad \text{and} \quad v = \dot{v}_0 t \ln(\xi R / a) , \quad \text{at } r = R \text{ and } 0 \leq \theta \leq \pi/2 , \qquad (7)$$

where

$$\dot{v}_0 = \frac{2P(1 - v^2)}{\pi t E} \quad \text{and} \quad \xi = 2\exp\left[-\frac{1}{2(1 - v)}\right] .$$

Here t is the time. Symmetry across $x = 0$ and across $y = 0$ requires

$$\tau_{xy} = u = 0 \quad \text{at } x = 0 \text{ and } 0 \leq y < R \qquad (8)$$

and

$$\tau_{xy} = v = 0 \quad \text{at } 0 < x < a \text{ and } y = 0 . \qquad (9)$$

Finally traction free crack surfaces require

$$\tau_{xy} = \sigma_y = 0 \quad \text{at } a \leq x < R \text{ and } y = 0 . \qquad (10)$$

The ratio R/a is put to 40. This is believed to be sufficiently large to establish the requirements for a boundary layer solution. A few calculations with smaller radii are made to check that these requirements are obtained.

3 Numerical Model

The commercial code ABAQUS is used for the numerical calculations. To obtain rapid convergence the finite element model employs an initially finite root-radius, $r = 0.2a$ with its center at $x = 1.2a$ and $y = 0$. This is assumed to be acceptable whereas the radius grows several times the original notch radius [7].

The body selected for the calculations is covered by a mesh containing 481 nodes and 432 plane strain isoparametric bilinear displacement based constant pressure elements (see Fig. 2b). Full integration is used. A large deformation theory is assumed. During the analyses the elements become distorted. To avoid extreme side-to-side aspect ratios the original near-tip element shape is chosen so that an element experiences an extending strain along the side originally short and a compressive strain along the side originally long.

4 Experiments

Notched specimens are rapidly loaded in tension using a set-up shown in Fig. 3a. According to manufacturer, material data for the test specimen are $E = 206$ GPa $\sigma_Y = 216$ MPa and $v = 0.3$. A loading rod (diameter 27 mm, length 1500 mm) is impacted at its far end by a projectile (diameter 27 mm, length 150 mm) from an air gun causing a compressive elastic wave in the loading rod. As the wave reflects, the end of the loading rod impacts a yoke which is rapidly accelerated. The head of the T-shaped specimen (Fig. 3b) is fitted to the yoke and thus a transient tensile force is applied to the specimen. At the other end of the specimen a 500 mm long measuring rod of steel is attached with means of an adhesive but- and overlap joint. This rod has the same cross-sectional dimensions as the small end of the specimen (3×5 mm^2) and the joint is very small. Undistorted wave propagation down the measuring rod is therefore assumed and the strain history in the wave represents the force history at the tensile fracture in the notch. At a position 100 mm from the joint a strain gauge with an active length of 3 mm is attached to each side of the rod and the pair is coupled so that bending strains are suppressed. The bridge unbalance signal is fed to a strain amplifier (Measurement Group model 2210). At a position just below the head of the specimen a target for a non contacting displacement transducer (Zimmer 100D) is attached. Both these equipments have wide band characteristics with 3 dB limits of 100 and 400 kHz respectively.

The analog strain and displacement signals were recorded by a digital transient recorder (Lucas Datalab DL6034) at a rate of $2,5 \times 10^6$ sample per second and the digital data were then transferred (using the software Labview) to a computer.

5 Results

The objective is to study the energy dissipation in the ligaments and its dependence on the ligament extension rate, $\dot{v}_0$. Calculations are performed for extension rates from $\dot{v}_0 = 0.0387\,\dot{\gamma}_0 a$ to $80\,\dot{\gamma}_0 a$. The lowest results were computed primarily to correlate with the experimental results.

Fig. 4 shows the extent of the plastic zone at the extension rate $\dot{v}_0 = 1545\,\dot{\gamma}_0 a$. At low loads plastic deformation is confined to the notch bottom. As the plastic zone is

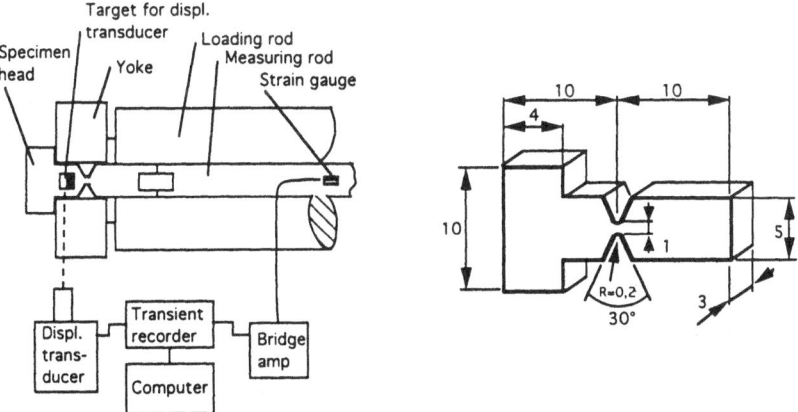

Fig. 3. a) The experimental setup and b) the test specimen

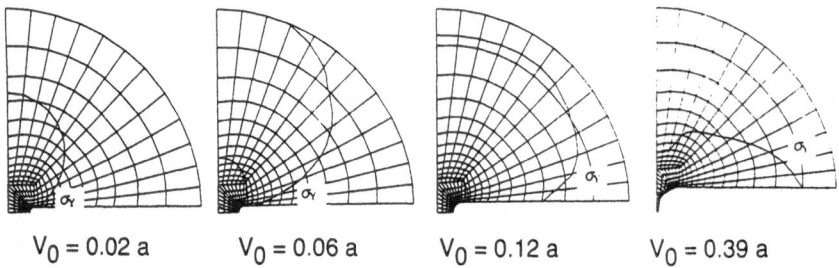

$V_0 = 0.02$ a $V_0 = 0.06$ a $V_0 = 0.12$ a $V_0 = 0.39$ a

Fig. 4. Plastic zone sizes at different stages of ligament extension.

growing with increasing displacements, it develops a circular shape. The maximum height is around $10a$ and is obtained at $v_o = 0.041a$. After a peak load is reached, the plastic zone assumes a different shape. A part immediately above the center of the ligament becomes elastic. Finally, when the calculation is interrupted at a very low load the plastic zone is still considerable (see Fig. 4d). This is assumed to be due to residual strains and will, thus, presumably remain even after the ligament has disappeared. The maximum heights for other extension rates vary from around $6a$ at the lowest rate to around $15a$ at the highest rate.

Reaction forces can be studied in Fig. 5a as functions of time for a few of the lower ligament extension rates. The results for higher rates are equal in shape but have a higher peak value. A maximum load is reached approximately at the instant when the plastic zone reaches its maximum height. At the subsequent decreasing loads the energy consumption is neglected when the ligament width is reduced to less than 1/5 of the original width. At this point the load is around 1/5 of the maximum load.

Of six experiments two were successful. Fig. 5b shows these experiments together with

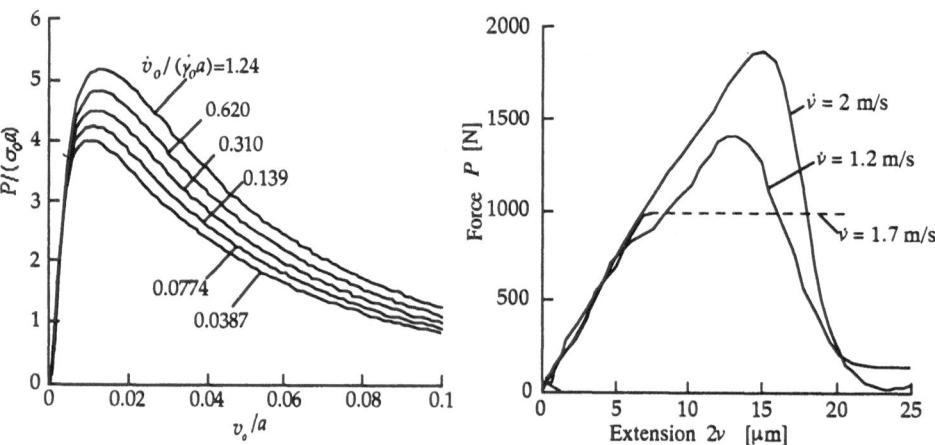

Fig. 5. Reaction forces for different ligament extension rates as functions of displacement. a) Finite element results and b) experimental results

one that failed after a short registration of an initial elastic phase. By comparing the initial elastic slopes in Figs 4b with the ones in Fig. 4a a much more rapid decrease in load is observed after the peak load is passed. The surfaces of the ruptured ligaments revealed little thinning of the ligament before final rupture. Growth and coalescence of voids are assume to have occurred. However, the initial plastic deformation causing the peak load is believed to be possible to correlate with the numerical result. The nucleation of voids requires large plastic straining. However, the comparison showed large differences in stress levels (50%) and, thus, the analysis is limited to comparing the shape of the curves up to maximum load.

The total energy dissipation during a completed tearing of a ligament at different extension rates, $\dot{v}_o$, is displayed in Fig. 6. The plastically dissipated energy is increasing with increasing ligament extension rates $\dot{v}_o$. This gradient is particularly high at low ligament extension rates.

6 Discussion

To relate the order of magnitude of the dissipated energy in the ligaments to the fracture energy, a few assumptions have to be made. To illustrate the significance of the rate sensitivity of ligament extension during fast fracture the following was assumed: An area coverage of the crack surface of 10%, $\dot{\gamma}_o = 7000$ s^{-1}, $E = 206$ GPa, $\sigma_0 = 216$ MPa and $a = 5$ μm.

Based of a strip yield solution [8, 9] with the length, $(\pi/8)GE/\sigma_Y^2$, of the yielded zone and the crack tip opening displacement G/σ_Y an average crack surface displacement rate can be calculated. This suggests that the crack tip speed as an average is around $(\pi/8)E/\sigma_Y = 400$ times larger than the rate of displacements of the crack surfaces. Thus, at a crack tip speed of 300 m/s $\dot{v}_o$ may be put to 0.75 m/s. Fig. 6 gives $G_{lig} = 9\sigma_0 a$ = 9720 N/m. With ligaments covering 10% of the crack surface $G_{lig} = 972$ N/m. The corresponding stress intensity factor $K_{Ic} = (G_{lig} E)^{1/2} = 14$ MPa m$^{1/2}$. Considering cleavage fracture this is not a small value.

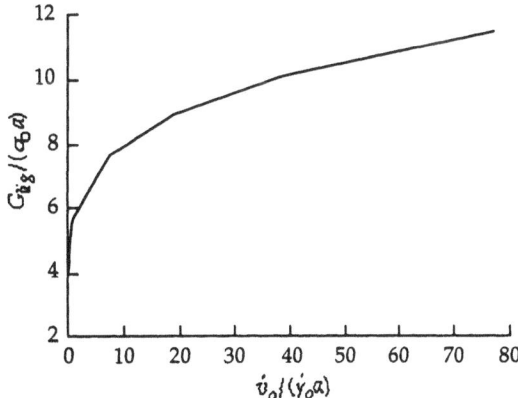

Fig. 6. Energy dissipation as function of ligament extension rate.

120

The increasing energy dissipation with increasing speeds will stabilize the crack tip speed provided that the decreasing energy dissipated in the plastic zone is compensated for. Without making any definite assumptions about the energy dissipated in the plastic zone it is believed that it is of the same order of magnitude as G_{lig} and that the ligaments in many cases are responsible for low arrest speeds and the difficulties at experiments to obtain crack tip speeds of the order of the Rayleigh wave speed.

References

[1] K. K. Lo, Dynamic Crack Tip Fields in Rate Sensitive Solids, *Journal of the Mechanics and Physics of Solids*, Vol. 31, 1983, pp. 123-306.

[2] B. Brickstad, A Viscoplastic Analysis of Rapid Crack Propagation Experiments in Steel, *Journal of the Mechanics and Physics of Solids*, Vol. 31, 1983, pp. 307-327.

[3] L. B. Freund and J. Hutchinson, High-Strain Rate Crack Growth in Rate-Dependent Plastic Solids, *Journal of the Mechanics and Physics of Solids*, Vol. 33, 1985, pp. 169-191.

[4] L. E. Malvern, Plastic Wave Propagation in a Bar of Material Exhibiting a Strain Rate Effect, *Quarterly of Applied Mathematics*, Vol. 8, 1951, pp. 405-411.

[5] P. Perzyna, The costitutive equations for rate sensitive plastic materials, *Quarterly of Applied Mathematics*, Vol. 20, 1963, pp. 321-332.

[6] N. I. Muskhelishvili, *Some Basic Problems of the Mathematical Theory of Elasticity*, P. Noordhoff Ltd., Groningen, Holland, 1953.

[7] R. M. McMeeking, Finite Deformation Analysis of Crack-tip Opening in Elastic-plastic Materials and Implications for Fracture, *Journal of the Mechanics and Physics of Solids*, Vol. 25, pp. 1977, 357-381.

[8] D. S. Dugdale, Yielding of Steel Sheets containing slits, *Journal of the Mechanics and Physics of Solids*, Vol. 8, 1960, pp. 100-108.

[9] P. Ståhle and L. B. Freund, *Process Region Characteristics During Fast Crack Growth -- An Examination of Wide Plate Crack Arrest Experiment*, Division for Solid Mechanics Report, Brown University, Providence, RI 02912, USA, 1990.

[10] M.S. J. Hashmi, *Journal of Strain Analyses*, Vol. 27 1980, pp. 273–283.

Velocity Dependent Dynamic Fracture Toughness of Araldite B Simultaneously Determined by Caustic and Photoelastic Methods

Kiyoshi Takahashi[1], and Mamoru Kido[2]

1 Research Institute for Applied Mechanics, Kyushu University, Kasuga-shi, Fukuoka 816, JAPAN
2 Presently, Central Research Laboratories, Mitsubishi Chemicals Co., Yokohama-shi, Kanagawa 226, JAPAN

Summary

Using the method of caustics and the photoelastic method, dual focus high speed photography was attempted to simultaneously evaluate values of dynamic toughness K_{ID} for Araldite B. Specimen geometries were so chosen that fracture acceleration and deceleration occurred in one specimen to study the dependence of K_{ID} on the crack velocity $\dot{a}$. It is concluded that curves of $K_{ID}(\dot{a})$ determined by both methods exhibit hysteresis similarly with higher K_{ID} values in the deceleration area than those in the acceleration area.

Keywords: Dynamic fracture, dynamic fracture toughness, dynamic stress intensity factor, high speed photography, caustic method, photoelastic method, epoxy

1. Introduction

Dependence of the dynamic stress intensity factor K_{Id} or the dynamic fracture toughness K_{ID} during crack propagation has experimentally been studied by various investigators using mostly caustic or photoelastic methods [1] . Whether a unique relationship exists between K_{Id} (or K_{ID}) and the crack velocity $\dot{a}$ has been one of the concerns in those studies. Using the caustic method Takahashi and Arakawa [2] showed that the K_{Id} - $\dot{a}$ curve had a hysteresis for an accelerated and then decelerated crack and that values of K_{Id} for an accelerated crack were smaller than those of a decelerated crack at the same crack velocity. This non-unique dependence of K_{Id} on $\dot{a}$ seems to be related with various problems associated with dynamic fracture.

Among the problems are the appropriateness of the assumption $K_{Id} = K_{ID}$, the effect of viscoelasticity, non-localization of the damage zone at the crack tip [3] , the effect of higher order terms of the series expansion to represent the stress distribution around the crack, i. e., the effect of specimen geometry, crack blunting, the effect of reflected waves from specimen edges, the effect of emitted waves from the moving crack, and validity limit of the optical method of caustics or photoelastisity for the deter-mination of K_{Id}. Some of those problems are related with each other.

The non-unique dependence was examined for its physical meaning by studying fracture phenomena such as roughness development [4, 5] and crack branching [6]. Qualitatively speaking, roughness λ on fracture surfaces of PMMA, epoxy (Araldite D) and polyester (Homalite 100) exhibited also the hysteresis with curves of λ - $\dot{a}$, similarly to K_{Id} - $\dot{a}$ [4]. A study on crack branching in Araldite D and Homalite 911 (CR-39) showed that neither $\dot{a}$ nor K_{Id} had their limiting values for the onset of the crack branching. Instead, a quantity which was introduced in Ref. [4]

$$R^{*} \cdot \dot{a} = \frac{K_{ID}^{2} - K_{IA}^{2}}{E} \cdot \dot{a}$$ (1)

had its limiting value for the onset of branching in each material, where K_{IA} is the dynamic stress intensity factor at crack arrest [6]. It is noted that the crack branching took place even during crack deceleration when K_{Id} was still increasing and that, if we take the parameter of eq. (1), the relationship between this parameter and the roughness λ indicates a good uniqueness through acceleration and deceleration procedures [4]. This parameter has a unit,

$$\frac{\text{energy}}{\text{length} \cdot \text{time}}$$ (2),

and is considered to physically represent fracture energy per unit crack width per unit time during dynamic fracturing. Thus, it has been postulated that the parameter shown by eq. (1) should be responsible as an influential parameter for dynamic fracture phenomena. Those experimental results may have more or less a relationship with the non-unique nature of K_{Id} ($\dot{a}$), although they do not explicitly explain it.

On the other hand, Dally, Agarwal and Sanford reported their experimental result on photoelastic determination of K_{ID} ($\dot{a}$) for Homalite 100 suggesting that the relationship between K_{ID} and $\dot{a}$ is almost unique within an experimental error.

As far as the experimental methods are concerned for determining K_{Id}, one of the problems has been to what extent values from the caustic methods are well correlated to those from the photoelastic method with respect to the velocity dependence. This problem was studied by several investigators. Nigam and Shukla [8] performed experiments using specimens of Homalite 100 and reported that values of K_{ID} from the photoelastic method were averagely by 20~30% larger than those from the caustic method. Taudou and Ravi-Chander [9] showed also for Homalite 100 that peaks of maximum and minimum values of K_{ID} from both methods appear almost similarly during crack propagation. In each of the studies above, two kinds of the optical experiments were performed separately for different specimens, where an effect of fracture phenomena scatter among specimens was unavoidable. Yang et al [10], on the other hand, attempted simultaneous evaluation of K_{ID} for Araldite D by both optical methods using a technique of dual-focus photography [6], and showed that values from the photoelastic method provides values larger by 20~25% than those from the caustic method. In this experiment, because intensity of the spark light sources was insufficient, a monochromatic filter was not used, therefore, isochromatic fringes were not clear enough. Afterwards, the light sources were intensified for its strength by 400%, so that a monochromatic filter could be employed in the photoelastic measurement. In the present study, using this modified camera, a simultaneous determination was attempted for an epoxy resin, Araldite B, to evaluate values of $K_{ID}(\dot{a})$.

2. Experimental

2.1 Specimens and loading jig

Commercially available epoxy plates of Araldite B were used for specimens in the present experiment. Figure 1 shows the specimen geometry presently employed to cause acceleration and deceleration twice in one crack propagation procedure. The tensile loads were biaxially applied with a set of V-jig as shown in Fig. 2. The left hand side of the specimen was considered as the first SEC-CT specimen and the right hand the second SEC-CT specimen.

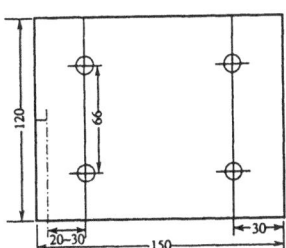

Fig. 1 Geometry of tested specimens.

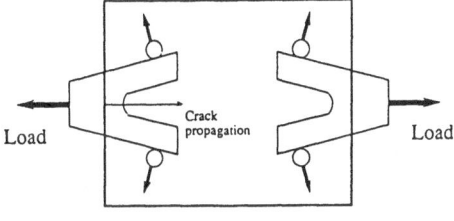

Fig. 2 A schematic drawing of a jig for tensile loading to cause doubly accelerated cracking.

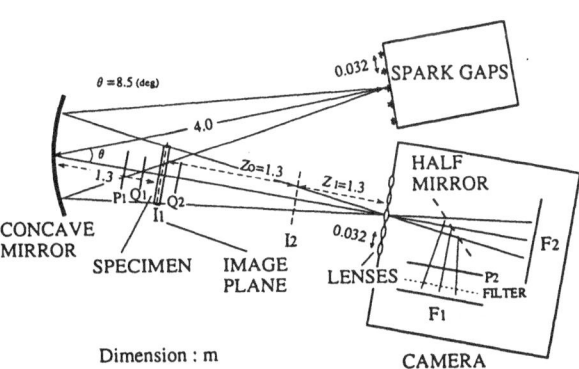

Fig. 3 Optical setup for simultaneous caustic and photoelastic high speed photography.

2.2 Dual focus high speed photography and determination of K_{ID}

Figure 3 illustrates the optical setup for simultaneous high speed photography using photoelastic and caustic methods. A half mirror in the Cranz-Schardin camera box splited light rays coming from 30 lenses into two film plates F_1 and F_2. The F_1 and F_2 could be focused independently on corresponding image planes. Thus, dual focus photography was made possible for photoelastic and caustic images on the plane I_1 and I_2, respectively. For the photoelastic setup, filtering plates P_1 and P_2, and 1/4 wave plates Q_1 and Q_2 were placed as shown in the figure. To obtain monochromatic light a filter (Edmund Co. No. 871) was used at a position between F_1 and P_2. For the determination of K_{ID}, the Irwin method [11] was used similarly to the previous study [10].

3. Results and discussion

3.1 Fracture velocity measurement

Pre-tests were performed for Araldite B as well as for PMMA plates to see fracture velocity change in a doubled SEC-CT specimen described above. Figure 4 presents fracture velocity changes in two PMMA specimens. It is shown that the first peak velocity is achieved for the cracks at a position which is located ca.10mm behind the first loading axis (dotted line). After the deceleration for a

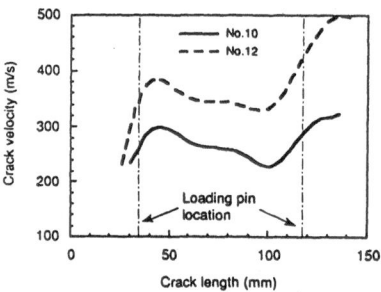

Fig. 4 Examples of crack velocity change in PMMA specimens.

while the second acceleration occurred at a position $15 \sim 20$mm ahead of the second loading axis. The second peak velocity peak was obtained at the final stage of the crack propagation. Figure 5 shows experimental results for an Araldite B specimen. The a - $\dot{a}$ and $\dot{a}$ - t relationships are shown in (a) and (b), respectively. It is noted that the a -$\dot{a}$ curve is different from the one for PMMA in that the third velocity peak, whose existence is slightly recognized in the curves for PMMA, is more pronounced at a position between the two loading axes, and that the position of the first velocity peak shifts closer to the first loading axis. This fracture behavior, different from that of PMMA, is considered to be related

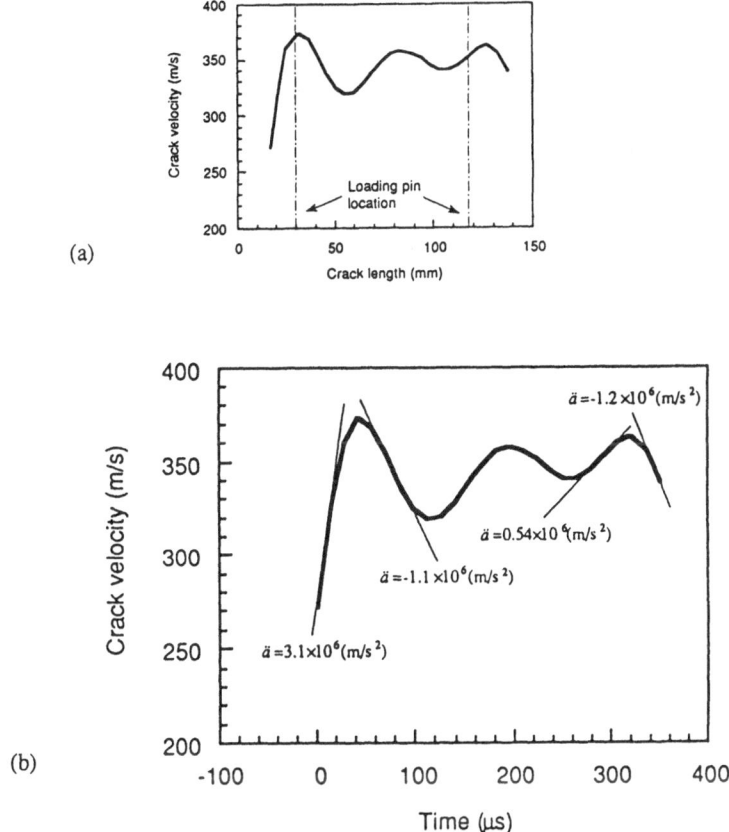

Fig. 5 An example of crack velocity change in an Araldite B specimen.

with brittleness of the material. Values of acceleration and deceleration were obtained for the first and the last peaks as shown in Fig. 5(b). The acceleration of the first peak is fairly higher than that of the second and third ones.

A set of caustic and photoelastic images is represented in Fig. 6 for an Araldite B specimen. Although the fringes are still not good enough in quality, the quantitative measurement was possible. The initial curve radius r_o of the caustics were in a range $0.9 < r_o / h < 1.3$, where h is the specimen thickness 5mm. On the other hand, the redius r of the isochromatic fringe taken for the K_{ID} evaluation was in a range $1.5 < r/h < 3.0$. The fringe pattern direction was changed to the opposite at a time 210 μ s when the load of the second axis became dominant.

Fig. 7 (a) and (b) represents the K_{ID}- $\dot{a}$ curves obtained by the caustic and the photoelastic method, respectively. Each of the numbers in the figures corresponds to the frame number in the film. The

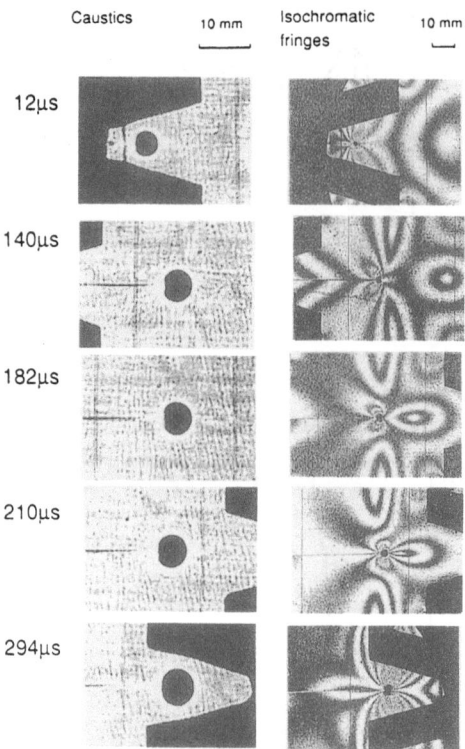

Fig. 6 Simultaneously taken pictures of caustic and photoelastic images
for a crack in an Araldite B specimen.

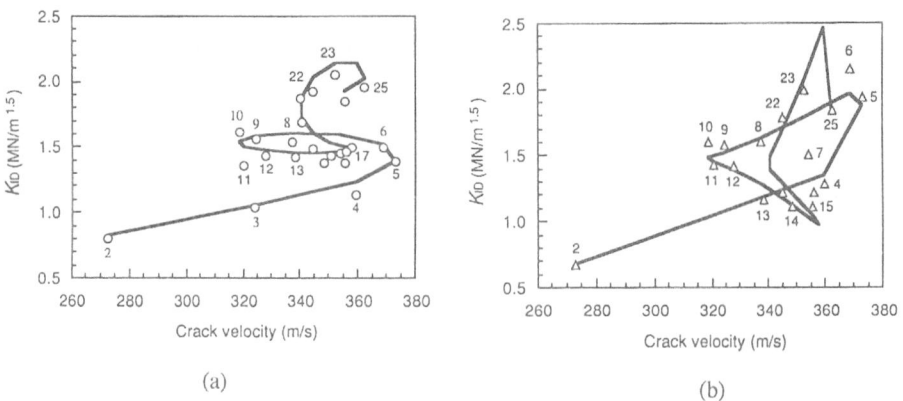

(a)

(b)

Fig.7 K_{ID} - $\dot{a}$ relationships for an Araldite B specimen in Fig. 6 obtained
by caustic (a) and photoelastic (b) methods.

change of K_{ID} is complicated particularly in the latter half of the fracture event reflecting the complex loading condition. However, if we observe the first half, *i. e.*, the first acceleration and deceleration procedure, both (a) and (b) in Fig. 7 indicate that values of K_{ID} are larger in the deceleration range, *i.e.*, from No.5 to No.10. This is important because the photoelastic method also resulted the hysteresis in the K_{ID} - $\dot{a}$ curve similarly to the caustic method.

Values of K_{ID} (C), evaluated by the caustic method, are compared in Fig.8 with those by the photoelasic method (K_{ID}(P)). The curves in the figure were obtained through a fitting technique using the 9th order polynominals [2] . Data from the photoelastic method had larger scatter, which was

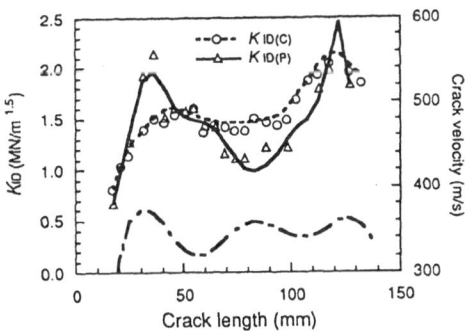

Fig. 8 Comparison of simultaneously determined K_{ID}, $\dot{a}$ vs. a relationships for an Araldite B specimen.

caused partly by the nature of the photoelastic determination of K_{ID}. The measurement of r involves ambiguity because of the broadness of the fringe. Even if the scatter is reckoned with, we see in Fig.8 that the peak of K_{ID} (P) comes later than that of the crack velocity for the first cycle of acceleration and deceleraiton, which is the most important result in the present experiment. The peak of K_{ID} (C) seems to come slightly later than that of K_{ID}(P). As far as the difference between values of K_{ID} (C) and K_{ID} (P) is concerned, we see roughly a difference of 30% at the peak position where values of K_{ID} (P) are larger. In contrast, values of K_{ID} (C) were by similar percentage larger in the middle portion of the specimen.

4. Conclusions

Using a modified Cranz- Schardin camera, dual focus high speed photography was attempted to simultaneously evaluate K_{ID}($\dot{a}$) for an epoxy, Araldite B, by the caustic (K_{ID}(C)) and photoelastic (K_{ID} (P)) methods. The following conclusions were obtained:
(1) Not only the caustic method, but also the photoelastic method gave K_{ID} - $\dot{a}$ curves which exhibited a hysteresis similarly. Values of K_{ID} were lager in the acceleration area than in the deceleration area at the same crack velocity.

(2) Values of K_{ID}(P) were by ca. 30% larger than those of K_{ID} (C) near the first peak of $\dot{a}$, whereas the relation is reversed in the middle portion of the specimen.

References

[1] See, for example, *Dynamic Fracture,* eds. M. L.Williams and W. G. Knauss
     (Martinus Nijhoff Pub., Dordrecht, 1985).

[2] K. Takahashi and K. Arakawa, *Experimental Mechanics,* Vol. 27, 1987, pp. 195-200.

[3] K. B. Broberg, *in this Proceedings.*

[4] K. Arakawa and K. Takahashi, *Int. J. Fracture,* Vol. 48, 1991, pp. 103-114.

[5] D. Hull, *J. Materials Sci.,* to appear.

[6] K. Arakawa and K. Takahashi, *Int. J. Fracture,* Vol. 48, 1991, pp.245-254.

[7] J. W. Dally, R. K. Agarwal and R. J. Sanford, *Experimental Mechanics,* Vol. 30, 1990,
     pp.177-183.

[8] H. Nigam and A. Shukla, *Experimental Mechanics,* Vol. 28, 1988, pp. 123-135.

[9] C. Taudou and K. Ravi-Chander, *Experimental Mechanics,* Vol. 32, 1992, pp. 203-210.

[10] H. T. Young, Y. Sakurada, T. Mada and K. Takahashi, *Bulletin of Research Institute for
     Applied Mechanics, Kyushu University,* No. 68, 1989, pp. 543-550.

[11] G. R. Irwin, Proceedings of Society of Experimental Stress Analysis, Vol. 16, 1958,
     pp.93.

Computer Simulation of Structural and Mechanical After-Effects of Shock Waves Induced by Impulse Laser in Solids

A.B.Volyntsev and A.N.Shilov

Physics of Metals Dept., Perm State Univ., 15 Bukirev Str., GSP, Perm, 614600, RUSSIA

Summary

Athermal dislocation climb and polarization of dislocation associations are proposed to be a reason of a new mechanism of solids softening during plane shock impact compression. This process is investigated by means of the computer simulation within the framework of continuous dislocation distribution.

Keywords: Dislocation, computer simulation, shock impact compression, internal stresses, X-ray laser.

1 Introduction

Compression in laser induced shock waves may achieve more than 100 GPa. Athermal dislocation climb that is possible in a front shock zone may induce the spatial polarization of dislocation associations, because edge dislocations with different signs climb in opposite directions (see Figure 1). The estimation of a critical compression for the beginning of this process gives the value from 10 to 100 GPa. Such polarization will inevitably entail the formation of high internal stresses. If shock wave propagates in loaded materials (in case of tension or shear external loading), the internal stresses mentioned above will be superposed on external stresses and induce an additional dislocation motion. Such sinergetic process may result in significant reduction of the yield point of metals. This is a new way of solids plastification during shock loading proposed for the first time by the authors in [1], where it was investigated by means of computer experiment with calculations for X-ray laser parameters. The idea of a broad-scale Strategic Defense Initiative was a powerful inducement for the whole spectrum of scientific explorations in the field of mechanical behaviour of different materials under the influence of powerful lasers. In this connection the behaviour of solids during and after X-ray laser influence seems to be one of the urgent problems. From the practical point of view it defines the effectiveness of the most probable cosmic weapon component [2]. Shock impact impulse is the main destroying factor of X-ray laser as well as of other powerful impulse lasers. On the other hand it is undoubtedly one of the fundamental problems of physics of solids further we will analyse some new aspects connected with the mechanism proposed in [1] for X-ray laser parameters.

130

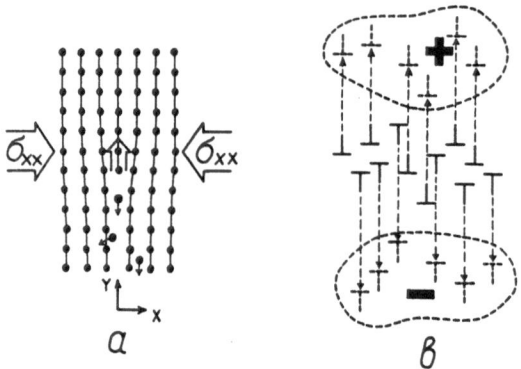

Figure 1: Behaviour of edge dislocations during powerful plane compression along one axis: a - athermal climb and interstitial atoms generation; b - polarization of dislocation charge in dislocation clots

2 Model

The model proposed in [3] for the investigation of the large dislocation ensembles dynamics is used in the research under consideration. General model assumptions include the following main points:

1. Continuous distribution of parallel dislocations stretched along z-axis is considered. The distribution is defined by the function of surface density $\rho(x, y)$ in the plane normal to the dislocation lines (two dimensional model). Only edge dislocations are under consideration.

2. The time evolution of the dislocation density is described by the two-dimensional continuity equations with or without functions of dislocation sources and sinks. These equations are used for the density of the positive ρ^+ and negative ρ^- dislocations and for the dislocation charge $\rho = \rho^+ + \rho^-$, which is the main cause of the long-range internal stresses in the model. Different algorithms are proposed for the calculation of these stresses.

3. The substructure and the microstresses caused by it, and therefore ρ, ρ^+ and ρ^- are supposed to be periodic. It is not a very strong model restriction because simulation region may include structure elements with different sizes and forms.

4. The dislocation forest and dislocation kinks are regarded as the main obstacles for the dislocation glide.

5. Activation energy for dislocation motion process depends upon the sum of internal and external stresses. With the total stresses increasing the thermally activated dislocation motion transforms gradually into athermal process, and the dislocation movement velocity asymptotically approaches its maximum value that is sound velocity.

6. Phonon scattering and the effective dislocation mass are taken into account for rapid dislocation movement.

7. The plastic deformation velocity is taken to be proportional to the integral current of positive and negative dislocations.

Two different versions of the dislocation interaction in the large spatial ensembles are used in the research. The first one is the same as in [3] and is based on the integration of whole internal stresses and the corresponding interaction forces for infinite periodic continuous dislocation medium on the base of the mathematical formalism of Fourier series.

Therefore dislocation interaction may be calculated on the very high level of precision but there is no any possibility to take into account the effects connected with the limited velocity of the elastic interaction propagation (that is, sound velocity). The influence of this effects may be considerable for the fast processes structure transformation during shock impact compression.

To take into account the really limited dislocation interaction propagation the second version of the interaction calculation is used. In this case the influence of some little dislocation ensemble element with dislocation charge density $\rho(\xi, \eta)$ on the dislocation situated in some other point (x,y) for the time moment t is completely defined by the value of the density $\rho(\xi, \eta)$ but for the time equal to $t - t'$, where t' is the time necessary for sound moving from point (ξ, η) to point (x,y).

The whole interaction force (x - component for example) is calculated according to:

$$F_x(x, y, t) = \int\limits_{y-l_y/2}^{y+l_y/2} \int\limits_{x-l_x/2}^{x+l_x/2} f_x \left[\rho(\xi, \eta, t - t') \right] d\xi \, d\eta,$$

where f_x is the function defining dislocation interaction (see for example [4]), $t' = [(x - \xi)^2 + (y - \eta)^2]^{1/2} / V_s$, V_s sound velocity, l_x and l_y are structural periods along the x- and y-axes, respectively.

The integral is taken for two-dimensional area with the point (x,y) under consideration situated in the centre of the integral field that is equal to one structural period along both dimensions. Thus the second version takes into account the limit velocity of dislocation interaction propagation but the calculation of this interaction is performed for each dislocation only for the surroundings within one structural period. The expansion of the surrounding field entails very intensive increasing of computer resources that are necessary for modelling procedure because this field with corresponding dislocation distribution must be taken into account for the previous time periods and all this information is saved in the computer memory and is used during the simulation process.

For further descriptions and discussions we will use denotations V1 and V2 for the first and the second versions of dislocation interaction calculation respectively.

3 Computer Experiments

3.1 V1-Calculations

3.1.1 Dislocation Clots

Figure 2 shows the initial (before loading) distribution of the positive dislocations density ρ^+ within one substructure period ($l_x = l_y = 10^{-4} cm$). The distribution of negative dislocations ρ^- is the same. So the initial dislocation charge and long range internal stresses are absent. Dislocation clots really existing in natural deformed metals are simulated here. The density of dislocation forest is uniform and equal to 10^{11} cm^{-2}.

Testing for the conditions of constant deformation ε_{xy} is being performed here. Starting external shear stress before compression shock $\sigma_{xy} = 600 MPa$, that is equal to about 1/2 yield point for material under consideration. The temperature of computer experiment is equal to 293K.

Plane compression along the x-axis (50 GPa) is simulated here. The distribution of positive dislocations for some time after the beginning of the compression impulse is shown in Figure 3. Compression entails the fragmentation of the structure and induces the

132

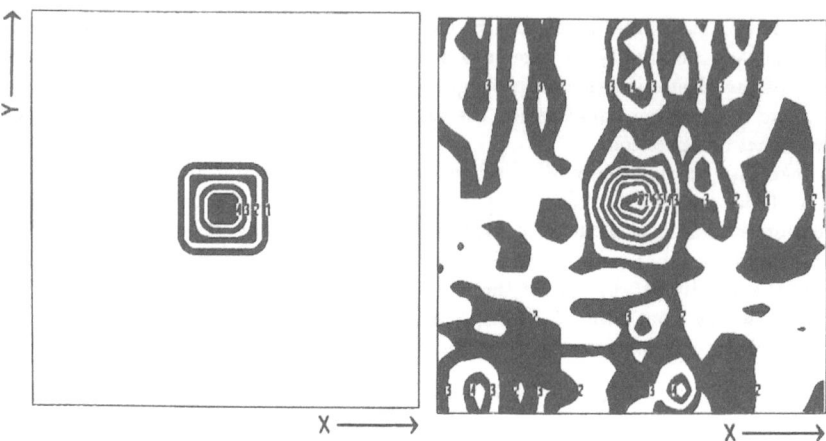

Figure 2: Initial distribution of the positive dislocations ρ^+ (in units of $10^{11} cm^{-2}$) within a period of the dislocation clots structure: 1 - 1.00, 2 - 3.25, 3 - 5.50, 4 - 7.75; the maximum value in the centre of the clot is equal to 10.0

Figure 3: Clots structure for $t = 1.38 \times 10^{-10} s$ after the beginning of compression impulse (V1 - calculations). The density of the positive dislocations ρ^+ is represented in units of $10^{11} cm^{-2}$: 1 - 0.0, 2 - 0.73, 3 - 1.47, 4 - 2.20, 5 - 2.93, 6 - 3.67, 7 - 4.40, 8 - 5.13

appearance of high internal microstresses (see Figure 4). The tendency of the dislocation system to reduce (as it is possible) the level of these microstresses entails the dynamic polyganization. It may be observed in Figure 3 in the form of alternating "ditches" and "banks" of dislocation density that are positioned along the y-axis. These "ditches" and "banks" represent low angle tilt boundaries.

The intensive movement of dislocations results in the increasing of plastic deformation ε^p_{xy}. For the conditions of the computer experiment when $\varepsilon_{xy} = \varepsilon^e_{xy} + \varepsilon^p_{xy} = const$, where $\varepsilon^e_{xy} = G\sigma_{xy}$ is the elastic deformation, the increasing ε^p_{xy} component leads to the reduction of ε^e_{xy} and, consequently, of σ_{xy} - that corresponds to the relaxation of external macrostresses (see Figure 5). These stresses fall to zero very quickly and the material can not stand any load in such conditions. Due to the existence of the effective mass fast dislocations pass their equilibrium positions and oscillations of the dislocation system are beginning. This process is reflected by internal and external stresses oscillations (see Figure 5). The oscillations of dislocation ensemble is an additional factor of structure instability which promotes to an almost absolute loss of solidity of the metal even without any shape changing of the specimen. The amplitude of internal microstresses amounted to 4 GPa and overcame even the theoretical limit of durability.

Special calculation were performed for the estimation of a long-term consequences of the shock impact compression.

The initial conditions of the new computer experiment were quite the same as for the previous one but the material was not initially loaded by shear stress. Plane compression was also the same. Structure evolution processes were qualitatively similar to the previous calculation. Compression lingered during 0.5 ns - the possible period of X-ray laser

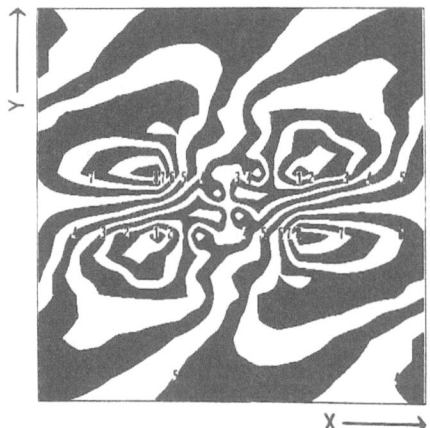 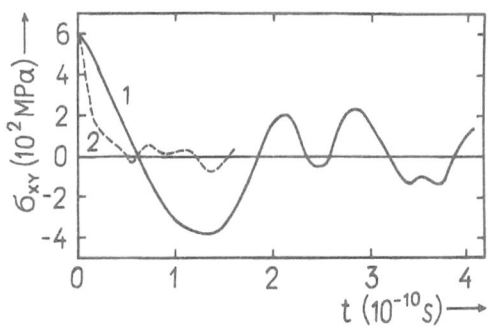

Figure 4: Distribution of internal stress σ_{ixy} for $t = 0.25 \times 10^{-10}s$ after the beginning of compression impulse (in units of $10^3 MPa$ for clots structure): 1 - -1.63, 2 - -1.21, 3 - - 0.80, 4 - - 0.38, 5 - +0.04, 6 - 0.46, 7 - 0.88, 8 - 1.29

Figure 5: Time dependence of macrostress necessary to maintain constant Specimen deformation (V1 - calculations): 1 - clots structure, 2 - cellular structure

impulse. After that the material was free from any external influences during 10 minutes. Then simulation of testing for $\dot{\varepsilon}_{xy} = const$ was performed ($\dot{\varepsilon}_{xy} = 10^{-2}s^{-1}$). Quite similar computer experiment was executed for the control material which was not exposed under the compression impulse.

The main feature of the previously compressed material behaviour is the absence of elastic deformation interval on $\sigma(\varepsilon)$ curve. The plastic deformation for it takes place from the very beginning. The plastic flow for noncompressed control material begins as usual from some yield point.

The reason of such behaviour for the previously compressed material is the local residual internal microstresses which may induce the dislocation movement in some structure regions even for a quite low level of external stress if the internal and external stresses coincide. It is a sinergetic effect. So it is impossible for a compressed material to bear any load without plastic deformation for a long time.

3.1.2 Cellular Structure

The computer experiment quite similar to 3.1.1 was performed for cellular structure with formless dislocation subboundaries. The unital distribution of the positive dislocation density $| \rho^+ |$ for this case is shown in Figure 6 ($\rho^+ =| \rho^- |$, so the initial dislocation charge is absent). It is the result of the numerical treatment of some real electron microscope photo for the deformed aluminum specimen with pictorially observed deformation cells. The shock compression entails the fragmentation of the structure (see Figure 7) and induces the appearance of high level internal stresses and oscillation of dislocation system as in

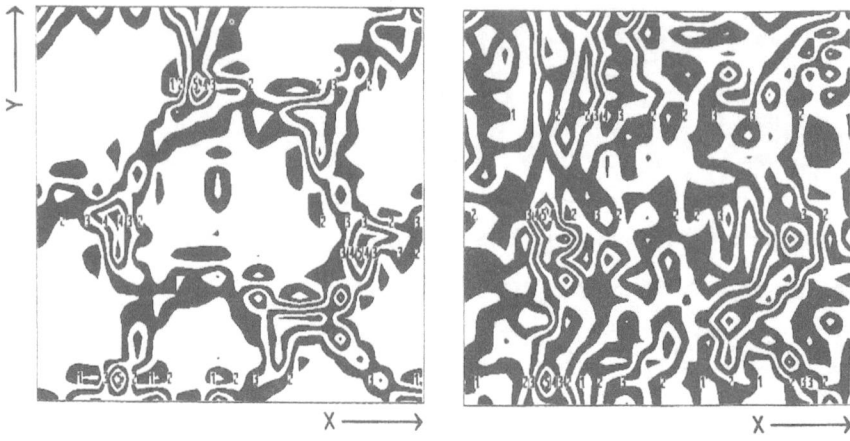

Figure 6: Initial distribution of the positive dislocations ρ^+ (in units of $10^{11}cm^{-2}$) within a period of cellular structure: 1 - 0.0, 2 - 0.52, 3 - 3.12, - 5.71, 5 - 8.31

Figure 7: Cellular structure for $t = 1.6 \times 10^{-10}s$ after the beginning of compression impulse. The density of the positive dislocations ρ^+ is represented in units of $10^{11}cm^{-2}$: 1 - 0.0, 2 - 1.38, 3 - 2.76, 4 - 4.15, 5 - 5.53

case of the computer experiment in 3.1.1. The material can not endure any mechanical load, but the amplitude of macrostress oscillations now is lower then for the more regular structure in 3.1.1 (compare curves 1 and 2 in Figure 5). It is the result of some statistical averaging for the structure which consists of cells with different sizes and forms.

This computer experiment shows that the model assumption of periodic structure is not a very strong model restriction because the simulation region may include different elements as it was performed. So it is possible to take into account statistical variations of structure components and at the same time to preserve the mathematical formalism of Fourier series.

The reduction of macrostresses at the first stage of the process is realized quicker for the cellular structure then for computer experiment described in 3.1.1 (compare curves 1 and 2 in Figure 5). It is due to the less average size of the structure elements for the new computer experiment (now there are several cells in the same simulated region - $10^{-4} \times 10^{-4} cm^2$).

3.2   V2 - Calculations

The initial structure is supposed to be the same as in 3.1.1 but dislocation density in the background between clots is ten times more than it was taken for the first clots structure. So dislocation density in the centre of each clot is now only two times greater than the background density. The new version of calculations occurs to be more sensible from the mathematical point of view to the degree of dislocation density spatial ununiformity. V2 is less stable in the comparison with V1 and for the estimation of the main physical peculiarities of the dislocation system behaviour according to V2 - calculations rather a more smooth initial dislocation distribution was taken.

The behaviour of dislocation system according to V2 at the first stages of the evolution process is qualitatively the same as it was described for V1 in 3.1.1: the external stress goes to zero, passes it and vibrations of the dislocation ensemble with corresponding oscillations of internal and external stresses are beginning (see Figure 8). But the long-term evolution stages according to V2 - calculations have some peculiarity in the comparison with V1. The amplitude of external stress oscillations gradually increase (see Figure 8) so not only internal but also external stresses overcome the theoretical level of durability. Another difference is that the degree of dislocation density inhomogeneity increases too as compared with gradual decreasing according to V1 (see Figure 9 and compare it with Figure 3). This peculiarity may be caused by the piling up of dislocations in front zones of internal stresses waves when dislocations situated in the forward periphery of the stress wave front with comparatively low level of internal stresses have not time to make a way for the fast dislocations moving from the inner regions of the stress wave with a high level of internal stresses.

So the structure evolution according to V2 - calculations may be characterized by more intensive processes of structure instability development in comparison with V1.

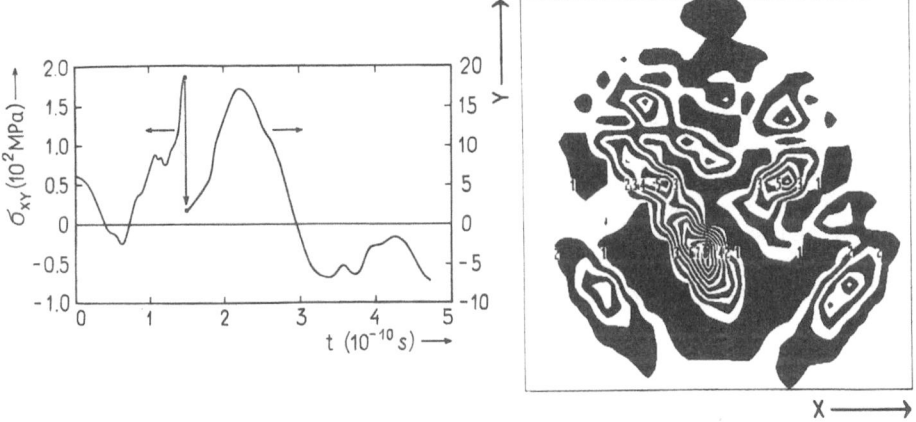

Figure 8: Time dependence of macrostress necessary to maintain constant specimen deformation for clots structure (V2 - calculations)

Figure 9: Clots structure for $t = 0.83 \times 10^{10}s$ after the beginning of compression impulse (V2 - calculations). The density of the positive dislocation ρ^+ is represented in units of $10^{11}cm^{-2}$: 1 - 0.0, 2 - 5.33, 3 - 10.7, 4 - 16.0, 5 - 21.3, 6 - 26.7, 7 - 32.0, 8 - 37.3

4 Discussion and Conclusions

Despite of some structural differences and external stress behaviour peculiarities according to V2 - calculations the main result is the same as it is according to V1: the polarization of dislocation clots during compression shock may lead to the absolute loss of solidity of metal construction units even without their shape changing. This effect is sufficiently durable thanks to the residual internal microstresses and slightly depends on initial dislocation structure. Moreover, the taking into account of the limit velocity of dislocation interaction propagation makes it possible to consider the new mechanism of solids plastification even more effectively in comparison with this factor omission during calculation. Special analysis showed that taking into account of the new mechanism of solids plastification may entail reduction of the energy deadline, normally defined for the destruction of the ballistic missiles casing and their supporting constructions from $20\ kJ/cm^2$ to $1\ kJ/cm^2$. Similar processes but for another time and spatial scales may take place in the deep-lying strata of the Earth or other planets.

Acknowledgements

The publication of some former secret results presented in the article is possible only thanks to the help of the Vice-President of the Russian Academy of Sciences, Director of Atomic Energy Institute, E.P.Velikhov to whom our grateful acknowledgements are made. The authors would like to express their indebtedness to Perm Subsidiary of Mosbusinessbank for the support of investigations. The authors also wish to express their thanks to I.C.Utrobina and G.S.Dvinyaninova for the help in the preparation of the manuscript.

References

[1] A.B.Volyntsev, A.N.Shilov, A New Way of Solids Plastification During Shock Loading, *Doklady Akademii Nauk (Russia).* Vol. 328, N 6, 1993, pp. 691 - 693.

[2] E.P. Velikhov, R.Z.Sagdeev and A.A.Kokoshin, *Cosmic Weapons: Dilemma of Security,* Publishing House: Mir, Moscow 1986.

[3] A.B.Volyntsev, Computer Modelling of the Dynamics of Space Dislocation Ensembles, *Phys. stat. sol. (b),* Vol. 165, 1991, pp. 343 - 354.

[4] J.P.Hirth, J.Lothe, *Theory of Dislocations,* McGraw-Hill Publ. Co., New York/London 1972.

Nonlinear Viscoelastic Constitutive Relations and Nonlinear Viscoelastic Wave Propagation for Polymers at High Strain Rates

Lili WANG (Li-Lih WANG)[1,2], Dejin HUANG[1], Su GAN[1]

1 Mechanics and Materials Sci. Res. Centre, Ningbo Univ., Ningbo 315211, CHINA
2 Dept. of Modern Mechanics, Univ. of Sci. and Tech. of China, Hefei 230026, CHINA

Summary

A Series of experimental and theoretical investigation at high strain rates revealed that the nonlinear viscoelastic behaviour of polymers and the related composites are well described by the so-called Zhu-Wang-Tang (ZWT) nonlinear viscoelastic constitutive equation. The impulsive reponse of ZWT materials consists of a rate-independent nonlinear elastic response and a higt frequency linear viscoelastic response. The dispersion and attenuation of nonlinear viscoelastic waves mainly depend on the effective nonlinearity and the high frequency relaxation time θ_2. An "effective influence distance/time" is defined to characterize the wave propagation range where θ_2 dominates the impact relaxation process.

Keywords: Nonlinear viscoelasticity, constitutive relation, polymer, wave propagation, high strain rate.

1 Introduction

It is well-recognized that inertia effects and strain-rate effects are two key-points what distinguish impact dynamics from static mechanics[1-3]. Inertia effects have led to the study of wave propagation in various forms, and strain-rate effects have promoted the study of all kinds of rate-dependent constitutive relations and failure criteria under high/very high strain rates.

One of the main difficulties in the study of those effects is that inertia effects (or wave effects) and strian-rate effects are, in practice, usually coupled together. On the one hand, no wave propagation can be analysed without knowing the corresonding dynamic constitutive relation of material, and consequently the basic characteristics of wave propagation inevitably depend on the strain-rate dependence of mechanical behaviour of materials. On the other hand, in the study of rate-dependent constitutive relations and failure criteria of materials, wave propagation effects can not be neglected as in quasi-static tests and thus must be taken into consideration. This is particularly true for polymers and polymer-matrix-composites, which are more susceptible to loading rate (or strain-rate) and are known as visco-elastic/plastic materials.

Two kinds of experimental techniques, the split Hopkinson bar technique(SHBT) and the wave propagation technique (WPT), have been extensively accepted to investigate the rate-dependent constitutive relations of materials at high strain rate, all taking account of wave propagation effects. In the SHBT proposed originally by Kolsky[4], two long elastic loading bars are used in which the strain rate effects can be approximately neglected, and the sandwiched specimen is short enough so that the wave effects herein can be approximately neglected. Thus the strain-rate effects and the wave effects are cleverly un-coupled, and the rate-dependent response of specimen is determined separately. However, little knowledge of the coupled character or inter-fluence between wave and rate-effects can be revealed from the SHBT tests.

In the wave propagation technique (WPT), on the other hand, the rate-dependent constitutive relation is deduced directly from the measurements and analyses of wave propagation signals themselves, as a solution of the so-called inverse problem. The coupled charater and inter-fluence between wave and rate-effects has been implicitly taken into account, and thus the obtained constitutive relation should be more suitable for predicting the wave propagation in the test material. However, several constitutive models may apparently all fit in with the same wave propagation data, if the experiments are not well designed. Thus, it is important to study more deeply the essential distinction of the wave propagation character for the materials with different constitutive models or parameters.

In the recent 15 years, a series of experimental and theoretical investigation on the rate-dependent constitutive relations for a variety of plastics and polymer-matrix-composites at finite deformation and in a wide range of strain-rates from 10^{-4} to 10^{-3} s^{-1} were performed by Zhu, Wang and their co-workers, and a nonlinear viscoelastic constitutive equation was proposed, which is simple for engineering application and is called Zhu-Wang-Tang (ZWT) equation. In the present paper the basic character of ZWT equation is analysed, and the dependence of the nonlinear viscoelastic wave propagation on the ZWT equation is studied for a typical thermoplastics, polymethyl methacrylate (PMMA), and a typical thermoset resin, epoxy.

2 ZWT Nonlinear Viscoelastic Constitutive Equation

The experimental investigation by Zhu, Wang and their co-workers for a variety of plastics, including polymethyl methacrylate (PMMA), polycarbonate (PC), polyamide (PA or Nylon), acrylonitrile-butadiene-styrene (ABS), polybutylene terephthalate (PBT), epoxide and phenolic thermoset plastics, and the related composites at strain rates from 10^{-4} to 10^3 s^{-1} showed that the nonlinear viscoelastic behaviour in uniaxial state for all the polymers studied can be well described by the following Zhu-Wang-Tang(ZWT) nonlinear viscoelastic constitutive equation[5-11].

$$\sigma = \sigma_e(\varepsilon) + E_1 \int_0^t \dot{\varepsilon}(\tau) \exp(-\frac{t-\tau}{\theta_1})d\tau + E_2 \int_0^t \dot{\varepsilon}(\tau) \exp(-\frac{t-\tau}{\theta_2})d\tau \tag{1}$$

where σ denotes the stress, ε the strain, $\dot{\varepsilon}$ the strain rate, t the time, and the first term $\sigma_e(\varepsilon)$ describes the nonlinear elastic equilibrium response; the next integral term describes the linear viscoelastic response for low strain rates, in which E_1 and θ_1 are the elastic constant and relaxation time, respectively, of the corresponding Maxwell element I; and the last integral term describes the linear viscoelastic response for high strain rates, in which E_2 and θ_2 are the elastic constant and relaxation time, respectively, of the corresponding Maxwell element II.

Moreover, it was experimentally revealed that the nonlinear elastic response $\sigma_e(\varepsilon)$ can be well described either by the following power polynominal form[5-9]

$$\sigma_e(\varepsilon) = E_0\varepsilon + \alpha\varepsilon^2 + \beta\varepsilon^3 \tag{2a}$$

where E_0 denotes the initial elastic modulus and α, β are nonlinear elastic constants, or by the following exponential form[10]

$$\sigma_e(\varepsilon) = \sigma_m[1 - \exp(-\sum_{i=1}^n (m\varepsilon)^i / i)] \tag{2b}$$

where σ_m denotes the asymtotic maximium, m the ratio of E_0 and σ_m, and the positive integer n is a material parameter charaterizing the initial linearity.

Theoretically, the ZWT equation can be deduced from the Green-Rivlin multiple-integral equation[14], or, alternatviely, from the Coleman-Noll finite linear viscoelastic theory[15,16]. But both of them are very difficult to deal with for engineering application. Experimental and theoretical investigation also revealed that the nonlinear viscoelastic behaviour of polymer-matrix-composites at high strain rates can be well described by the ZWT equation modified by a rate-independent coefficient which is related to the re-inforced fibre or particles[7,11].

The typical parameters in Eq. (1) and (2a), which were determined by SHBT for epoxy, PMMA and PC, are listed in Table 1,

Table 1 The typical nonlinear viscoelastic paramenters experimentally determined[5,7,9,10]

	Epoxy	PMMA-1	PMMA-2	PMMA-3	PC
ρ_0, kg/m³	1200	1190	1190	1190	1200
E_0, GPa	1.96	2.05	2.19	2.95	2.20
α, GPa	4.12	4.71	4.55	10.9	23
β, GPa	-181	-233	-199	-96.4	-52
E_1, GPa	1.47	0.897	0.949	0.832	0.10
θ_1, s	157	15.3	13.8	7.33	470
E_2, GPa	3.43	3.07	3.98	5.24	0.73
θ_2, μs	8.57	95.4	67.4	40.5	140

It is evident from Eqs. (1), (2) and Table 1 that for ZWT materials, namely, for the materials which behave in accordance with ZWT nonlinear viscoelastic constitutive relation, the following constitutive characteristics exist.

a) The physical or constitutive nonlinearity only comes from the pure elastic equilibrium response $\sigma_e(\varepsilon)$, while all the rate/time dependent responses are essentially linear and consequently can be still described by linear viscoelastic elements even when the apparent, overall response is nonlinear. Such a constitutive nonlinearity may be termed the "rate-independent nonlinearity" — a kind of weak nonlinearity. Then it provides a possibility of generalizing what has been established in linear viscoelasticity to the rate-dependent response of ZWT materials without substantial difficulties.

b) As shown by the experimental measurements in Table 1, generally, θ_1 is of the order of 10^0 to 10^2 s, while θ_2 is of the order of 10^{-6} to 10^{-4} s, namely, θ_1 is of the order of 10^5 to 10^7 times higher than θ_2. This means that each of them exerts its influence in its own dominant range of strain rates, since an analysis of relaxation process for a Maxwell element has shown that the "effective influence domain" for any relaxation time θ_j is about 4.5 order of magnitude in both strain-rate and time scale[6].

c) Consequently, under quasi-static loading conditions, where the time-scale is of the order of 10^0 to 10^2 s, the high frequency Maxwell element with relaxation time of the order of 10^2 to 10^0 μs has already relaxed even at the beginning of loading. Equation (1) then reduces to

$$\sigma = \sigma_e(\varepsilon) + E_1 \int_0^t \dot{\varepsilon}(\tau) \exp(-\frac{t-\tau}{\theta_1}) d\tau \tag{3}$$

This means that it is unable to determine the constitutive parameters for describing high velocity deformation, such as θ_2 and E_2, by any quasi-static test, but by high strain rate tests, and vice versa.

d) On the contrary, under impact loading conditions where the time-scale is of the order of 10^0 to 10^2 micro-second, the low frequency Maxwell element with relaxation time θ_1 of the order of 10^0 to 10^2 s does not have enough time to relax until the end of loading, and consequently reduces to a single spring element with an elastic constant E_1. Equation (1) then reduces to

$$\sigma = \sigma_e(\varepsilon) + E_1\varepsilon + E_2 \int_0^t \dot{\varepsilon}(\tau) \exp(-\frac{t-\tau}{\theta_2}) d\tau \tag{4a}$$

or, in an equivalent differetial form

$$\frac{\partial \sigma}{\partial t} + \frac{\sigma}{\theta_2} = \left[\sigma'_{eff}(\varepsilon) + E_2 \right] \frac{\partial \varepsilon}{\partial t} + \frac{\sigma_{eff}(\varepsilon)}{\theta_2} \tag{4b}$$

where $\sigma_{eff}(\varepsilon)$ is the effective pure nonlinear elastic response

$$\sigma_{eff}(\varepsilon) = \sigma_e(\varepsilon) + E_1\varepsilon, \qquad \sigma'_{eff}(\varepsilon) = \frac{d\sigma_{eff}(\varepsilon)}{d\varepsilon} = \frac{d\sigma_e(\varepsilon)}{d\varepsilon} + E_1 \tag{5}$$

e) It can be seen from Table 1 that the ratio of α/E_0 is of the order of 10^0 to 10, and β/E_0 is of 10 to 10^2. This means that nonlinearity should be taken into account if $\varepsilon > 1\%$, and it can be approximately neglected if $\varepsilon < 1\%$. In the latter case, $\sigma_e(\varepsilon)$ reduces to $E_0\varepsilon$, and Eq. (1) reduces to a linear viscoelastic relation. Correspondingly, Eq.(4) for high strain rates reduces to

$$\sigma = E_{eff}\varepsilon + E_2 \int_0^t \dot{\varepsilon}(\tau) \exp(-\frac{t-\tau}{\theta_2}) d\tau, \qquad E_{eff} = E_0 + E_1 \tag{6a}$$

or equivalently

$$\frac{\partial \sigma}{\partial t} + \frac{\sigma}{\theta_2} = \left(E_{eff} + E_2 \right) \frac{\partial \varepsilon}{\partial t} + \frac{E_{eff}\varepsilon}{\theta_2} \tag{6b}$$

Note that the only viscoelastic term in the nonlinear viscoelastic equation (4) is the same as in the linear viscoelastic equation (6). This means that the constitutive parameters for high strain rate response, θ_2 and E_2, determined at linear situation (e.g. $\varepsilon < 1\%$) should still be valid for nonlinear situation (e.g. $\varepsilon > 1\%$). Based on this principle, a simple experimental method was proposed in [13].

However, it is worthwhile to notice that in the case of large deformation, not only constitutive and geometrical nonlinearity but also damage evolution should be taken into account. A damage modified nonlinear ZWT equation was proposed in [10].

3 Nonlinear Viscoelastic Waves in Bars of ZWT Materials

The governing equations for longitudinal nonlinear viscoelastic waves propagating in a thin bar of ZWT materials are constituted by the following motion equation,

$$\rho_0 \frac{\partial v}{\partial t} = \frac{\partial \sigma}{\partial x} \tag{7}$$

continuity equation

$$\frac{\partial v}{\partial x} = \frac{\partial \varepsilon}{\partial t} \tag{8}$$

and the ZWT nonlinear viscoelastic constitutive equation at high strain rates, Eq. (4), where v denotes the particle velocity.

By means of the well-known characteristics method[2], the above partial differential equations are equivalent to three sets of ordinary differential equations, each set consisting of a characteristics equation and a corresponding compatibility condition along the characteristics. The first two sets of characteristics and their compatibility equations are

$$dx = \pm C_v dt \tag{9a}$$

$$dv = \pm \frac{1}{\rho_0 C_v} d\sigma \pm \frac{\sigma - \sigma_{eff}(\varepsilon)}{\rho_0 C_v \theta_2} dt = \pm \frac{1}{\rho_0 C_v} d\sigma + \left[\frac{\sigma - \sigma_{eff}(\varepsilon)}{(\sigma'_{eff} + E_2)\theta_2} \right] dx \tag{9b}$$

where the positive sign is for rightward waves, the negative sign for leftward waves, and C_v the wave velocity along the characteristics

$$C_v = \sqrt{\frac{1}{\rho_0} \frac{d\sigma_{eff}}{d\varepsilon}} = \sqrt{\frac{\sigma'_e(\varepsilon) + E_1 + E_2}{\rho_0}} \tag{10}$$

The third set corresponds to the relation along the particle motion locus and consists of

$$dx = 0, \tag{11a}$$

$$d\varepsilon - \frac{d\sigma}{\sigma'_{eff} + E_2} - \frac{\sigma - \sigma_{eff}(\varepsilon)}{\sigma'_{eff} + E_2} \frac{dt}{\theta_2} = 0. \tag{11b}$$

It is worthwhile to note from the compatibility condition Eq.(9b) and (11b) that the high frequency relaxation time θ_2 always appears in the terms with dt or dx, and in a form of $(\sigma - \sigma_{eff}) dt / \theta_2$ or $(\sigma - \sigma_{eff}) dx / (C_v \theta_2)$, the dispersion and attenuation of viscoelastic waves are, in fact, described by those terms, and thus mainly depend on θ_2 and the stress difference between the overall viscoelastic stress and the pure elastic equilibrium stress, $(\sigma - \sigma_{eff})$, namely, the stress relaxation process.

By using the numerical characteristics method mentioned above, the nonlinear viscoelastic wave propagation in a semi-infinite polymer bars can be computationally simulated. In the following, the analyses and comparison will be made between two typical polymers, namely, a thermoplastic one, polymethylmethacrylate (PMMA), and a thermoset one, epoxy, to see how the nonlinear viscoelastic wave propagation character depends on the costitutive character. For PMMA, the constitutive parameters used in the simulation are $\sigma_m = 91.8$MPa, $m=22.3$, $n=4$, $E_1=0.897$GPa, $\theta_1 = 15.3$s, $E_2=3.07$GPa, $\theta_2=95.4$ μ s, $\rho_0 =1190$kg/m³. For a constant velocity impact boundary condition, the typical results of computational simulation are given in Fig. 1, including stress profiles (Fig.1a), strain profiles (Fig.1b) and particle velocity profiles (Fig.1c) at the different distances from the impact boundary ($x=0$).

142

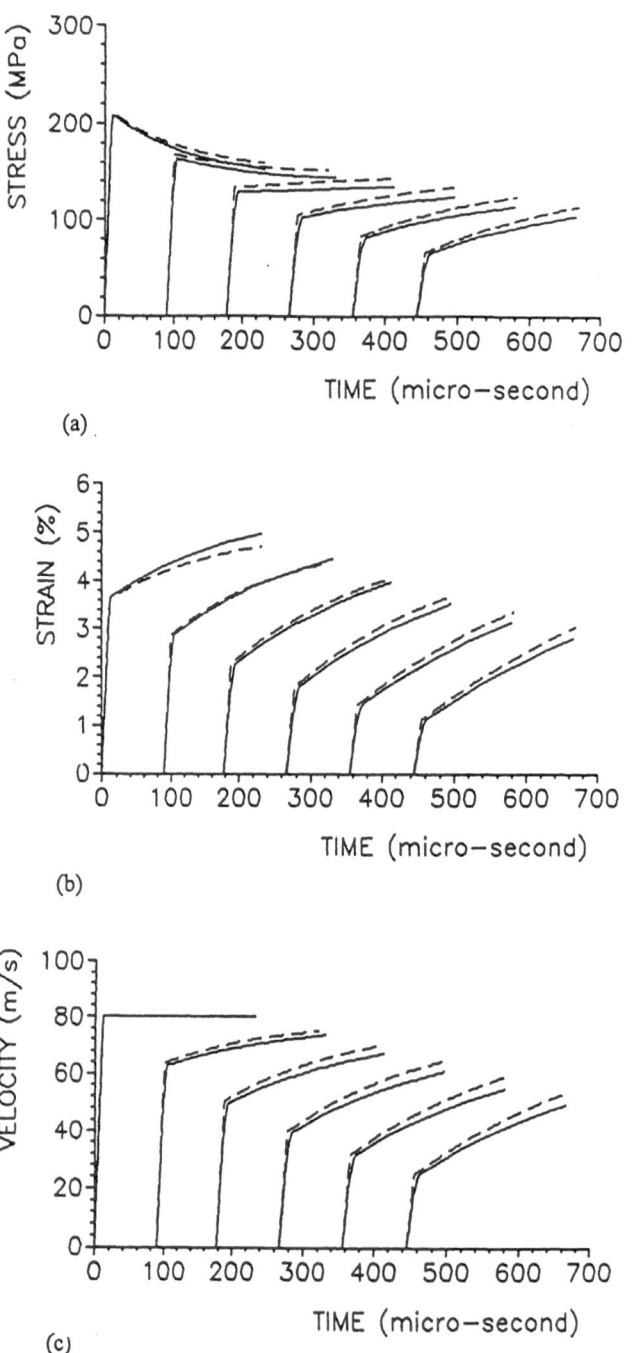

Fig.1 Nonlinear viscoelastic waves in PMMA bar at constant velocity (80m/s) impact condition, showing (a) stress, (b)strain, (c) particle velocity, at x=0, 0.2, 0.4 0.6, 0.8, 1.0m,respectively.

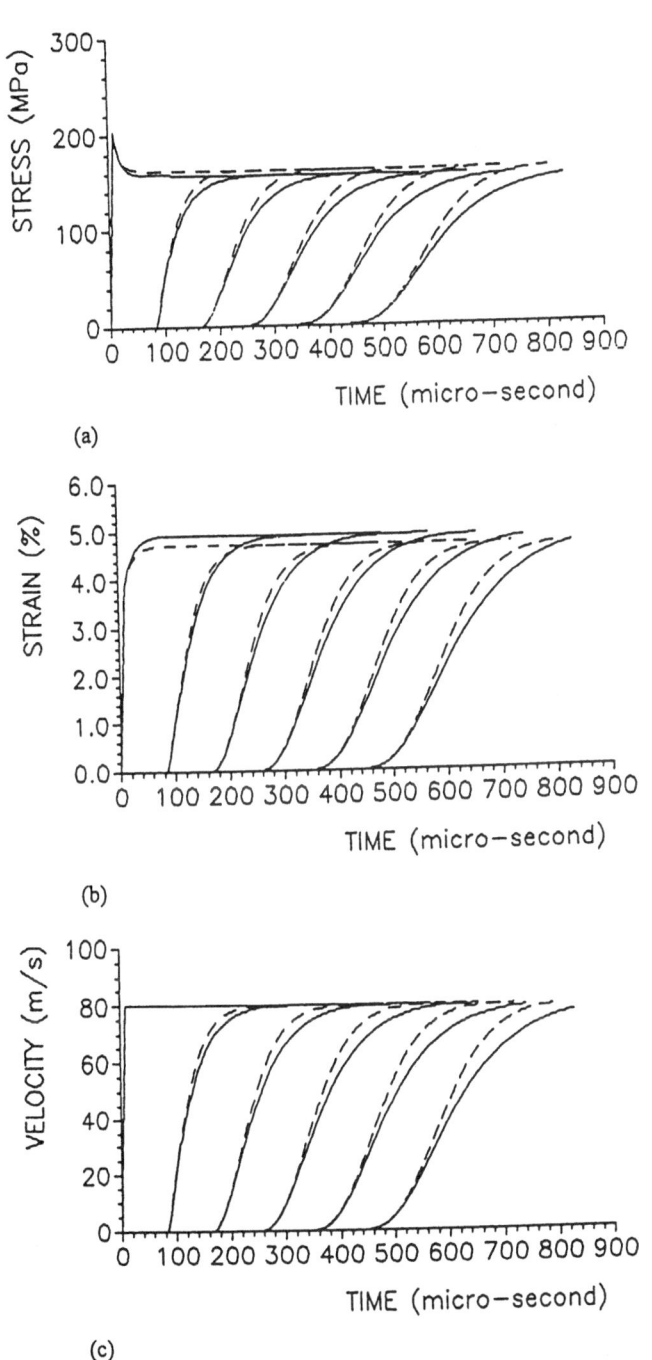

Fig.2 Nonlinear viscoelastic waves in epoxy bar at constant velocity (80m/s) impact condition, showing (a) stress, (b)strain, (c) particle velocity, at x=0, 0.2, 0.4 0.6, 0.8, 1.0m,respectively.

For epoxy, the constitutive parameters used in the simulation are σ_m=97.5MPa, m=20.1,n=4, E_1=1.47GPa, θ_1=157s, E_2=3.43GPa, θ_2=8.574 μ s, ρ_0=1200kg/m³. The typical results for a constant velocity impact boundary condition are given in Fig. 2.

Some common characteristics can be seen from both of Fig. 1 and Fig.2. First, in contrast to linear elastic waves, no proportional relation or no simple equivalency exists between the viscoelastic stress wave, strain wave and particle velocity wave. One of the most distinct characters of viscoelastic waves is that for the position near and at the impact boundary, when the viscoelastic stress decreases with time, showing a "stress relaxation" character, the viscoelastic strain on the contrary increases with time, showing a "strain creep" character. It implies that the viscoelastic dynamic stress at a given position cannot be determined directly from an experimentally measured dynamic strain signal at that positioin by simply multiplying with the apparent elastic modulus.

Next, all viscoelastic waves attenuate with distance x, whichever expression form (σ, ε or v) is in terms of. Comparing with the results in Fig.1 and Fig.2, it can be seen that the viscoelastic waves in a PMMA bar, of which the θ_2 is higher, show a weaker attenuation. This result is consistent with the theoretical analysis according to Eq. (9b) and (11b), since the viscoelastic wave attenuation is mainly dominated by the term of $[\sigma - \sigma_{eff}(\varepsilon)]/[(\sigma'_{eff}+E_2)\theta_2]$, or in other words, becomes weaker with increasing θ_2.

Moreover, with increasing propagation distance, the stress profile of viscoelastic wave is gradually transformed from the "relaxation" shape (i.e. decreasing with time) into the "creep" shape (i.e. increasing with time). Such a transformation of stress profile character is a reflection of the viscoelastic relaxation process. In fact, as mentioned above, the relaxation response of ZWT material at high strain rates is mainly characterised by θ_2, and the "effective influence domain" of θ_2 in terms of time t, as pointed out in [6], is $10^{-2} < t/\theta_2 < 10^{2.3}$. Then, an "effective influence time" $t_{eff} = \theta_2$, or equivalently, an "effective influence distance" $X_{eff}=C_v\theta_2$ can be defined to characterise the time and the propagation distance, respectively, when θ_2 effectively dominates the relaxation process.

According to the constitutive parameters for computational simulation mentioned above, X_{eff}=0.214m for PMMA, while X_{eff}=0.0205m for epoxy.

Thus, for epoxy, it becomes easy to understand why the "stress relaxation profile" only appears at x=0 (impact boundary), but not at $x > 0.2$m. In a similar case, it should be emphasized that the measure positions must be correctly designed by taking account of the "effective influence distance" X_{eff}, whatever wave propagation method is used to determine the viscoelastic constitutive relation.

The results for linear approximation, namely, by using the linear viscoelastic constitutive relation Eq.(4), instead of the nonlinear viscoelastic constitutive relation Eq. (4), are also given in both Fig. 1 and Fig. 2, and drawn by dot lines. It can be seen that the constitutive nonlinearity (Eq. 2) exerts a marked influence on both the wave amplitude and the wave profile shape, especially on the wave front part, and thus the nonlinear effects cannot be neglected in the case of high-velocity large deformation, even if the constitutive nonlinearity is rate-independent. However, if the strain is small enough, e. g. ε <1%, in the present examples, the linear approximation can be accepted since it coincides well with the nonlinear one.

4 Conclusion

Experimental investigation of a variety of plastics, both thermoplastic and thermoset, and the related polymer-matrix-composites at strain-rates from 10^4 to 10^3 s^{-1} revealed that their nonlinear viscoelastic behaviour are well described by the ZWT nonlinear viscoelastic constitutive equation, Eqs. (1) and (2). It consists of three parts: a nonlinear elastic equilibrium response, a low strain-rate linear viscoelastic response and a high strain-rate linear viscoelastic response, and consequently its nonlinearity is rate-independent.

Under high strain rates, the constitutive parameters which describe the impulsive response of ZWT materials are attributed to the effective nonlinear elastic constants (E_0, E_1, α, β, or E_1 σ_m, m, n), which can be determined from quasi-static tests, the high frequency elastic constant E_2 and the high frequency relaxation time θ_2. The last two parameters should be determined from high strain rate tests.

The dispersion and attenuation of nonlinear viscoelastic waves are mainly influenced by the effective nonlinearity, $\sigma_{eff}(\varepsilon)$, and high frequency relaxation time, θ_2. It is found that only in a certain range of wave propagation, in terms of either propagation distance or time, the θ_2 effectively dominates the relaxation process. An "effective influence distance (or time)" is thus defined.

Acknowledgements

The research was partly supported by the Natural Science Foundation of Zhejiang.

References

[1] H. Kolsky. *Stress Waves in Solids*, Clarendon Press, Oxford, 1953.

[2] Lili Wang. *Foundations of Stress Waves*, National Defense Industry Press, Beijing, 1985.

[3] J. A. Zukas, T. Nicholas, H. Swift, L. B. Greszczuk, D. R. Curran. *Impact Dynamics*, John Wiley & Sons, Inc., New York, 1982.

[4] H. Kolsky, An Investigation of the Mechanical Properties of Materials at Very High Rates of Loading, *Proc. Phys. Soc.*, Vol. B62, 1949, p.676.

[5] Zhiping Tang, Lanqiao Tian, Chao-Hsiang Chu, Lili Wang, Mechanical Behaviour of Epoxy Resin under High Strain Rates, *Proc. 2nd Nat. Conf. Explosive Mechanics*, 1981, p.4-1-2.

[6] Chao-Hsiang Chu, Lili Wang, Daben Xu, A Nonlinear Thermo-viscoelastic Constitutive Equation for Thermoset Plastics at High Strain Rates, Ed. Wei-Chang Qian, *Proc. Int. Conf. Nonlinear Mechanics*, 1985, p.92.

[7] Liming Yang, Chao-Hsiang Chu, Lili Wang, Effects of Short Glass Fibre Reinforcement on Nonlinear Viscoelastic Behaviour of Polycarbonate, *Explosion and Shock Waves*, Vol.6, 1986, p.1

[8] Zhaoxiang Zhu, Daben Xu, Lili Wang, Thermo-Viscoelastics Constitutive Equation and Time-Temperature Equivalence of Epoxy Resin at High Strain Rates, *Jour. Ningbo University (Natural Sci. & Engng. Edition)*, Vol.1, 1988, p.58.

[9] Lili Wang, Xixiong Zhu, Shaochu Shi, Su Gan, Hesheng Bao, An Impact Dynamics Investigation on Some Problems of Birds Striking the Windshields of High Speed Aircraft, *Chinese Jour. Aeronautics*, Vol.5, 1992, p.205.

[10] Fenghua Zhou, Lili Wang, Shisheng Hu, A Damage Modified Nonlinear Viscoelastic Constitutive Relation and Failure Criterion of PMMA at High Strain Rates, *Explosion and Shock Waves*, Vol.12, 1992, p.333.

[11] Lili Wang, Liming Yang, A Class of Nonlinear Viscoelastic Constitutive Relation of Solid Polymeric Materials, Eds. Lili Wang, Tongxi Yu, Yongchi Li, *Progress in Impact Dynamics*, The Press of China University of Science and Technology, Hefei, 1992, p.88.

[12] Lili Wang, K. Labibes, Z. Azari, G. Pluvinage, Generalization of Split Hopkinson Bar Technique to Use Viscoelastic Bars, *Int. Jour. Impact Engng.*, Vol.15, 1994, p.669.

[13] K. Labebis, Lili Wang, G. Pluvinage, On Determining the Viscoelastic Constitutive Equation of Polymers at High Strain-Rates, *DYMAT Jour.*, Vol. 1, 1994, p. 135.

[14] A. E. Green, R. S. Rivlin, The Mechanics of Nonlinear Materials with Memory, Part 1, *Arch. Rat. Mech. Anal.*, Vol. 1957, p.1.

[15] B. D. Coleman, W. Noll, An Approximation Theorem for Functionals with Applications in Continum Mechanics, *Arch. Rat. Mech. Anal.*, Vol. 6, p.355.

[16] B. D. Coleman, W. Noll, Foundations of Linear Viscoelasticity, *Rev. Mod. Phys.*, Vol.33, 1961, p.239.

Dynamic Measurement of Elastic Moduli for Composite Materials Using Disk Specimens

Yoshiaki Yamauchi[1], Motohiro Nakano[1], Keizo Kishida[1], Takashi Hashimoto[1] and Yuji Sogabe[2]

1 Department of Precision Science and Technology, Osaka University. 2-1 Yamada-Oka.
   Suita 565, JAPAN
2 Department of Mechanical Engineering, Faculty of Engineering. Ehime University.
   3 Bunkyo-cho, Matsuyama. Ehime 790, JAPAN

Summary

We propose a new method to measure orthotropic elastic moduli for composite materials. In this method. the strain histories of a disk specimen subjected to impact loading in various directions are measured in the experiments. These strains are quantitatively compared in the frequency domain with the numerical results of the dynamic finite element analyses and orthotropic elastic moduli are estimated. Using this method. all elastic moduli may be obtained with only one specimen and one test apparatus. We tried to measure elastic moduli for a carbon fiber/epoxy composite material. As a result, it was found that the orthotropic elastic moduli E_1, E_2, G_{12} could be measured easily within acceptable accuracy. but this method was suitable for measuring the Poisson's ratio ν_{12}.

Keywords: Composite materials, elastic moduli, impact loading. natural vibration. FFT

1 Introduction

Recently composite materials have been widely used for structures in the fields of the aeronautical and space sciences, where the requirements for the lightweight are stronger. In order to make the most effective use of composite materials, first of all we must estimate the mechanical properties for them as precisely as possible.

In general, to measure elastic moduli for orthotropic materials such as composite laminates, it is necessary to prepare different specimens with several kinds of shapes according to the elastic modulus to be measured, and then the several experiments must be conducted using different test apparatuses.[1-3]

In this study, we try to develop a new method to measure elastic moduli conveniently by using a disk specimen loaded by the elastic wave. The advantage of using the disk specimen is that the angle between the direction of the loading and that of the reinforcement can be easily varied. For orthotropic materials, since the velocities of the elastic waves depend on the direction in which the waves propagate, various natural vibrations can be excited by the impact load. The frequencies of these vibrations depend strongly on the elastic moduli of the specimen.

Using modal analyses, these natural vibrations were investigated. Then. the dynamic finite element analyses with a direct time integration scheme were conducted. The strain histories obtained in the experiments were compared with the numerical results in the time domain. Moreover, by analyzing the behavior of the specimen in the frequency domain with fast Fourier transform (FFT), we tried to estimate the elastic moduli for the present material precisely.

147

2 Specimens and Experimental Apparatus

In the present study. a typical CF (carbon fiber)/epoxy composite material was examined. Unidirectional prepregs TR340J-125S (Mitsubishi Rayon Co.. Ltd.) were laid up and the UD panel were processed in an autoclave according to the manufacture's recommended cure cycle. Disk specimens were machined from this panel. The diameter of the disk specimen D was 50mm. the thickness h 5.8mm.

The one-point impact test apparatus as shown in Fig. 1 was used for the experiments. A compressive wave was generated by the longitudinal impact of the striker. and propagated along the input bar. Part of the compressive wave was transmitted through the disk specimen. The impact load $P(t)$ was calculated from the incident and reflected wave measured with the strain gages on the input bar according to the one-dimensional theory for the elastic wave propagation. The length of the striker was 200 mm and the duration time of the impact load was about 80μs. The strain histories of the specimen were measured with strain gages as shown in Fig. 2. In order to eliminate the effect of bending as far as possible. these gages were set on both sides of the disk specimen and connected in series.

The material principal axis system is defined as 1-2 axes. as shown in Fig. 2. θ is the angle between the axis 1 (the direction of the reinforcement) and the direction of loading. The advantage of this method is that various natural vibrations can be exiting by varying the angle θ. In this study, we selected 0.30.45,60 and 90 degrees for the angle θ.

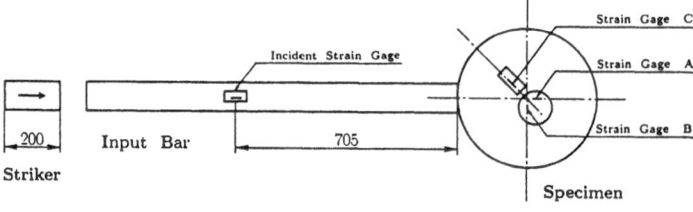

Unit : [mm]

Figure 1: One-point impact apparatus using disk specimen

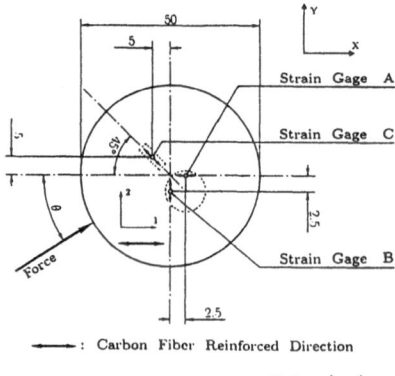

⟶ : Carbon Fiber Reinforced Direction

Unit : [mm]

Figure 2: Geometry of disk specimen and positions of strain gages

3 Numerical analyses

For orthotropic materials, the velocities of the elastic waves depend on the direction in which the waves propagate. These elastic waves reflect from the free boundary and form a very complicated stress field in the specimen.

It is difficult to analyze this complicated stress state mathematically. In this study. we analyzed this stress state numerically using the dynamic two-dimensional finite element method with a direct time integration scheme. The finite element mesh is presented in Fig. 3. The isoparametric plane stress elements with four nodes were employed. The element number was 1200. the node number 1241. The time integration scheme was Newmark-β method with $\beta = 1/4$. the time increment was $\Delta t = 0.1\mu$s. The elastic moduli measured with static tensile loading tests were used for initial numerical analyses as tentative values. These moduli are shown in Table 1.

Table 1. Mechanical properties used for initial numerical analyses

Young's modulus		Shear modulus	Poisson's ratio	Density	Volume fraction
E_1 [GPa]	E_2 [GPa]	G_{12} [GPa]	ν_{12}	ρ [kg/m^3]	V_f [%]
115	8.01	3.84	0.306	1.52×10^3	57.3

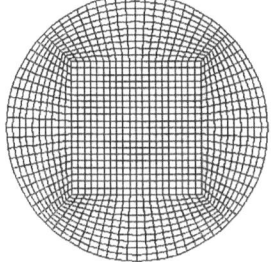
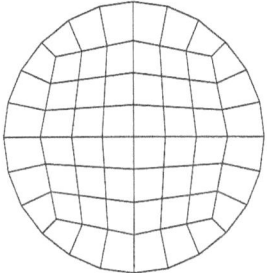

Figure 3: Finite element mesh for direct   Figure 4: Finite element mesh for model analysis
time integral analysis

4 Modal analyses

In order to investigate distinctive mode shapes and frequencies. modal analyses using a simple model was conducted. We analyzed in-plane deformation modes only using a FEM code. The element number was 60, the node number 73. The finite element mesh is presented in Fig. 4. The elastic moduli shown in Table 1 were used.

According to the analyses, the lowest frequency is about $f_a = 24.4$ kHz and this mode shape is, as shown in Fig. 5(a). a deformation that the disk specimen is elongated and compressed along the direction inclined $\pm45°$ from the material principal axis. Since the shear modulus G_{12} is lower than the other moduli. it seems that shear deformations were easily caused.

About $f_b = 26.5$ kHz . a characteristic deformation mode that the specimen vibrates in the direction of the principle axis 2 can be seen. as shown in Fig. 5(b). This vibration is closely related to the elastic modulus E_2 perpendicular to the reinforcement.

The natural vibration mode about $f_c = 93.4$ kHz are shown in Fig. 5(c). A deformation that the disk vibrate in the direction along the reinforcements can be seen in this mode. This deformation strongly depends on the elastic modulus E_1.

Moreover, one parameter among the elastic moduli E_1, E_2, G_{12} was intentionally varied. We investigated how the three modes above behaved. The results are shown in Table 2.

Table 2.   Variation of frequencies with change of elastic moduli

(unit:kHz)

frequency	initial	$E_1 \times 0.9$	$E_1 \times 1.1$	$E_2 \times 0.9$	$E_2 \times 1.1$	$G_{12} \times 0.9$	$G_{12} \times 1.1$
f_a	24.4	24.4	24.4	24.3	24.4	<u>23.1</u>	<u>25.5</u>
f_b	26.5	26.5	26.6	<u>25.3</u>	<u>27.7</u>	26.5	26.7
f_c	93.4	<u>89.4</u>	<u>97.2</u>	93.0	93.7	93.0	93.8

It was found that each mode intensely depends on the specified elastic modulus, while it has little relation to the other moduli. According to quantitative consideration, the following relations were obtained.

$$f_a \propto \sqrt{G_{12}} , \quad \frac{\partial f_a}{\partial E_1} \approx 0 , \quad \frac{\partial f_a}{\partial E_2} \approx 0 \qquad (1)$$

$$f_b \propto \sqrt{E_2} , \quad \frac{\partial f_b}{\partial E_1} \approx 0 , \quad \frac{\partial f_b}{\partial G_{12}} \approx 0 \qquad (2)$$

$$f_c \propto \sqrt{E_1} , \quad \frac{\partial f_c}{\partial E_2} \approx 0 , \quad \frac{\partial f_c}{\partial G_{12}} \approx 0 \qquad (3)$$

The velocity of the elastic wave is in proportion to the square root of the elastic modulus. This has a close relation to the fact that the specified natural frequencies above are in proportion to the square root of the specified elastic moduli.

As for the Poisson's ratio ν_{12}, modal analyses were performed in the same way, but it was found that there was no natural mode that depended on ν_{12}

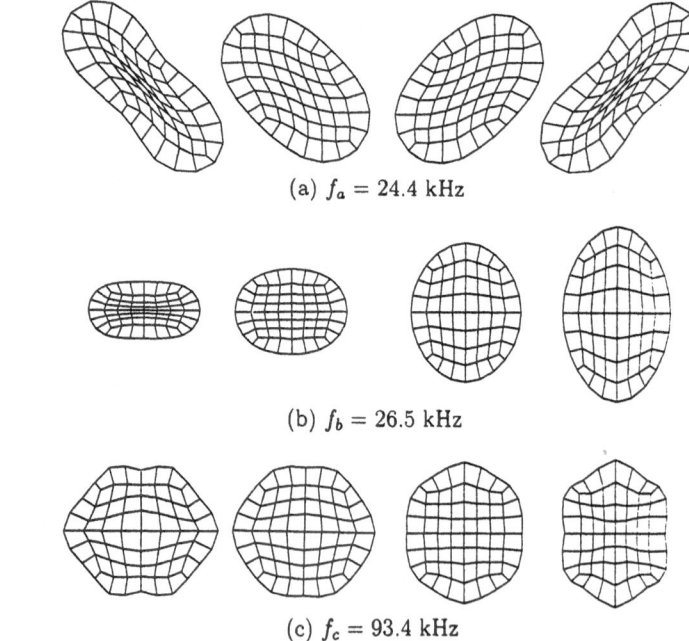

(a) $f_a = 24.4$ kHz

(b) $f_b = 26.5$ kHz

(c) $f_c = 93.4$ kHz

Figure 5: Mode shapes in specified frequencies

5 Result and Discussion

The strain histories measured with the experiment with $\theta = 0°$ are shown in Fig. 6. Output from the gage C that set on the direction inclined 45° from the reinforcement is very large. This means that, at the first stage of the impact loading. a large shear deformation occurred. In the experiments that $\theta = 30, 45, 60$ and 90 degrees. output from the gage C is comparatively large, but the gage A set on the direction along the reinforcement is insensitive.

We performed the dynamic finite element analyses using typical impact loads $P(t)$ obtained with the experiments. When $\theta = 0°$, the numerical results are shown in Fig. 7. The vibrations of low frequencies are similar to those in Fig. 6. The strains are damped comparatively soon in the experiments, but the vibrations continue for a long time in the results of FE analyses. The cause of this is that the present FEM code did not include the mechanism of the energy dissipation. In the other cases that $\theta = 30, 45, 60$ and 90 degrees. the numerical results are also similar to the experiments in the vibration of the low frequencies except that there is no damping in the strain waves.

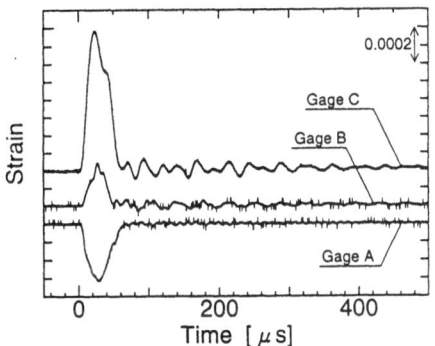

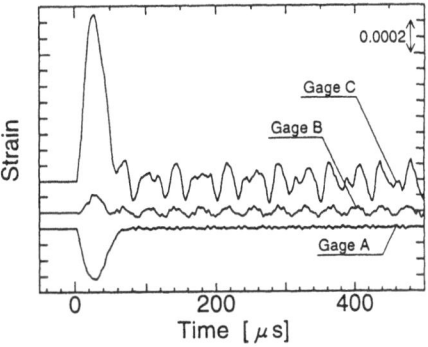

Figure 6: Strain histories measured in experiment, when $\theta = 0°$

Figure 7: Strain histories calculated in FE analysis, when $\theta = 0°$

In order to investigate these strain waves in detail, these waves were resolved into the components of frequencies using fast Fourier transform (FFT). The number of the strain data N was $2^{14}=16384$, which were sampled every $0.1\mu s$. After correcting the zero points by the averaging of 2^{14} points, components of the high frequencies that caused by cutting the waves off were eliminated through the Hanning window. The frequency increment between two successive points of the spectrum was $\Delta f = 1/N\Delta t = 610$ Hz.

The results of the gage A when $\theta = 0°$ and the gage C when $\theta = 45°$ are shown in Fig. 8. where part of the low frequencies is magnified. The frequencies are normalized by the representative frequency f_0. which is given by.

$$f_0 = \frac{1}{2D} \cdot \sqrt{\frac{E_1}{\rho}} \tag{4}$$

In the present study, $f_0 = 87.0$ kHz. The vertical axis is the logarithm for the magnitude. This is the absolute value of the complex frequency domain waveform.

There are many peaks in both experimental and numerical results. The peaks in the numerical results are sharper than those in the experiments. Most of these peaks have correspondence between the experiments and the analyses. We can see the small peak at $\hat{f} = 0.077$ in the curves of some experiments. There was no peak in the results of the FE analyses, where only in-plane deformations were considered. In modal analyses where out-plane deformations were permitted, a natural vibration mode was observed about this frequency. Therefore, this peak seems to be the effect of the out-plane bending.

On the basic of the modal analyses, we pay to our attention to the three peaks, which are correspond to the three natural deformation modes as shown in Fig. 5. In Fig. 8(a), the vibrations corresponding to f_c, f_b appear, but the vibration of f_a cannot be seen in the numerical analyses. On the other hand, in Fig. 8(b), although the peak of f_c is not clear, two peaks can be seen in the numerical analysis, which correspond to f_a, f_b respectively. These two frequencies are so close that they cannot be distinguished in the experiments.

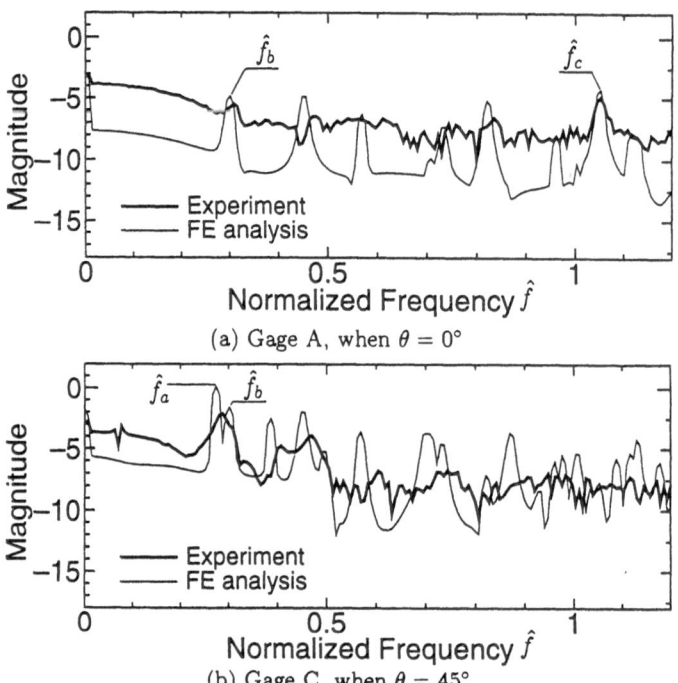

(a) Gage A, when $\theta = 0°$

(b) Gage C, when $\theta = 45°$

Figure 8: Comparison in frequency domain between spectra measured and calculated using initial elastic moduli

In these three modes, each natural frequency is approximately in proportion to the square root of the specified elastic modulus, while it has little relation to the other moduli. Therefore, comparing these frequencies obtained by the experiments with the numerical results, the elastic moduli can be corrected as shown in Table 3.

Table 3. Comparison of frequencies and correction of elastic moduli

	Experimental	Numerical		Initial	Corrected
$\hat{f}_a$	0.288	0.273	G_{12} [GPa]	3.84	4.24
$\hat{f}_b$	0.309	0.302	E_2 [GPa]	8.01	8.39
$\hat{f}_c$	1.053	1.053	E_1 [GPa]	115	115

Moreover, using these corrected elastic moduli, the dynamic finite element analyses were carried out again. The strain histories at the gage A when $\theta = 0°$. the gage C when $\theta = 45°$ are shown in Fig. 9, which are compared with the experiment results. Both results almost agree except the effect of the damping.

The effect of the damping is comparatively small in the strain measured with the gage A. but the strains measured with the gages B.C is considerably affected by it. It is said that carbon fiber/epoxy composites have no property of the viscosity along the direction of the reinforcement and the viscosity originates in the property of the matrix. The experimental results agrees with it. In order to investigate the viscoelastic property for composite materials. we must consider the propagation of the strain waves in the disk specimen using FFT and determine its complex compliance.[4]

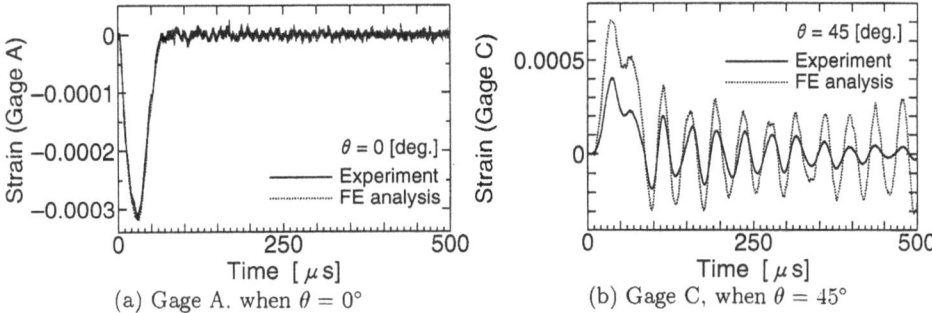

(a) Gage A. when $\theta = 0°$ (b) Gage C, when $\theta = 45°$

Figure 9: Comparison between strain histories measured
and calculated using corrected elastic moduli

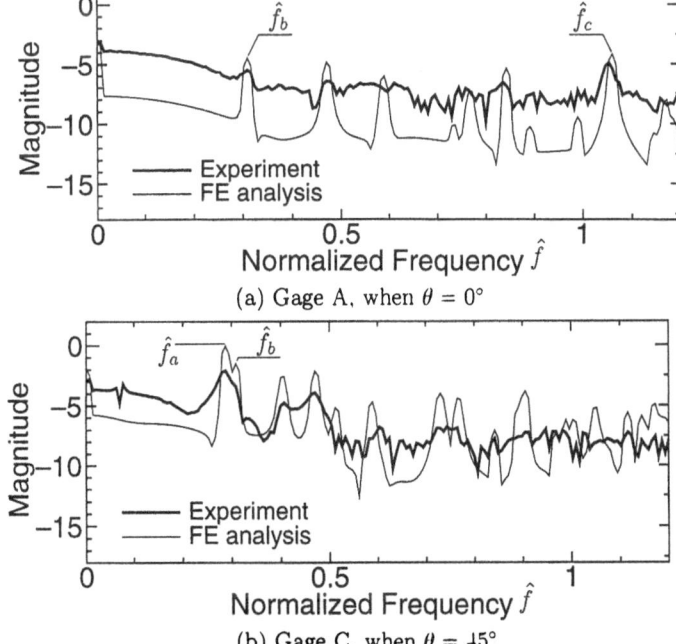

(a) Gage A, when $\theta = 0°$

(b) Gage C, when $\theta = 45°$

Figure 10: Comparison in frequency domain between spectra
measured and calculated using corrected elastic moduli

The comparisons in the frequency domain are shown in Fig. 10. Until the normalized frequency $\hat{f} \sim 0.7$, the frequencies of the peaks agree very well. As $\hat{f}$ is larger. $\hat{f}$ become overestimated in the numerical results. Particularly, although E_1 was not changed in the recalculation. $\hat{f}_c$ was estimated a little largely. Since the value $\frac{\partial f_c}{\partial E_2}$, $\frac{\partial f_c}{\partial G_{12}}$ is small. but not negligible. as shown in Table 2, the influence of E_2, G_{12} appears on $\hat{f}_c$. The effect on the value of E_1 is less than 1.5 %. Since E_1 hardly affects the value of $\hat{f}_a, \hat{f}_b$, if the recalculation are conducted after correcting E_1, it seems that we can get a better results.

Compared with the frequency increment Δf. the frequencies $f_a. f_b$ is not so large. Therefore. the accuracy is reduced in this region. In order to improve the frequency resolution. the duration of the time domain waveform $T = N\Delta t$ must be increased. In the experiments. however, since the damping is intensive. it is meaningless to increase T. To improve the accuracy, we must perform the comparison of the spectrum in more frequencies than three.

6 Conclusions

Elastic moduli for a CF/epoxy composite material was measured using a new method. The strain histories of a disk specimens subjected to impact loading in various directions were measured with the experiments. The dynamic finite element analyses using the measured impact load were performed. Based on modal analyses, these results were quantitatively compared in the specified frequencies and the elastic moduli were obtained.

As a result, the following conclusions are obtained. Using only one specimens and one test apparatus, we can measure the orthotropic elastic moduli E_1, E_2, G_{12} within acceptable engineering accuracy. Particularly. since shear deformations are easily caused by impact loading. the shear modulus G_{12} can be estimated with simple specimen and test apparatus. On the other hand, it is difficult to measure the Poisson's ratio ν_{12} with this method. Further improvement are required to determine this value.

After this, considering the propagation of the strain waves in the disk specimen using FFT. the viscoelastic properties for composite materials can be estimated by applying this method.

Acknowledgment – The authors would like to express their appreciation to the former graduate student Mr. T.Okabe for their technical assistance in the programming of the finite element analyses code. We wish to thank Arisawa Factory Co.. Ltd. for the help in preparing the test specimens.

Reference

[1] JIS K 7073 Testing Method for Tensile Properties of Carbon Fiber Reinforced Plastics. 1988

[2] JIS K 7076 Testing Method for Compressive Properties of Carbon Fiber Reinforced Plastics, 1991

[3] JIS K 7079 Testing Method for In-Plane Shear Property of Carbon Fiber Reinforced Plastics by ±45° Tension Method and Pairs of Rails Method. 1991

[4] Yuji.Sogabe, Masayuki.Tsuzuki and Keizo.Kishida. Effect of Fiber Orientation Angle on Viscoelastic Properties of CFRP Subjected to Impact Loads. *Journal of the Society of Material Science*. Vol.42. 1993, pp.536-541

Dynamic Yield Loci of A Porous Visco-plastic Material by Using A lower Bound Approach

Zhuping Huang[1], Xi Yuan[2], Ren Wang[1] and Jianguo Ning[3]

1. Dept. of Mechanics, Peking University, Beijing, 100871, China
2. Dept. of Civil Eng. , North Carolina State University, Raleigh, NC 27695, U. S. A.
3. Dept. of Mech. Eng. , Beijing Institute of Technology, Beijing, 100081, China

Summary

Based on the study of a representative volume element (RVE) of porous visco-plastic material, a void growth model is proposed in this paper. The constitutive relation of the matrix material is in the over stress form given by previous authors. A lower bound approach by constructing a statically admissible stress field in the RVE is employed, and the corresponding dynamic yield loci of this porous material are obtained. The dynamic yield loci derived from both the lower bound approach in the present paper and the upper bound approach in Ref. [5] are compared for different values of porosity f and strain rate parameter m. They are rather close.

Keywords: Dynamic yield loci, void growth, visco-plastic material, representative volume element, lower bound approach

1 Introduction

The process of void nucleation, growth and eventual coalescence is known to be a principal micromechanism of fracture in ductile metals. The evolution law of the above process may be modelled through either macroscopic or microscopic consideration. From the latter point of view, the microvoid evolution can be related explicitly to the density of local defects in the material. This not only gives a microscopic interpretation on the damage mechanism, but also requires less material parameters to be determined by experiment.

There are several void growth models in the literature which are based on the study of a representative volume element (RVE) of a porous material[1 − 6]. However, in all these models, only the upper bound approach was employed, in which different matrix material properties were assumed and different kinematically admissible velocity fields were considered. In this paper, we employ a lower bound formulation to construct a statically admissible stress field in the RVE. An approximate macroscopic strain rate potential together with its corresponding constitutive relation in a porous visco-plastic material are developed. From this constitutive relation, dynamic yield loci at constant equivalent strain rates $\dot{E}_a =$ $(2\dot{E} : \dot{E} /3)^{1/2}$ are obtained. In order to estimate the accuracy of the present model, comparisons between the results of the present model and those from an upper bound approach, given by Sun and Huang [5], are made. It can be seen that the difference between these results is not so large for the void volume fraction $f = 0. 05$ and $f = 0. 1$. Hence it may be concluded that the dynamic yield loci either from the upper bound approach proposed by Sun and Huang or from the lower bound approach in the present paper may be used to derive the macroscopic constitutive relation of porous visco-plastic materials.

2   Macroscopic Strain Rate Potential

Let us take a hollow sphere as a RVE. Denote the volumes of the matrix and the void by V_m and V_v, respectively. The total volume is $V = V_m + V_v$. Denote the macroscopic stress and the macroscopic strain rate acting on the RVE by Σ and $\dot{E}$, respectively, while the corresponding microscopic ones in the matrix are σ and $\dot{\varepsilon}$, respectively. When the inertial effect is neglected, the following equations should be satisfied by a statically admissible stress field in the matrix:

$$\sigma \cdot \nabla = 0 \qquad \text{(in } V_m)$$
$$\sigma \cdot n = 0 \qquad \text{(on } \partial V_v) \qquad (1)$$
$$\sigma \cdot n = \Sigma \cdot n \qquad \text{(on } \partial V)$$

So, we have

$$\Sigma = \frac{1}{V} \int_{V_m} \sigma \, \mathrm{d}v = \frac{1}{2V} \int_{\partial V} (\sigma \cdot n \otimes x + x \otimes n \cdot \sigma) \, ds . \qquad (2)$$

Suppose that the matrix material is incompressible and the constitutive relation of the matrix is given in the overstress formulation proposed by Perzyna. Then we have

$$\dot{\varepsilon} = \frac{\partial \psi(\sigma)}{\partial \sigma} = \frac{3\dot{\varepsilon}_0}{2\sigma_e}\left(\frac{\sigma_e}{\sigma_0}-1\right)^m S, \qquad \text{(when } \sigma_e \geqslant \sigma_0) \qquad (3)$$

and the corresponding microscopic strain rate potential $\psi(\sigma)$ and the microscopic stress potential $\varphi(\dot{\varepsilon})$ in the matrix are:

$$\psi(\sigma) = \frac{\sigma_0 \dot{\varepsilon}_0}{(m+1)}\left[\frac{\sigma_e}{\sigma_0}-1\right]^{m+1}, \qquad \text{(when } \sigma_e \geqslant \sigma_0) \qquad (4)$$

and

$$\varphi(\dot{\varepsilon}) = \frac{\sigma_0 \dot{\varepsilon}_0}{(n+1)}\left[\left(\frac{\dot{\varepsilon}_e}{\dot{\varepsilon}_0}\right)^{n+1} + (n+1)\frac{\dot{\varepsilon}_e}{\dot{\varepsilon}_0}\right], \qquad (5)$$

where σ_e and $\dot{\varepsilon}_e$ are the effective stress and the effective strain rate, respectively, σ_0, $\dot{\varepsilon}_0$ and $m = \frac{1}{n}$ ($\geqslant 1$) are material parameters.

Noting that $\psi(\sigma)$ and $\varphi(\dot{\varepsilon})$ are convex functions of their variable, we have the following inequality:

$$\varphi(\dot{\varepsilon}^{(2)}) \geqslant \dot{\varepsilon}^{(2)} : \sigma^{(1)} - \psi(\sigma^{(1)}), \qquad (6)$$

where $(\sigma^{(1)}, \dot{\varepsilon}^{(1)})$ and $(\sigma^{(2)}, \dot{\varepsilon}^{(2)})$ are two arbitrarily specified stress-strain rate states. Now let $(\sigma^{(2)}, \dot{\varepsilon}^{(2)})$ be the actual stress-strain rate field which satisfies all the field equations and boundary conditions corresponding to $\dot{E}^{(2)}$. Let $\sigma^{(1)}$ be any statically admissible microscopic stress field σ satisfying eqs. (1), then by integrating inequality (6), we obtain

$$\Phi \geqslant \dot{E}^{(2)} : \Sigma - \Psi(\Sigma, \sigma), \qquad (7)$$

where $\Phi = \frac{1}{V}\int_{V_-} \varphi(\dot{\varepsilon}^{(2)}) \, dv,$ \qquad (8)

and $\Psi(\Sigma, \sigma) = \frac{1}{V}\int_{V_-} \psi(\sigma) \, dv.$ \qquad (9)

For a given Σ, the corresponding macroscopic strain rate potential $\Psi(\Sigma)$ is by minimizing the functional $\Psi(\Sigma, \sigma)$ among all statically admissible stress fields satisfying eqs. (1), i. e. ,

$$\Psi(\Sigma) = \min_{\sigma} \Psi(\Sigma, \sigma). \tag{10}$$

Hence, when the actual macroscopic strain rate $\dot{E}^{(2)}$ is denoted by $\dot{E}$, the macroscopic constitutive relation may be expressed by

$$\dot{E} = \frac{\partial \Psi(\Sigma)}{\partial \Sigma} \tag{11}$$

3. Construction of a Statically Admissible Stress Field

Suppose the RVE is subjected to an axial-symmetric deformation and the non-zero components of the macroscopic stress Σ are such that

$$\Sigma_{33} \geqq \Sigma_{11} = \Sigma_{22}$$

(the case of $\Sigma_{33} < \Sigma_{11} = \Sigma_{22}$ can be discussed similarly). The dilatant part and the deviatoric part of Σ may be denoted by

$$\Sigma_m = \frac{1}{3}(2\Sigma_{11} + \Sigma_{33})$$

and $\Sigma_e = (\frac{3}{2}\Sigma : \Sigma)^{1/2} = \Sigma_{33} - \Sigma_{11}$, respectively. The components of the microscopic stress field in the spherical polar coordinate (r, θ, φ) may be written as $(\sigma_r, \sigma_\theta, \sigma_\varphi, \sigma_{r\theta})$. This stress field would satisfy the equlibrium equation by introducing two stress functions $M(r, \theta)$ and $N(r, \theta)$ such that:

$$\sigma_r = \frac{1}{r^2 \sin\theta} \cdot \frac{\partial^2(M-N)}{\partial \theta^2},$$

$$\sigma_\theta = -\frac{1}{\sin\theta}\left(\frac{\partial^2 N}{\partial r^2} + \frac{2}{r}\frac{\partial N}{\partial r}\right) + J,$$

$$\sigma_\varphi = \frac{1}{\sin\theta}\left(\frac{\partial^2 N}{\partial r^2} + \frac{2}{r}\frac{\partial N}{\partial r} + \frac{1}{r}\frac{\partial^3 M}{\partial r \partial \theta^2}\right) - J, \tag{12}$$

$$\sigma_{r\theta} = \frac{1}{r\sin\theta}\frac{\partial^2 N}{\partial r \partial \theta},$$

where $J = \frac{1}{\sin^2\theta}\int\left(\frac{\partial^2 N}{\partial r^2} + \frac{2}{r}\frac{\partial N}{\partial r} + \frac{1}{r}\frac{\partial^3 M}{\partial r \partial \theta^2}\right)\cos\theta\, d\theta.$

Now the macroscopic stress Σ is additively decomposed into two parts. The first one corresponds to Σ_m:

$$\Sigma_{11}^{(1)} = \Sigma_{22}^{(1)} = \Sigma_{33}^{(1)} = \Sigma_m, \qquad \Sigma_{ij}^{(1)} = 0 \quad (\text{when } i \neq j),$$

and the second one corresponds to Σ_e:

$$\Sigma_{11}^{(2)} = \Sigma_{22}^{(2)} = -\frac{\Sigma_e}{3}, \quad \Sigma_{33}^{(2)} = \frac{2}{3}\Sigma_e, \qquad \Sigma_{ij}^{(2)} = 0 \quad (\text{when } i \neq j).$$

For the dilatant part Σ_m, the corresponding microscopic stress field may be easily obtained since the problem is spherically symmetric, which gives

$$M^{(1)} = -r^2 \sin\theta \left[\frac{2m}{3}(\sigma_\theta(a) - \sigma_0)\left(1 - \left(\frac{a}{r}\right)^{3/m}\right) + 2\sigma_0 ln\frac{r}{a} \right],$$

$$N^{(1)} = 0, \tag{13}$$

and

$$\sigma_r^{(1)} = \frac{2m}{3}(\sigma_\theta(a) - \sigma_0)(1 - x^{-3/m}) + 2\sigma_0 lnx,$$

$$\sigma_\theta^{(1)} = \sigma_\varphi^{(1)} = \sigma_r^{(1)} + (\sigma_\theta(a) - \sigma_0)x^{-3/m} + \sigma_0, \tag{14}$$

$$\sigma_{r\theta}^{(1)} = 0,$$

where $\sigma_\theta(a) = \dfrac{3}{2m}(\Sigma_m - 2\sigma_0 ln\beta)/(1 - \beta^{-3/m}) + \sigma_0,$

a and b are inner and outer radii of the RVE respectively. f is the porosity of the material, $x = \dfrac{r}{a}$ and $\beta = \dfrac{b}{a} = f^{-1/3}$.

For the deviatoric part Σ_e, the problem becomes much more difficult, and a procedure to minimize the functional $\Psi(\Sigma, \sigma)$ should be performed. As a first order approximation, we shall make use of an approximate microscopic stress field which corresponds to a solution of a linear viscous hollow sphere problem. Hence we may take

$$M^{(2)} = \left[\left(\frac{7+6\upsilon}{2}\right)\mathscr{A}_1 r^4 - \frac{1}{3}\mathscr{A}_2 r^2 - \left(\frac{1-2\upsilon}{3}\right)\mathscr{A}_3 r^{-1} - \frac{8}{3}\mathscr{A}_4 r^{-3} \right] sin^3\theta,$$

and

$$N^{(2)} = \left[\left(\frac{7+2\upsilon}{2}\right)\mathscr{A}_1 r^4 - A_2 r^2 + 4(1+\upsilon)\mathscr{A}_3 r^{-1} + \frac{16}{3}\mathscr{A}_4 r^{-3} \right] sin^3\theta, \tag{15}$$

where υ is a parameter to be determined by minimizing $\Psi(\Sigma, \sigma)$. From eq. (15) and eq. (12), we obtain

$$\sigma_r^{(2)} = [3\upsilon\mathscr{A}_1 x^2 + \mathscr{A}_2 - 2(5-\upsilon)\mathscr{A}_3 x^{-3} - 12\mathscr{A}_4 x^{-5}](1 + 3cos2\theta),$$

$$\sigma_\theta^{(2)} = [7(2+v)\mathscr{A}_1 x^2 - \mathscr{A}_2 + (1-2v)\mathscr{A}_3 x^{-3} + 7\mathscr{A}_4 x^{-5}](1+3\cos2\theta)$$
$$+ [-2(7-4v)\mathscr{A}_1 x^2 + 2\mathscr{A}_2 + 4(1-2v)\mathscr{A}_3 x^{-3} - 4\mathscr{A}_4 x^{-5}],$$
$$\sigma_\varphi^{(2)} = [(7+11v)\mathscr{A}_1 x^2 + 3(1-2v)\mathscr{A}_3 x^{-3} + 5\mathscr{A}_4 x^{-5}](1+3\cos2\theta)$$
$$+ [2(7-4v)\mathscr{A}_1 x^2 - 2\mathscr{A}_2 - 4(1-2v)\mathscr{A}_3 x^{-3} + 4\mathscr{A}_4 x^{-5}],$$
$$\sigma_{r\theta}^{(2)} = 3[(7+2v)\mathscr{A}_1 x^2 - \mathscr{A}_2 - 2(1+v)\mathscr{A}_3 x^{-3} - 8\mathscr{A}_4 x^{-5}]\sin2\theta \qquad (16)$$

where $\mathscr{A}_1 = 5(\beta^5 - \beta^3)\Sigma_e/\Omega$

$$\mathscr{A}_2 = [(49-25v^2)\beta^{10} + 126\beta^5 + 25(v^2-7)\beta^3]\Sigma_e/6\Omega,$$
$$\mathscr{A}_3 = 5(7+5v)(\beta^{10} - \beta^3)\Sigma_e/12\Omega,$$
$$\mathscr{A}_4 = -(7+5v)(\beta^{10} - \beta^5)\Sigma_e/4\Omega \qquad (17)$$

and

$$\Omega = [(49-25v^2)\beta^{10} + 25(v^2-7)\beta^7 + 252\beta^5 + 25(v^2-7)\beta^3 + (49-25v^2)].$$

Thus a statically admissible microscopic stress field is constructed which is the summation of the above two stress fields: $\sigma^{(1)} + \sigma^{(2)}$. This field not only satisfies the equilibrium equation, but also satisfies the boundary conditions on the inner and the outer surfaces of the RVE.

4. Numerical Calculations and Discussions

As a first order approximation, the microscopic strain rate potential $\psi(\sigma)$ may be calculated from eq. (4) and the statically admissible microscopic stress field given in section 3. Hence from eqs. (9) and (10), the minimized value of the macroscopic strain rate potential Ψ may be obtained by taking $v=0.5$. Hence the corresponding macroscopic constitutive relation eq. (11) can be expressed as follows:

$$\dot{E}_{kk} = \frac{\partial\Psi}{\partial\Sigma_m} = \frac{1}{V}\int_{v_a} \frac{3}{2}\frac{\dot{\varepsilon}_0}{\sigma_e}\left(\frac{\varepsilon_e}{\sigma_0} - 1\right)^m \frac{\partial\sigma^{(1)}}{\partial\Sigma_m} : S\, dv,$$

$$\dot{E}_e = \frac{\partial\Psi}{\partial\Sigma_e} = \frac{1}{V}\int_{v_a} \frac{3}{2}\frac{\dot{\varepsilon}_0}{\sigma_e}\left(\frac{\varepsilon_e}{\sigma_0} - 1\right)^m \frac{\partial\sigma^{(2)}}{\partial\Sigma_e} : S\, dv, \qquad (18)$$

where S is the deviatoric part of the stress field $\sigma = \sigma^{(1)} + \sigma^{(2)}$. It can be proved that when $\Sigma_e = 0$, we have $\dot{E}_e = 0$, and eq. (18) reduces to

$$\left(\frac{\dot{E}_{kk}}{\dot{\varepsilon}_0}\right) = \frac{3}{2}\left[\frac{3\Sigma_m + 2\sigma_0 \ln f}{2m\sigma_0(1 - f^{\frac{1}{m}})}\right]^m \cdot f \qquad (19)$$

This result is exactly the same with that given by Sun and Huang [5], i. e. , when the RVE is subjected to a spherically symmetric tension, the present result derived from a lower bound approach coincides with that derived from a upper bound approach. Dynamic yield loci for constant $\dot{E}_a = \left(\dot{E}_e^2 + \frac{2}{9} \dot{E}_{kk}^2 \right)^{1/2}$ may be obtained from eq. (18) by numerical calculations. In order to estimate the accuracy of the present result, dynamic yield loci from both the lower bound approach in this paper and the upper bound approach in [5] are depicted in $\Sigma_m/Y_0 \sim \Sigma_e/Y_0$

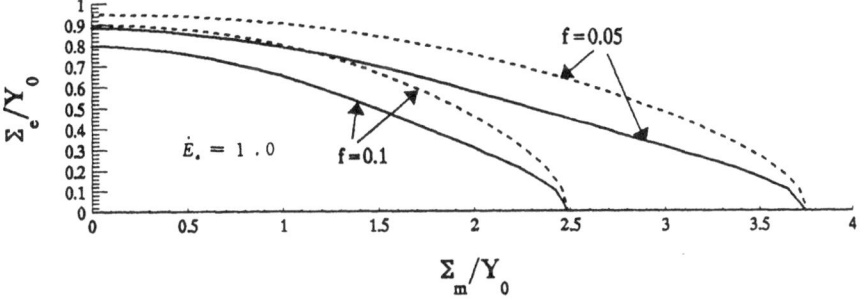

Fig. 1: Dynamic yield loci for different porosity f with rate-sensitive exponent $m=2$.
————present result, ----calculated from the upper bound approach in [5].

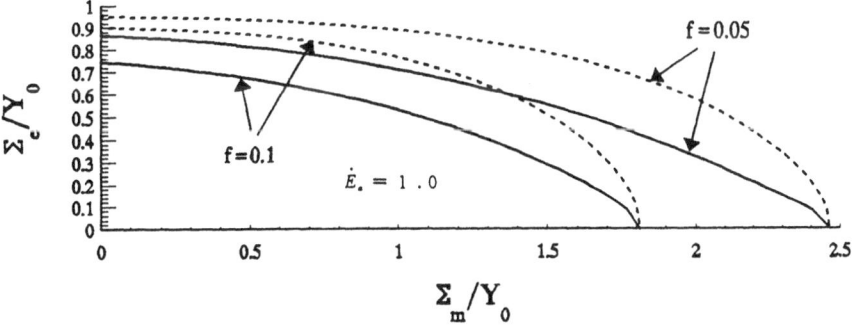

Fig. 2: Dynamic yield loci for different porosity f with rate-sensitive exponent $m=5$.
————present result, ----calculated from the upper bound approach in [5].

162

coordinate as shown in Fig. 1 and Fig. 2, where $Y_0 = \sigma_0 \left[\left(\frac{\dot{E}_a}{\dot{\varepsilon}_0} \right)^n + 1 \right]$. It is seen that the differences between these results for same values of $f = 0.05$ and 0.1 with $m = 2$ and 5 are not so large. By inference, the same is true for other values of f and m between these limits. Since the actual dynamic yield locus should be located between these two bounds, the dynamic void growth model either from the upper bound approach proposed by Sun and Huang, or from the lower bound approach in the present paper may be used to derive the macroscopic constitutive relation of porous visco-plastic materials.

Acknowledgements——This study was supported by the National Natural Science Foundation of China. The authors are also grateful to Mr. H. L. Li and Mr. Y. Liu for their helpful discussions on this paper.

References

[1]A. L. Gurson, Continuum Theory of Ductile Rupture by Void Nucleation and Growth: Part I——Yield Criterion and Flow Rules for Porous Ductile Media, *J. Eng. Mater. Tech.*, *Trans. ASME*, Vol. 99, 1977, pp. 2—15.

[2]B. Budiansky, J. W. Hutchinson, S. Slutsky, Void Growth and Collapse in Viscous Solids, in *Mechanics of Solids*——*The R. Hill 60th Anniversary Volume*, (H. G. Hopkins and M. J. Sewell eds.), Pergamon Press, 1982, pp. 13—45.

[3]J. M. Duva, J. W. Hutchinson, Constitutive Potentials for Dilutely Voided Nonlinear Materials, *Mech. Mater.*, Vol. 3, 1984, pp. 41—54.

[4]P. Gilormini, C. Licht, P. Suquet, Growth of Voids in a Ductile Matrix: A Review, *Arch. Mech.*, Vol. 40, 1988, pp. 43—80.

[5]L. Z. Sun, Z. P. Huang, Dynamic Void Growth in Rate-Sensitive Plastic Solids, *Int. J. Plasticity*, Vol. 8, 1992, pp. 903—924.

[6] Z. P. Huang, L. M. Yang, K. L. Pan, Dynamic Void Growth Models in Ductile Materials, *Proceedings of IUTAM Symp. on Impact Dynamics*, Peking University Press, 1994, pp. 310—322.

On the Inelastic Failure Criterion for Structures Subjected to Large Dynamic Loads

Norman Jones[1] and Jilin Yu[2]

[1]Impact Research Centre, Department of Mechanical Engineering, The University of Liverpool, Liverpool L69 3BX, U.K.
[2]Department of Modern Mechanics, University of Science and Technology of China, Hefei, Anhui 230026, P.R. China

Summary

This article is concerned with the response of structural members which are subjected to impact loads causing large inelastic deformations and material failure. The literature published on the dynamic inelastic failure of structural members is reviewed from which a need for additional theoretical and experimental studies emerges clearly.

Some experimental results for clamped mild steel beams under impact loading are simulated numerically using the finite-element code ABAQUS with plane stress elements. The critical conditions of the beams during the response are revealed and various possible failure criteria are examined and discussed by comparing the numerical predictions with the experimental results.

Keywords: dynamic inelastic failure, structural impact, beams, finite-element calculations.

1   Introduction

This article is concerned with the response of structural members which are subjected to impact loads causing large inelastic deformations and failure. Engineers and designers in many industries require information on the maximum amount of energy which can be absorbed for a specified deformation. For example, the deformations of passenger cars are designed in a range of accident scenarios to maintain a survivable volume with acceleration levels which avoid human injury [1]. In other practical situations, the maximum amount of energy which can be absorbed prior to a material failure is required. This information would be necessary in order to assess the integrity of LNG tankers involved in a collision, for example. Many other safety calculations and hazard assessments involving impact, dynamic pressure or explosive loads would require information on structural impact.

Many authors [1-7] have examined the behaviour of structures subjected to impact loads which produce large ductile deformations. These methods of analysis may be used to predict the impact energy absorbed by a structure and the associated magnitude of the permanent deformations when assuming that the material has an unlimited ductility. No information is obtained, therefore, on the structural integrity. However, this aspect is important for many engineering problems since designers need to assess the integrity of critical components which could be breached in various accident scenarios involving dynamic events.

A survey on the dynamic inelastic failure of beams was presented in Reference [8]. The beams were made from ductile materials, which could be modelled as rigid, perfectly plastic, and they were subjected either to a uniformly distributed impulsive velocity, as an idealisation of an explosion, or

163

to a mass impact to idealise a dropped object loading. Thus, large plastic strains could be produced and the possibility of material rupture was studied for sufficiently large dynamic loads.

It was observed that the different dynamic loadings examined in Reference [8] may cause the development of different failure modes. The simplest failure modes were associated with a uniformly distributed impulsive velocity loading. A rigid-plastic method of analysis for this particular problem [6] shows that membrane forces as well as bending moments must be retained in the basic equations for the response of axially restrained beams subjected to large dynamic loads which cause transverse displacements exceeding the beam thickness, approximately. This is known as a Mode I response where the dynamic energy is absorbed plastically without material failure.

If the external impulse is severe enough then the large strains which are developed at the supports of an axially restrained beam, would cause rupture of the material which is known as a Mode II failure. At still higher impulsive velocities, the influence of transverse shear forces dominates the response, and failure is more localised and occurs due to excessive transverse shear displacements (Mode III).

It is important to note that a Mode III transverse shear failure is more likely to occur in dynamically loaded beams than in similar statically loaded beams. Thus, even a beam with a solid rectangular cross-section and a large length-to-thickness ratio can suffer a transverse shear failure under a dynamic loading (Mode III response), as observed by Menkes and Opat [9] and analysed in Reference [10]. This effect is even more pronounced for beams with open cross-sections.

Paradoxically, it is observed in Reference [8] that despite the lower impact velocities of the mass impact case, the failure behaviour is much more complex than the impulsive loading case. The Mode II and III failure modes discussed above for an impulsive loading also occur for a mass impact loading. However, it transpires that other more complex failure modes may develop. For example, a striker may cause an indentation on the struck surface of a beam, which, if sufficiently severe, may lead to failure. The impact velocity of a strike near to the support of a beam might not be sufficiently severe to cause a transverse shear failure (Mode III) but could distort severely a beam and cause rupture due to the combined effect of transverse shear force, membrane force and bending moment.

It is evident [8] that the dynamic inelastic rupture of beams and other structures is an extremely complex phenomenon and that there is a pressing need for the development of a reliable criterion which can be used in theoretical methods, numerical schemes and computer codes in order to predict the onset of structural failure due to material rupture for hazard assessments and safety calculations throughout the field of engineering.

In an attempt to obtain a universal failure criterion, which could be used for a large class of dynamic structural problems, an energy density failure criterion was introduced in Reference [11] and discussed in Reference [12]. It is assumed that rupture occurs in a structure when the absorption of plastic work (per unit volume) reaches a critical value which contains the plastic work contribution related to all of the stress components.

The numerical predictions of the energy density failure criterion in Reference [11] for the dimensionless impulses at the transitions between a Mode I and Mode II failure and between a Mode II and Mode III failure compare favourably with the corresponding experimental results of Menkes and Opat [9] and the theoretical rigid, perfectly plastic predictions in Reference [10]. Many other theoretical predictions for the various parameters in an impulsively loaded beam are presented in References [11] and [12] to which an interested reader is referred. The energy density failure criterion predicted good agreement with the available experimental results on impulsively loaded beams and confirmed broadly the failure characteristics which were first observed using the elementary analysis developed in Reference [10].

The critical energy density failure criterion was also used in Reference [13] to examine the failure of fully clamped beams subjected to mass impact loads.

Some recent experimental and theoretical studies have been reported on the dynamic inelastic failure of fully clamped circular plates [14,15] and square plates [16] subjected to uniformly distributed impulsive loads. The influence of the boundary conditions on the magnitude of the critical impulse is shown to be very important, which, in fact, has already been observed for beams [17,18]. This complicates further the predictions of an energy density failure criterion but shows the importance of specifying carefully the exact details of the boundary conditions in both experimental tests and theoretical studies.

The entire area of structural failure due to material rupture is a very important one in engineering, but the present state of knowledge is incomplete from both experimental and theoretical viewpoints. The simple rigid-plastic methods of analysis have, in fact, been quite successful for predicting the failure of several structural problems which have been examined and they are particularly useful for preliminary design. The energy density failure criterion is also promising for predicting the failure of a broader range of structural problems but it is more difficult to use and is a function of the strain rate [19]. Eventually, it could be incorporated into any theoretical method or numerical scheme, but insufficient information is available currently to do this with confidence except for beams and frames. It is often assumed that rupture occurs when the equivalent strain in a structural member reaches the rupture strain recorded in a static uniaxial test. It is important to emphasise that, generally speaking, this assumption is incorrect [20] and that, moreover, the rupture strain is a complex function of the strain rate [21,22].

Finite-element methods have been used by several authors to simulate the dynamic inelastic response of structures [18,23]. For example, Clift et al. [23] observed that only the criterion based on the critical generalised plastic work density successfully predicted sites found experimentally in several metal forming operations. A numerical simulation of clamped aluminium alloy beams impacted transversely by a mass at different locations on the span, reported in Reference [18], showed that the mode II failure of the beams could be predicted by a criterion based on the maximum tensile strain, or equivalently, the overall rotation angle. These studies have shown the potential of combining numerical investigations with experimental tests to obtain the criterion which controls a structural failure.

This present study reports on part [24] of a systematic research programme on the dynamic inelastic failure of structures, and is a continuation of the previous careful experimental investigation reported in Reference [25] on clamped beams struck by a solid mass. The mechanical properties of the mild steel and aluminium alloy specimen materials were obtained from both quasi-static and dynamic uniaxial tensile tests. The mild steel beams were loaded at the mid-span and the one-quarter-span positions.

A numerical simulation of the experimental tests on the mild steel beams in Reference [25] is presented in Reference [24] and discussed briefly here. The actual experimentally determined material properties are used in the finite-element models and the global response and strain history are compared with the experimental data. Detailed information on the dynamic inelastic response of the beams related to structural failure is obtained. Through the combined study of the experimental investigation in Reference [25] and the present numerical simulation using ABAQUS, the reliability of both numerical and experimental results is confirmed and the critical conditions of the beams during the response are revealed. Various possible failure criteria are then examined by comparing the experimental and numerical results.

2   Numerical Results

Table 1 contains a comparison of the transverse displacements for dynamically loaded steel beams obtained numerically with the corresponding experimental data from Reference [25]. The agreement between the numerical results and the experimental data is quite reasonable, particularly when considering that the values of W_f were obtained with a limited temporal and spatial resolution of the high-speed photographs. The calculated deflection-time history for specimen SB07, from Reference [25], which was loaded dynamically at the one-quarter span position and was observed to be cracked and severely necked after a test, is compared with the experimental curves in Figure 1. Excellent agreement is observed

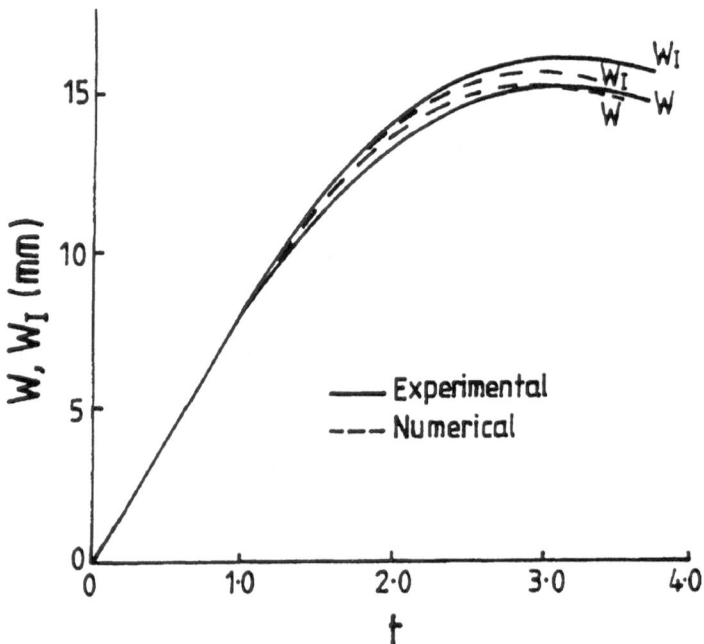

Figure 1: Deflection-time history for specimen SB07 from Reference [25]

The comparisons in Table 1 and Figure 1 together with others made in Reference [24] for strain-time histories from strain gauges and indentations underneath the striker, as well as quasi-static test results, confirm the accuracy and reliability of the numerical scheme. Thus, it is possible to explore many other parameters which might control failure with a degree of confidence. In particular, it is possible to examine various quantities in the numerical calculations at the critical impact energy input which is observed to cause failure in the experimental test beams in Reference [25]. These comparisons can be made without introducing any material failure criteria into the numerical scheme, and, therefore, observations are made without any prior prejudice towards any particular failure criterion.

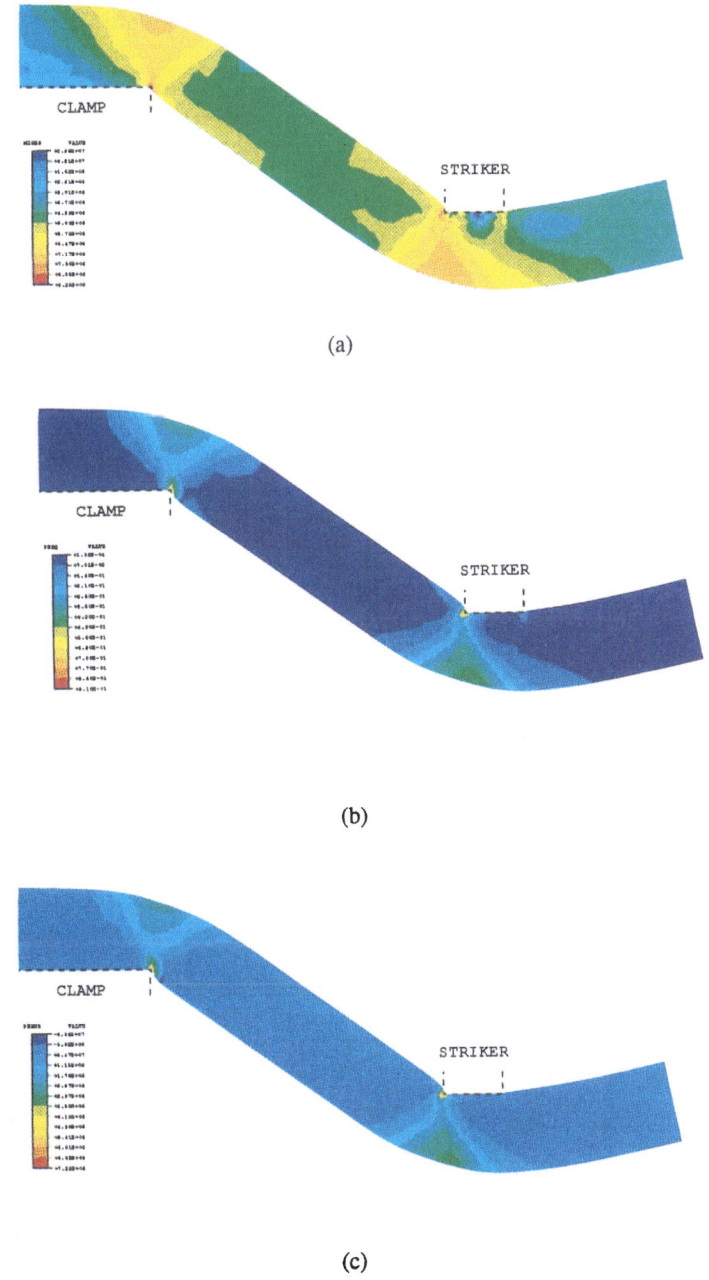

Figure 2: Distribution of (a) Mises stress, (b) equivalent plastic strain and (c) plastic strain energy density for specimen SBO7 at t = 2.86 ms

Table 1: Numerical and experimental results for the transverse displacements
of dynamically loaded mild steel beams

Specimen	Numerical	Experimental	Comments on Tests
SBO9	W_p = 20.56 mm	W_p = 20.9 mm	
SBO8	W_{max} = 21.26 mm	W_f = 21.8 mm	just broken
SBO6	W_{max} = 22.80 mm	W_f = 22.8 mm	broken
SBO7	W_p = 14.97 mm	W_p = 14.5 mm	crack and severe necking
SBO5	W_{max} = 16.03 mm	W_f = 15.0 mm	broken
SBO4	W_{max} = 17.34 mm	W_f = 16.0 mm	broken

3   Discussion

The maximum values of the tensile stress and strain are located on the lower surface underneath the striker, at the symmetric axis for the specimens impacted at the mid-span, or slightly towards the support for those impacted at the one-quarter span position, where necking or a tensile tearing failure occurs. On the other hand, the maximum values of the Mises stress, Tresca stress, shear strain, equivalent plastic strain and the plastic strain energy density occur on the upper surface, at, or near, the impact point where a shear failure occurs. The distributions of the Mises stress, equivalent plastic strain and plastic strain energy density for specimen SBO7 at the cessation of material inelastic flow are shown in Figure 2. It transpires that the equivalent plastic strain and the plastic strain energy density are concentrated near the impact point, in contrast with the smooth change along the axial direction found in the neck region of a tensile specimen, as shown in Reference [24], for example.

The distribution of the Tresca stress along the ridge of the Tresca stress contour across the beam thickness tends to be uniform at the later stage of the response except in a region very near to the impact point as shown in Figure 3. The angle of the ridge of the Tresca stress contour is coincident with the fracture or sliding surface observed in the dynamic tests reported in Reference [25], indicating that failure develops along the direction of the maximum shear stress.

The results of the numerical simulation are in excellent agreement with the experimental data reported in Reference [25] on the impact behaviour of mild steel beams, especially when considering the fact that there are no adjustable parameters in the specification of the material properties, which were obtained experimentally for the same material. Hence, the numerical results here and in

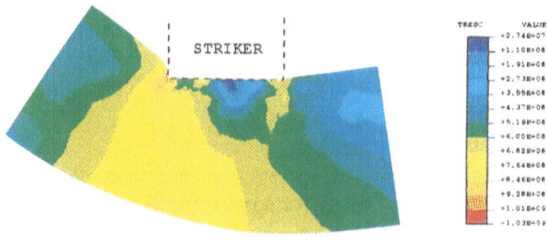

Figure 3: Contour of Tresca Stress for specimen SB07 at t = 2.86 ms

Reference [24] together with the experimental data in Reference [25] provide a reliable set of information for further analysis.

In comparison with the quasi-static loading cases, the stress wave effect is important during the early stage of the response for beams subjected to an impact loading even though the maximum impact velocity is only 8 m/s. The stress also increases in the dynamic cases for materials having a strain-rate hardening effect. Complex failure modes with the combined effects of tensile tearing and shearing, and possibly, geometrical instability, were obtained experimentally. However, the geometrical instability is delayed by the inertia effect in the dynamic cases.

Among the possible failure criteria discussed in Reference [24], the maximum membrane force in a beam cross-section appears to be the most promising for a tensile failure, while the maximum Tresca stress or Mises stress and the maximum plastic strain energy density are worthy of further study for a shear failure. The maximum Tresca stress, or shear strain at the centroidal axis, and the maximum yield index, or failure index based on an interaction yield surface, are promising parameters for a global failure criterion. Further investigations are required to confirm the reliability of these criteria, especially for predicting a shear failure since only one set of test results, i.e., the beams loaded at one quarter-span position, are related to shear failure in the experimental study reported in Reference [25].

4   Conclusions

The approach used in the present numerical study provides a critical way to evaluate the reliability and generality of a criterion for predicting the failure of structures subjected the dynamic loads which produce large inelastic strains. The key requirement is a complete set of reliable experimental data including both the structural response and the dynamic properties of the material. Unfortunately, there is a paucity of such data, especially those related to the rupture behaviour of materials under dynamic loading, including the influence of the strain-rate and the hydrostatic pressure. It is clear that further experimental studies and numerical investigations are necessary to achieve a better understanding of the phenomena of structural failure.

Acknowledgments

J. L. Yu wishes to express his gratitude to the Royal Society, the Chinese Academy of Sciences and the Impact Research Centre. Special thanks are also due to Mrs. M. White and Mr. M. Alves.

References

[1]   W. Johnson and A. G. Mamalis, *Crashworthiness of Vehicles*, Mechanical Engineering Publications, London, 1978.

[2]   W. Johnson and S. R. Reid, Metallic Energy Dissipating Systems, *Applied Mechanics Reviews*, Vol. 31, 1978, pp 277-288; also *Applied Mechanic Reviews Update*, 1986, pp. 315-319.

[3]   N. Jones and T. Wierzbicki (Eds.), *Structural Crashworthiness*, Butterworths, London, 1983.

[4]   T. Wierzbicki and N. Jones (Eds.), *Structural Failure*, John Wiley, New York, 1989.

[5]   N. Jones, Recent Studies on the Dynamic Plastic Behaviour of Structures, *Applied Mechanics Reviews*, Vol. 42, No. 4, 1989, pp. 95-115.

[6]   N. Jones, *Structural Impact*, Cambridge University Press, 1989.

[7]   N. Jones and T. Wierzbicki (Eds.), *Structural Crashworthiness and Failure*, Elsevier Applied Science, London and New York, 1993.

[8]   N. Jones, On the Dynamic Inelastic Failure of Beams, *Structural Failure*, T. Wierzbicki and N. Jones (Eds.), John Wiley, New York, pp. 133-159, 1989.

[9]   S. B. Menkes and H. J. Opat, Broken Beams, *Experimental Mechanics*, Vol. 13, 1973, pp. 480-486.

[10]   N. Jones, Plastic Failure of Ductile Beams Loaded Dynamically, *Trans. ASME, J. Engineering for Industry*, Vol. 98 No. B1, 1976, pp. 131-136.

[11]   W. Q. Shen and N. Jones, A Failure Criterion for Beams Under Impulsive Loading, *International Journal of Impact Engineering*, Vol. 12, No. 1, 1992, pp. 101-121 and Vol. 12, No. 2, 1992, p. 329.

[12]   N. Jones and W. Q. Shen, Criteria for the Inelastic Rupture of Ductile Metal Beams Subjected to Large Dynamic Loads, *Structural Crashworthiness and Failure*, N. Jones and T. Wierzbicki (Eds.), Elsevier Applied Science, London and New York, pp.95-130, 1993.

[13]   W. Q. Shen and N. Jones, Dynamic Plastic Response and Failure of a Clamped Beam Struck Transversely by a Mass, *International Journal of Solids and Structures*, Vol. 30, No. 12, 1993, pp. 1631-1648.

[14]   R. G. Teeling-Smith and G. N. Nurick, The Deformation and Tearing of Thin Circular Plates Subjected to Impulsive Loads, *International Journal of Impact Engineering*, Vol. 11, No. 1, 1991, pp. 77-91.

[15]   W. Q. Shen and N. Jones, Dynamic Response and Failure of Fully Clamped Circular Plates Under Impulsive Loading, *International Journal of Impact Engineering*, Vol. 13, No. 2, 1993, p. 259-278.

[16]   M. D. Olson, G. N. Nurick and J. R. Fagnan, Deformation and Rupture of Blast Loaded Square Plates - Predictions and Experiments, *International Journal of Impact Engineering*, Vol. 13, No. 2, 1993, pp. 279-291.

[17]   J. H. Liu and N. Jones, Experimental Investigation of Clamped Beams Struck Transversely by a Mass, *International Journal of Impact Engineering*, Vol. 6, No. 4, 1987, pp. 303-335.

[18]   J. Yu and N. Jones, Numerical Simulation of a Clamped Beam Under Impact Loading, *Computers and Structures*, Vol. 32, No. 2, 1989, pp. 281-293.

[19]   K. Kawata, I. Miyamoto and M. Itabashi, On the Effects of Alloy Components in the High Velocity Tensile Properties, *Impact Loading and Dynamic Behaviour of Materials: Proceedings of Impact '87, Bremen'*, Vol. 1, C. Y. Chiem, H.-D. Kunze and L. W. Meyer (Eds.), DGM Informationsgellschaft Verlag, pp. 349-356, 1988.

[20]   T. A. Duffey, Dynamic Rupture of Shells, *Structural Failure*, T. Wierzbicki and N Jones (Eds.), John Wiley and Sons, New York, pp. 161-192, 1989.

[21]   K. Kawata, S. Fukui, J. Seino and N. Takada, Some Analytical and Experimental Investigations on High Velocity Elongation of Sheet Materials by Tensile Shock, *IUTAM: Behaviour of Dense Media Under High Dynamic Pressure, Paris, Dunod*, 313-323, 1968.

[22]   K. Kawata, S. Hashimoto and K. Kurokawa, Analyses of High Velocity Tension of Bars of Finite Length of BCC and FCC Metals with Their Own Constitutive Equations, *High Velocity Deformation of Solids*, K. Kawata and J. Shioiri (Eds.), Springer Verlag, New York, pp. 1-15, 1977.

[23]   S. E. Clift, P. Hartley, C. E. N. Sturgess and G. W. Rowe, Further prediction in Plastic Deformation Processes, *International Journal of Mechanical Sciences*, Vol. 32, 1990, pp. 1-17.

[24]   Jilin Yu and N. Jones, Numerical Simulation for the Failure of Impact Loaded Steel Beams, Impact Research Centre Report No. IRC/127/95, Department of Mechanical Engineering, The University of Liverpool, 1995.

[25]   Jilin Yu and N. Jones, Further Experimental Investigations on the Failure of Clamped Beams Under Impact Loads, International Journal of Solids and Structures, Vol. 27, No. 9, 1991, pp. 1113-1137.

Projectile Impact on Rotating Plates

Werner Goldsmith and Xiofan Hou[1]

Department of Mechanical Engineering
University of California, Berkeley, CA, USA

Summary

A previous experimental and theoretical investigation involving normal impact of blunt-nosed cylinders on plates mounted on the end of a rotating beam was extended to higher target tangential speeds and greater striker velocities. In the new arrangement, annular targets were mounted on motor-driven rotating disks. A high-speed framing camera operating at about 55,000 pps and initial and terminal projectile velocity measuring units as primary data acquisition devices. Strikers consisted of 12.7 mm (0.5 in) diameter hard steel cylinders with either blunt or 60 conical tips and an aspect ratio of 3, fired from both pneumatic and powder guns with initial velocities ranging from 100 to 1000 m/s. No deformation of the striker was found in any of the tests.

Targets consisted of annular disks with an outer diameter of 610 mm (or 660 mm) and an inner diameter of 356 mm, composed of 6061-T6 aluminum, CR 1010 steel and polycarbonate (Lexan), with thicknesses of 1.588, 3.175, 6.350 and 9.525 mm. Tangential plate speeds at the impact point were 53.3, 93.3 and 133 m/s. Eighty-six runs were conducted.

The earlier model of the piercing process was improved by including transverse shear and utilizing the effective strain as the failure criterion in the initial, plugging, phase. Target speed was incorporated in the successive petaling stage for an improved specification of the strain in the crater, and the J integral was employed in the determination of the fracture energy.

A finite-element solution using DYNA-3D calculated the histories of the mass center position and velocity, and of the trajectory angle, as well as the crater length for various impact conditions. Histories of the total resistive force and target/striker cross sections were also obtained. Excellent correspondence was found between analytical and numerical predictions when compared to test results for the final striker velocity in the 3.175 mm thick aluminum targets, and good agreement for the trajectory angle and crater length. Somewhat greater divergence discrepancies existed at the same speed for mild steel targets of that thickness. Petals, but no plug are formed when a conical striker is used; the projectile rotation for this case is somewhat greater.

[1]Current address: Ford Motor Company, Detroit, MI

Key Words: Projectile Impact, Moving Targets, Plate Perforation

Introduction

While several recent publications have been concerned with the effect of jets striking moving targets, investigations of projectile impact on moving targets have been scarce. Perforation of thin 140 mm diameter steel and aluminum plates mounted on the ends of a rotating beam by rigid blunt cylinders have been examined [1, 2] where a phenomenological model of the perforation process was developed based on observations that indicated successive stages of plugging and petaling. Plugging consisted of the three consecutive phases of plastic wave propagation, common deceleration of striker and corresponding target material, and plate failure. Motion of the plate beyond the contact area was assumed to be governed by membrane action, permitting the use of a target motion model developed for non-perforation impact [3] to the present case with the caveat that failure was based on the ultimate tensile strain. Conservation of energy was applied during the petaling action with energy rates prescribed for curvature changes, including reversal, and stretching of the petal as well as crack propagation. Striker motion was treated by rigid-body dynamics. A computational solution using the HULL code, based on the test conditions of [1], was also found to be in good agreement with those experimental results [4].

The present investigation altered the test arrangement of [1] by using a continuous target surface rather than a finite sized plate and improved the earlier model [2] by: (a) Inclusion of shear deformation in the common motion stage, (b) Addition of transverse shear in the total resistive force during plugging, (c) An effective strain criterion governed plate failure, (d) Inclusion of the effect of target speed on petaling, and (e) Use of the J integral and a plastic tearing concept for calculating the fracture energy rate. In addition, a numerical investigation of the phenomenon was performed using the code DYNA-3D.

Experimental arrangement

The experimental setup is presented in Fig. 1. The projectiles had a diameter D = 12.7 mm, an overall length L = 38.1 mm and either blunt or 60° conical tips, with masses of 39.5 and 30.5 g, respectively. They were fired from a pneumatic gun at speeds of v_0 = 45-200 m/s and from a powder gun at velocities up to 1000 m/s. Circular annular targets with outside diameters of 610 (or 660) mm and central holes of 356 mm diameter consisted of 6061-T6 aluminum with a thickness ranging from h = 0.79 to 9.53 mm, cold-rolled CR 1010 steel with a thickness of h = 3.175 mm, and polycarbonate with thicknesses of h = 1.59, 3.175 and 6.35 mm. These plates were fastened to a disk attached to a target shaft with a mounted pulley that was driven by a belt from a corresponding pulley located on the shaft of the 15 hp (11.2 kw) 3-phase motor operating at a synchronous speed of 3500 Hz. The mechanical properties of the striker and the targets is shown in Table 1.

The motor size was determined using the requirement that the acceleration time to reach the operating speed should not exceed 10 seconds to prevent overheating of the driving unit. The size of the annular disks was established by stipulating a minimum safety factor of 8 for their stress levels due to centrifugal motion relative to the yield stress of the material. Unsymmetrical bearing loads due to target perforation were found to be insignificant.

The striker trajectory was so positioned that impact occurred at a point 254 mm from the center of rotation; the belt drive ratios of 1.442, 1 and 0.631 resulted

Table 1. Striker and Target Properties [5,6]

Material	Density, t, kg/m^3	Poisson's Ratio, μ	Young's Modulus, E, GPa	Dynamic Yield Stress, σ_Y, MPa Compression	Shear	Ultimate Tensile Strain, ϵ_u	Fracture Toughness K_{Ic} MN/m$^{3/2}$
Steel Striker	7,977	0.29	210	1,390	804		
6061-T6 Aluminum	2,708	0.33	71	295	190	0.2	45
CR 1010 Steel	7,670	0.33	210	310	220	0.4	140
Polycarbonate	1,200	0.4	2.3	62	86	1.2	3

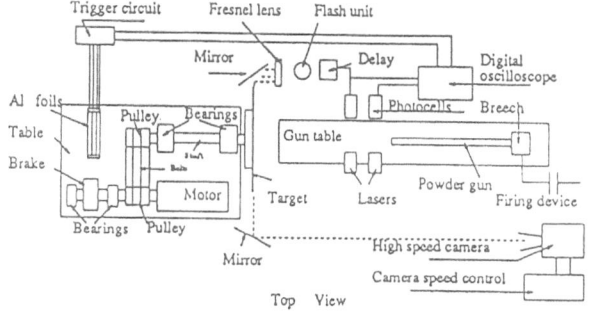

Top View

Fig. 1 Schematic of experimental arrangement

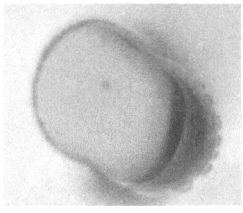

Fig. 2 Distal side of a 1.59 mm thick 6061-T6 Al plate: v_t = 133 m/s. 12.7 mm D striker, v_0 = 897 m/s. Uniform petaling, Type (i)

in linear speeds of the impact point of v_t = 133, 93.1 and 51 m/s.

Striker and target properties are listed in Table 1. The initial striker velocity was measured from signals produced by the interruption due to projectile passage of two parallel laser beams, 152 mm apart, directed against two photodiodes whose output was connected to an oscilloscope. When perforation occurred, the final velocity path was determined from the signals due to projectile contact with two sets of aluminum foil pairs that closed a circuit; the deviation of the striker trajectory was manifested by the tears in these foils. A rotating prism Beckman-Whitley WB2 framing camera operating through a rheostat at rates from 50,000-55,000 frames/sec monitored the whole event. Light, diffused by a Fresnel lens, was provided by a stroboscopic source with a 1.8 ms duration, triggered through a delay circuit from the interruption of the second beam of light.

Experimental results

a) 6061-T6 aluminum

The damage pattern produced above the ballistic limit by blunt strikers in 6061-T6 aluminum targets ranging in thickness from 1.59 to 9.53 mm consists of the formation of a plug accompanied by a deflection region with an extent from 2-4 times the projectile radius R; this is followed by petaling. The thinner the plate, the larger the deflection; the plug cross section is nearly the same as that of the striker, but its frontal surface is inclined due to target rotation. At higher speeds, the plugging mechanism is a secondary feature.

A crater is formed with a width slightly larger than D and a length that depends on both the initial striker and the target velocity. After initial piercing, a large bearing stress produced by the projectile side acts on the crater, causing subsequent plate failure in the form of petaling. Three types of petaling modes were observed:

   i) A regular, continuous lip with some small cracks at its edge at the edge of the crater, as shown in Fig. 2. This occurs in thin plates traveling at high velocities and struck by a very fast projectile.
ii) A more irregularly-edged lip, consisting of three major petals, as shown in Fig. 3 that exhibits two or more major cracks propagating outward in a combination of Mode I and Mode III fracture. This is observed in targets that are thin or of intermediate thickness struck by a projectile at a relative high velocity. Material is pushed to each side and bent opposite to the direction of motion, which may cause the front petal to break off.

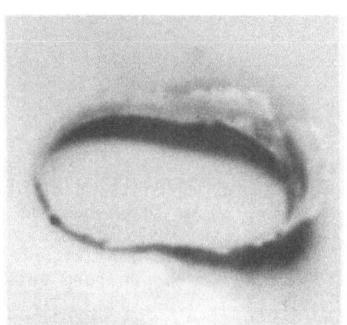

Fig. 3 Distal side of a 1.59 mm thick 6061-T6 Al plate: v_t = 133 m/s. 12.7 mm D blunt striker, v_0 = 887 m/s. Multiple petals, 2 major cracks, type (ii)

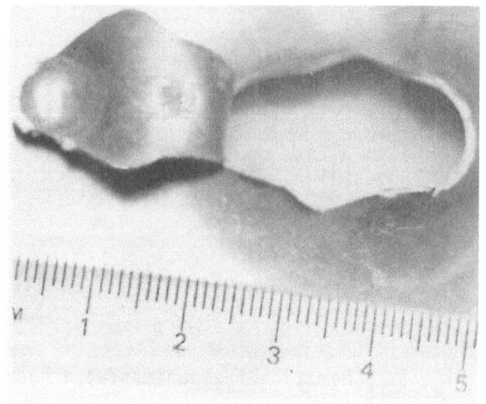

Fig. 4 Distal side view of a single petal of a 1.59 mm thick Al plate, v_t = 51 m/s. 12.7 mm D blunt striker. Single petal damage (iii) due to bending and tearing. v_0 = 148 m/s.

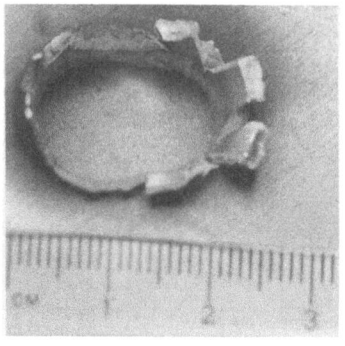

Fig. 5 Distal side view of petaling failure in a 3.18 mm thick 6061-T6 Al plate, v_t = 133 m/s, struck by a by a 60 cylindro-conical projectile of 12.7 mm D, v_0 = 1031 m/s

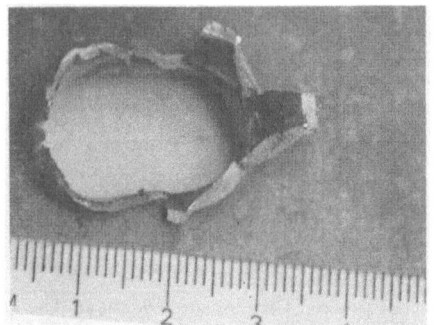

Fig. 6 Distal side petal failure in a 3.18 mm thick CR 1010 Steel Plate, v_t = 133 m/s, struck by a 60 cylindro-conical projectile of 12.7 mm D, v_0 = 495 m/s

iii) A major petal with the diameter of the striker and large curvature is formed, as shown in Fig. 4. The dual crack propagation occurs in mode III. This process takes place in thin targets for low target and striker velocities. A plug is not generated, and more pronounced petaling initiates immediately when a aluminum target is struck by a cylindro-conical projectile. The length of the crater is less than that for a blunt impact under comparable conditions; one or more petals may become detached in thicker specimens, as indicated in Fig. 5.

b) Cold-rolled 1010 steel

Blunt strikers produce a response in this type of target similar to that in aluminum, but the plate deflection is larger. A bulging deformation is also sometimes found in the front petal, indicating strong resistance to lateral striker motion. Cylindro-conical projectiles produce petaling due to tensile rather than shear stresses with deflections larger than for blunt strikers or for corresponding aluminum tests. Plate petaling failure is a combination of bending, mode I fracture due to a large circumferential stress, and mode III fracture due to compression and tearing. An example of this type of damage is shown in Fig. 6.

c) Polycarbonate

The plastic deformation in the plugging stage is much greater than for metallic targets tested under the same conditions due to the higher ultimate strain of the polymer. The penetration time is short in view of the low material strength. The heat generated at the interface further reduces the tensile strength of the target, producing adhesion to the striker of some of the separated target elements, while others pursue an independent trajectory. Except for a clearly discernible front petal, the remainder of the cohesive plate tends to return to its original shape. Elastic recovery and inward thermo-plastic flow reduce the width of the crater to values less than D. Similar damage is produced by both projectile tipes, with a slightly greater crater length found with the cylindrical projectile.

Analysis

a) General concepts

The model is an extension of the analysis of [10] and considers the normal impact of a blunt, undeformable striker on thin targets moving orthogonally to the projectile trajectory. The initial plugging process is modeled in terms of a stationary membrane which is then modified to account for transverse motion; this comprises the consecutive, but interrelated phases of plastic wave propagation -- the most dominant response to the impact -- and common motion of striker and target, and, finally, by target failure. This failure is based on the effective strain ϵ_e reaching the ultimate tensile strain ϵ_u, and includes the transverse shear strain $\epsilon_{rz} = \frac{1}{2}(\delta w/\delta r)$ as well as radial strain $\epsilon_{rr} = \frac{1}{2}(\delta w/\delta r)^2$, i.e.

$$\epsilon_e = \frac{\sqrt{2}}{3}[2\epsilon_{rr}^2 + \frac{3}{2}\epsilon_{rz}]^{\frac{1}{2}} = \epsilon_u \tag{1}$$

where r and z denote the radial and thickness directions of the target, respectively. The use of a coordinate system fixed in the target and moving at constant velocity v_0 requires its superposition in the opposite direction on the striker motion; transverse shear is included in the total force during plugging.

The material outside the crater is defined by a rigid/perfectly plastic solid while that within the contact area is regarded as elastic/perfectly plastic, both independent of strain rate and characterized by yield stress σ_Y. This permits plastic wave propagation with speed

$$c_p = (K/\rho_t) \tag{2}$$

where K is the bulk modulus and ρ the density of the target. Elastic deflections are ignored, unloading is not considered and the force acting on the striker boundary is regarded as uniformly distributed. Large target slopes occur only in the vicinity of the crater, primarily due to dishing, and hence a small-angle treatment for this slope is utilized. Observations indicate that the tumbling velocity of the striker during plugging is essentially constant.

The petaling process accounts for target motion and is represented using a Green strain tensor in terms of the upper bound theory of plasticity. Global plate deflections are neglected as are energy rates due to plate thinning, minor crack initiation and propagation, and other dissipative processes. The locally deformed plate region during this stage consists of a semi-ellipse expected on the basis of plastic wave propagation, suggesting the use of an elliptical coordinate system. The curvature in the radial direction, k_r, is considered to be uniform and constrained by ϵ_u; the energy rate during reverse petal bending is assumed to be that of initial flexure. The circumferential strain $\epsilon_{\theta\theta}$ is postulated to vary linearly from zero at the petal tip to its maximum value at the root which occurs at the extreme point of the major axis of the semi-ellipse. The contact force during petaling is normal to the projectile axis inasmuch as friction is neglected. The J integral was used to calculate the energy rate due to fracture.

b) Normal impact on stationary targets

The wave equation for a thin plate of thickness h composed of a rigid/per-fectly-plastic substance that includes shear effects can be expressed in terms of transverse displacement w (in the z-direction) by

$$\nabla^2 w + \left(\frac{h}{12 c_m}\right)^2 \nabla^2 \left(\frac{\delta^2 w}{\delta t^2}\right) = \frac{1}{c_m^2}\left(\frac{\delta^2 w}{\delta t^2}\right) \tag{3}$$

The plastic wave speed for the elementary theory, where shear effects are neglec-ted, is $c_m = (\sigma_Y/\rho)^{\frac{1}{2}}$. Separation of variables in Eq. (3) by functions of radius r and time t shows that the plastic wave speed c_p is here given by

$$c_p = c_m / [1 + \frac{h^2}{12}\left(\frac{2\pi}{\lambda}\right)^2] \tag{4}$$

with λ as the wave length of a harmonic component; when λ becomes very large, $c_p \rightarrow c_m$. The projectile is slowed by both the in-plane force $N_r = h\sigma_Y$ and by the shear force $Q_r = I_\theta(\delta^2\phi/\delta t^2) = I_\theta(\delta^3 w/\delta r \delta t^2)$ at the boundary which was not considered in the analysis of [2]. Term ϕ is the rotation angle relative to the x-axis and $I_\theta = \rho_t h^3/12$ is the moment of inertia per unit plate area.

The boundary conditions for the plate/projectile contact region, of radius r = R = ½D, and outside this contact zone are given by

$$2\pi R[I_\theta \frac{\delta^3 w}{\delta t^2 \delta r} + N_r \frac{\delta w}{\delta r}] = (m + m_p) \frac{\delta^2 w}{\delta t^2} \quad \text{at r = R} \quad \text{and}$$

$$w = 0 \quad \text{at r > r}_p = c_p t \tag{5}$$

Here, m and mp are the mass of the striker and plug, respectively. The duration of this phase is brief; the time of the plate traverse by the plastic wave, given by $t_1 = h/c_p$, is of the order of a few microseconds so that the common motion of plug and projectile ensues quickly. The initial conditions for this motion are:

$$v|_{t_1} = \frac{\delta w}{\delta t} = v_0[m/(m + m_p)] \quad \text{at r = R}$$
$$= 0 \qquad \qquad \qquad \qquad \text{r > R} \qquad \text{and } w|_{t_1} = 0, \; r \geq R \tag{6}$$

c) Thin plate penetration

i) Plastic Wave Propagation

Axial plastic wave propagation will always occur upon projectile/plate con-tact. Dilatation is considered to be governed by a linear elastic uniaxial stress-strain relation, while deviatoric stresses are determined from an elastic/perfectly plastic relation based on the von Mises yield criterion. This leads to a relation between normal stress σ_{zz} and strain ϵ_{zz} given by

$$\sigma_{zz} = \frac{2}{3}\sigma_Y + K\epsilon_{zz} \tag{7}$$

Plastic strains propagation in this direction with the speed given by Eq. (2), and the time for this wave to reach the distal face is $t_1 = h/c_p$ which is of the order of a microsecond for the test conditions employed here.

ii) Common Motion of Plug and Projectile

From conservation of momentum, the velocity of the entire plate within the contact region at time t_1 acquires velocity v_z given by the first of Eqs. (6).

d) Projectile penetration including target motion

i) Rotation of the moving coordinate system

Figure 7 depicts the projectile position in a reference system embedded in the

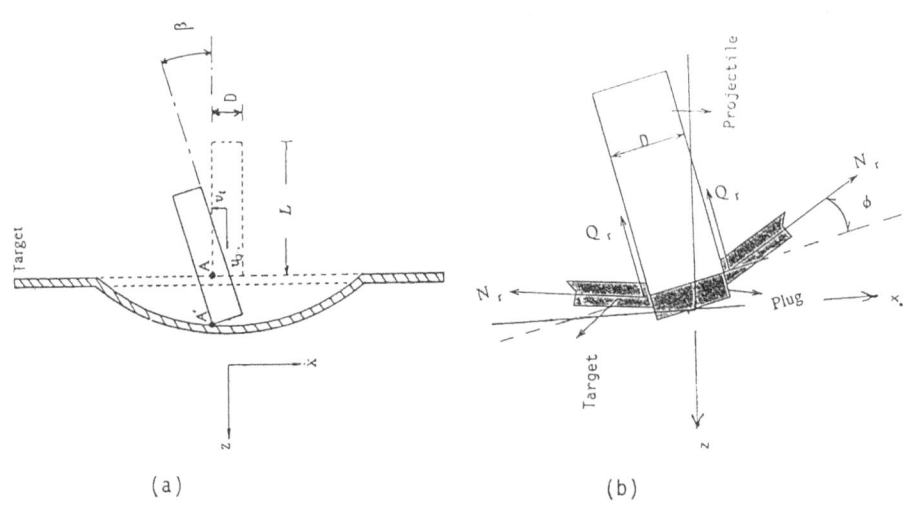

Fig. 7 Motion of Projectile and Target in Embedded Coordinate System
(a) First contact (b) Full engagement

target inclined at oblique angle β to the z-axis, (a), and after full engagement, (b). Neglecting the small gravitational component, combination of the linear and angular impulse-momentum equations at the end of the common motion motion at time $t = t_1 + t_2$ gives for the angular velocity of the striker

$$\left.\frac{\delta\beta}{\delta t}\right|_{t=t1 + t_2} = 1.5\ v_1/L \tag{8}$$

which is assumed to be constant. The oblique angle, β, i.e. the angle between the mass center velocity vector and the target normal is given by

$$\beta = 1.5(v_1/L)(t_1 + t_2) \tag{9}$$

   ii) Perforation

      The deceleration force acting on the striker during plugging is given by

$$F = 2\pi R[Q_r + N_r \sin \phi] \tag{10}$$

where $\sin \phi \approx \phi = \delta w/\delta r$ is the slope of the thin target (Fig. 7b). Failure is determined by Eq. (1) with initial penetration occurring at the point of the striker face opposite to the direction of target motion. The time t_2 is deter-mined from satisfaction of the condition of failure, Eq. (1). Using this value and Eq. (10), the velocity component in the direction z at the end of plugging, $v_z|_{t_1+ t_2}$, may be determined from the impulse-momentum equation

$$\int_0^{t_1+t_2} F\ dt = (m + m_p)(v_0 - v_z|_{t_1+ t_2}) \tag{11}$$

and the angular velocity and position of the cylinder are given by Eqs. (8) and (9). This system of equations in dimensionless form was solved by finite-dif-ference methods.

e) Petaling model

The petaling process is depicted by means of an energy rate balance, or

178

$$\sum_{i=1}^{2} \dot{W}_{i_t} + \dot{T}_b = F \cdot v_c \tag{12}$$

where $\dot{W}_{i_t}$ is the kinetic energy rate of the integral portion of the target, $\dot{T}_b$ is the kinetic energy rate of the petal, F is the contact force, and v_c is the contact velocity. Energy rates considered include (A) plastic deformations due to the compressive force, including that stored in the crater lip, (B) petal bending and (C) crack extension due to tearing and Mode I fracture. The crater and assumed plate deformation, symmetrical about axis x, is shown in Fig. 8.

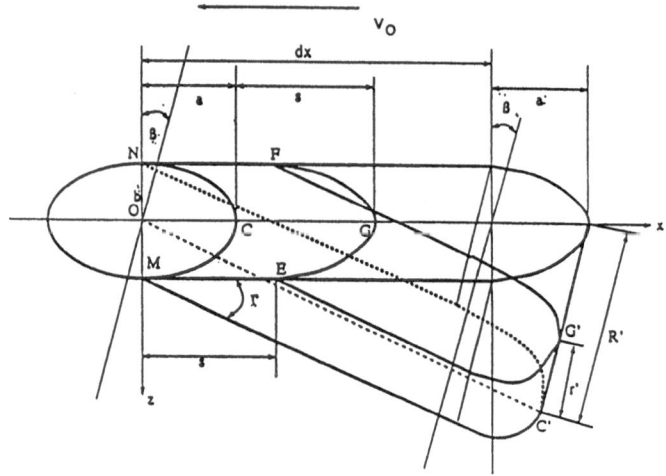

Fig. 8 Assumed Crater Shape and Plate Deformation during Petaling

   i) Plastic deformation energy rate

The contact zone is a semi-ellipse with with semi-minor axis b and semi-major axis $a = b/\cos \beta$. In time dt, during which β is assumed to be constant, an element of the plate of length dx is moved to the crater lip by the compressive force acting on this segment. It is further hypothesized that the lip material shown in Fig. 9 has yielded completely with $\sigma_{rr} = \sigma_{zz} = 0$ and $\sigma_{\theta\theta} = \sigma_Y$ (since the plate is thin and cracks occur in the petals) with the deformation of each semi-elliptical line element uniform. Curve MCN, in contact with the striker at time t_1, is transformed to curve MC'N at time $t_1 + dt$, while curve EGF moves to EG'F, with $C'G' = r'$ and $OG = s + a$, where s is the distance between crater centers at the top and bottom surface in the x-direction. The initial length L_0 of EGF becomes L_f for EG'F where

$$L_0 = \tfrac{1}{2}\pi(\tfrac{1}{2}[a^2 + b^2])^{\frac{1}{2}} \qquad \text{and } L_f = 2A'(dx = s) + B' \tag{13}$$

with $A' = (\cos \beta)/(\cos (\Gamma - \beta))$ and $B' = \tfrac{1}{2}\pi(\tfrac{1}{2}b^2 + \tfrac{1}{2}a^2A'^2)$ (14)

where A' and B' remain constant during dt as long as β remains invariant. Using the symmetric hole approximation of [7] for the present asymmetric crater, the maximum height of the lip H' and petal thickness h' are given as

$$H' = \tfrac{3}{4} dx \qquad \text{and } h' = 1.15h[r'/dx]^{\frac{1}{2}} \tag{15}$$

From Fig. 10 and Eqs. (15), it follows that

$$\Gamma = \tan^{-1}[(\cos \beta)/(\tfrac{4}{3}(1 + \tfrac{b}{dx}) - \sin \beta)] \tag{16}$$

resulting in the uniform strain

$$\epsilon_{\theta\theta} = \int_{L_0}^{L'} \frac{dL'}{L'} = \ln[\{(2A'(dx - s) + B'\}/L_0]$$ (17)

leading to the plastic energy rate of hole enlargement

(A) $$\dot{W}_1 = \int_0^{ds} 2b\sigma_Y \, \epsilon_{\theta\theta} \, \frac{ds}{dt} = 2bh\sigma_Y \, \ln[\frac{2A'dx + B'}{L_0}\dot{x}] \quad \text{with } b = R$$ (18)

ii) Petal bending energy rate

Petal bending ensues due two parallel cracks propagating outward from the crater that produce a cantilever beam of width w which is subjected to further contact pressure by the striker, as shown in Figs. 9a and 9b. Neglecting elastic components, the plastic moment necessary to bend a length dx of petal, of width w, to a radius of curvature is given by [8]

$$M_p = 2 \int_0^{\frac{1}{2}h} w\sigma z' \, dz' \quad \text{with strain } \epsilon = z'/K_0$$ (19)

A work-hardening law for the stress with constant n is used in the form

$$\sigma = \sigma_0^n$$ (20)

where n = 0 represents a perfectly-plastic material with $\sigma = \sigma_0$. The rate of plastic bending to produce a constant radius of curvature K_0 is then given by

(B) $$\dot{W}_2 = dx \int_0^{1/K_0} M_p \frac{d}{dt} (1/K) = \frac{1}{2}w\sigma_0\gamma\epsilon_u\dot{x}$$ (21)

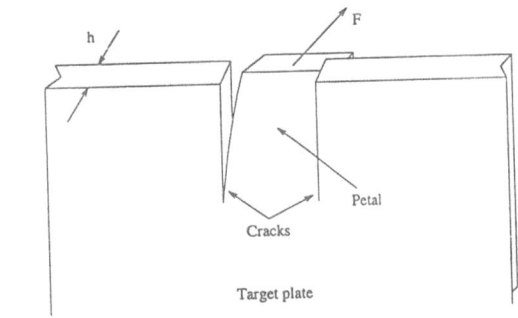

(a)

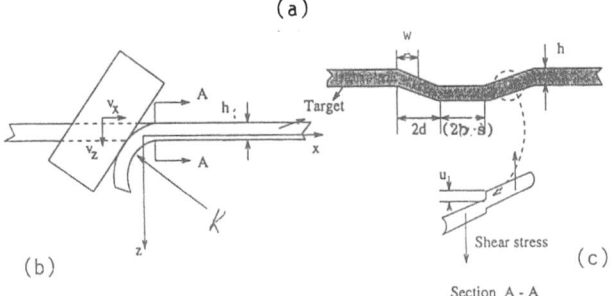

(b)   (c)

Section A - A

Fig. 9 Crack Propagation due to Bending and Tearing of the Petal
(a) Tearing Model (b) Petal Bending (c) Shearing of Petal and Shear Zone

iii) Crack propagation energy rate

T) Lip production and tearing

The propagation of the two cracks producing the petal probably occurs by a combination of tearing and circumferential stress action. It is assumed that the two mechanisms are not coupled so that the energy rates can be determined separately. Work must not only be expended to produce Mode III shear fracture in tearing, but must also create the transition region of width D at the edge of the petal shown in Fig. 9c. This work consists of both the energy of shearing along length dx, $dW^{(1)}_3$ and the energy of tearing, $dW^{(2)}_3$. The first component is evaluated in terms of the effective strain $\epsilon_e = \tau_0/\sqrt{3}$, where ϵ_0 is the maximum shear strain which is related to the assumed constant shear stress τ_0 by the Prandtl-Reuss relation

$$o = 3\bar{\epsilon}p \ \tau_0/\bar{o} \tag{22}$$

for pure shear. Here, the equivalent stress $\bar{o} = \sqrt{3} \ \tau_0$ and ϵp is the equivalent plastic strain. For an assumed perfectly plastic material, $n = 0$ in Eq. (20) and

$$o_0 = o_Y, \ \bar{\epsilon}p = {}_0/\sqrt{3} \ \text{ and } \ \tau_0 = o_Y/\sqrt{3} \tag{23}$$

Based on [14], the total incremental work for two damage mechanisms is

$$dW_3 = dW^{(1)}_3 + dW^{(2)}_3 = (W_3/h)dx = 2D \int_0^{\epsilon_e} o \ d\epsilon + \frac{1}{h} \int_0^h \tau_0(h - u)du \tag{24}$$

and for the two cracks need for petal production, the total energy rate is

$$T) \ \dot{W}_{3_T} = [4ho_Y \ {}_0D + o_Yh^2](\dot{x}/\sqrt{3}) \tag{25}$$

The shear zone half-width w and plate thickness h < 6.5 mm can be approximately related empirically from the data for steel of [9] by

$$w = 0.11h + 0.043 \ h^2, \text{ for mild and } h = 0.067w + 1.067w^2 \text{ for stainless steel.}$$

I) Mode I crack opening release rate

For a thin sheet, the crack opening displacement, δ, must be accomodated by plastic deformation over the thickness h so that

$$\epsilon = \frac{\delta}{h} \tag{26}$$

The energy release rate for Mode I cracks was specified in terms of a critical value of the average strain, ϵ_C, related through Eq. (26) to the critical value of the well-known J-integral for a mode I crack by the empirical relation [10]

$$J_{Ic} = \tfrac{1}{4}C_1(o_Y + o_u)\delta \tag{27}$$

The numerical parameter C_1 has the value 1.2 for plane stress and 1.6 for plane strain. The J-integral is a measure of the entire elasto-plastic stress-strain field surrounding the crack. If the crack driving force $_J$ (or energy release rate) is equated to J (valid only for purely linear elastic processes, but a reasonable approximation when plastic effects are not too large), the energy release rate that produces an increment dx in two cracks is given by

$$I) \ \dot{W}_{mode \ I} = C_1\epsilon_0(o_Y + o_u)h^2 \ \dot{x} \tag{28}$$

With the original hypothesis of independence of the crack propagation energy processes, the total release rate is the sum of Eqs. (25) and (28). However, it has been recognized [11] that this combination yields much too low an energy rate compared with experimental results, indicating greater dissipation due to possible interactions and other unknown mechanisms. A correction was applied to estimate the correct release energy rate, given in the form

$$\dot{W}_{3_T} = C_2(\dot{W}_3 + \dot{W}_{mode\ I}) \tag{29}$$

where the empirical constant $C_2 = 2$ provides reasonable agreement with data.

f) Rigid-body striker motion

As friction during petaling is neglected, the force $F = F_n$ acting at the contact point between striker and target is normal to the surface and uniformly distributed [10]. Its value is obtained from Eq. (12) as

$$F_n = (\sum_{i=1}^{3} \dot{W}_i)/v_{ct} = (\sum_{i=1}^{3} \dot{W}_i)/(v_{cx} \cos \beta + v_{cz} \sin \beta) \tag{30}$$

where v_{ct} is the transverse component of the projectile at the contact point with the coordinate system embedded in the moving target, as shown in Fig. 10. The line of action of the force can intersect the striker centerline at a distance r_1 (a) ahead, (b) at or (c) behind the mass center G. The equivalent force system at G consists of Eq. (30) and a moment given by $M_G = r_1 F_n$, where r_1 is positive for case (a), zero for case (b) and negative for case (c). The projectile rotation is defined in terms of the yaw angle α, the angle between the projectile axis and its velocity vector at G, the oblique angle β and the impact angle θ, formed by the axis of symmetry and the normal to the target, Fig. 10. Clearly

$$\alpha = \beta + \theta \tag{31}$$

The increment in angular velocity, $d\beta$, is obtained from the relation

$$d\dot{\beta} = (\frac{M_G}{I_G})\ dt \tag{32}$$

while the lift and drag forces, F_ℓ and F_D are given by

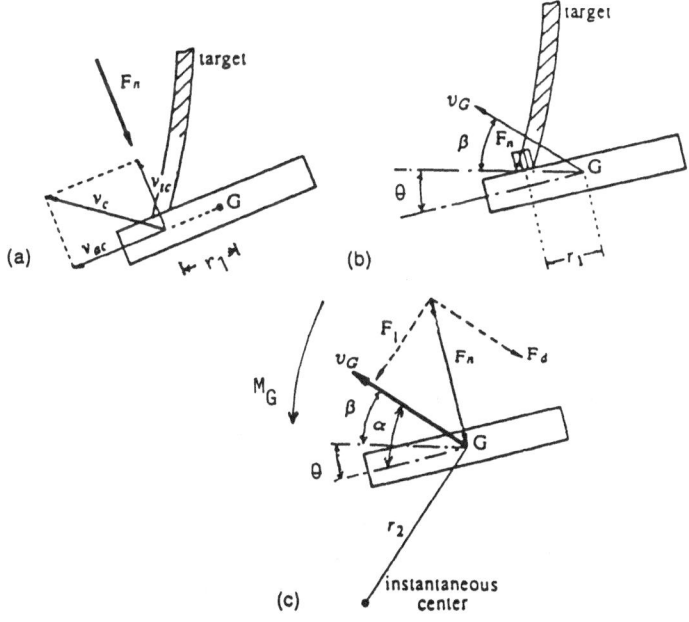

Fig. 10 Force System acting on the Projectile during Petaling (a) Forces at the Contact Point (b) Oblique and Impact Angle (c) Equivalent Force System

$$F_{\ell} = F_n \sin \alpha \qquad \text{and} \qquad F_D = F_n \cos \alpha \tag{33}$$

The change in translational motion of G in the moving coordinate system and the the change in the striker path, $d\alpha$, obtained by assuming a circular motion for a brief time interval dt are given by

$$dv_G = - \left(\frac{F_D}{m}\right) dt \qquad \text{and} \qquad d\beta = - (v_G/r^2) \, dt \tag{34}$$

where r_2, the instantaneous radius of this motion, is determined by

$$r_2 = mv_G^2/F_D \tag{35}$$

The mass center coordinates in the moving coordinate system in terms of the current yaw angle α are

$$x_G = x_{Go} + (v_G \cos \beta) \, dt \qquad \text{and} \qquad z_G = z_{Go} - (v_G \sin \beta) \, dt \tag{36}$$

and the same representation in the inertial system gives

$$\bar{x}_G = x_G \qquad \text{and} \qquad \bar{z}_G = z_{Go} = (v_t - v_G \sin \beta) \, dt \tag{37}$$

Comparison of test results with analysis and numerical model predictions

Figures 11 - 13 present a comparison of the experimental, analytical and numerical information, the latter from finite-element code DYNA-3D [12], from an investigation of the perforation of the aluminum and steel targets moving orthogonally to the initial striker trajectory with a speed of 93.1 m/s. Figure 11 shows that the final projectile velocity is linearly related to its striking velocity and is slightly overpredicted by the analytical model; this results from the neglect of certain minor energy absorbing mechanisms such as friction, use of plate theory assumptions, and failure to include strain rate effects.

The data for the final crater lengths shown in Fig. 12 are in excellent accord with the numerical analysis and conform reasonably well with the model, particularly a low impact speeds. The trend of the values is such that, at sufficiently high striker velocity, the crater configuration will approximate that obtained for a stationary target. The measured terminal impact angle is reasonably predicted by the phenomenological model, but is at considerable variance with the numerical results, especially at low striker speed, Fig. 13. It is believed that this is due to (i) the use of an erosion model where the failed material imposes no loads on the penetrator, and (ii) the discrepancy in the size of the actual target compared to its numerical counterpart, so that the actual momentum present, and hence the moment applied to the striker are smaller than actually extant. Contact durations decrease hyperbolically with initial velocity ranging from 150 us at $v_0 = 200$ m/s to 60 us at $v_0 = 850$ m/s.

Conclusions

A new experimental arrangement was devised to permit impact on annular plates moving orthogonal to the initial striker trajectory at both higher target and projectile speeds than reported earlier [9]. Concomitantly, a new analytic technique has been advanced, based on the earlier model [10] that includes transverse shear, employs effective strain as a failure criterion, includes the effect of target speed on petaling and provides a better delineation of the plastic petal deformation energy rate and of the fracture energy. A conical tip produces petaling without plugging as well as substantial striker rotation; substantial mushrooming was noted when a deformable projectile was simulated. The correlation of both analytical and numerical predictions with the test

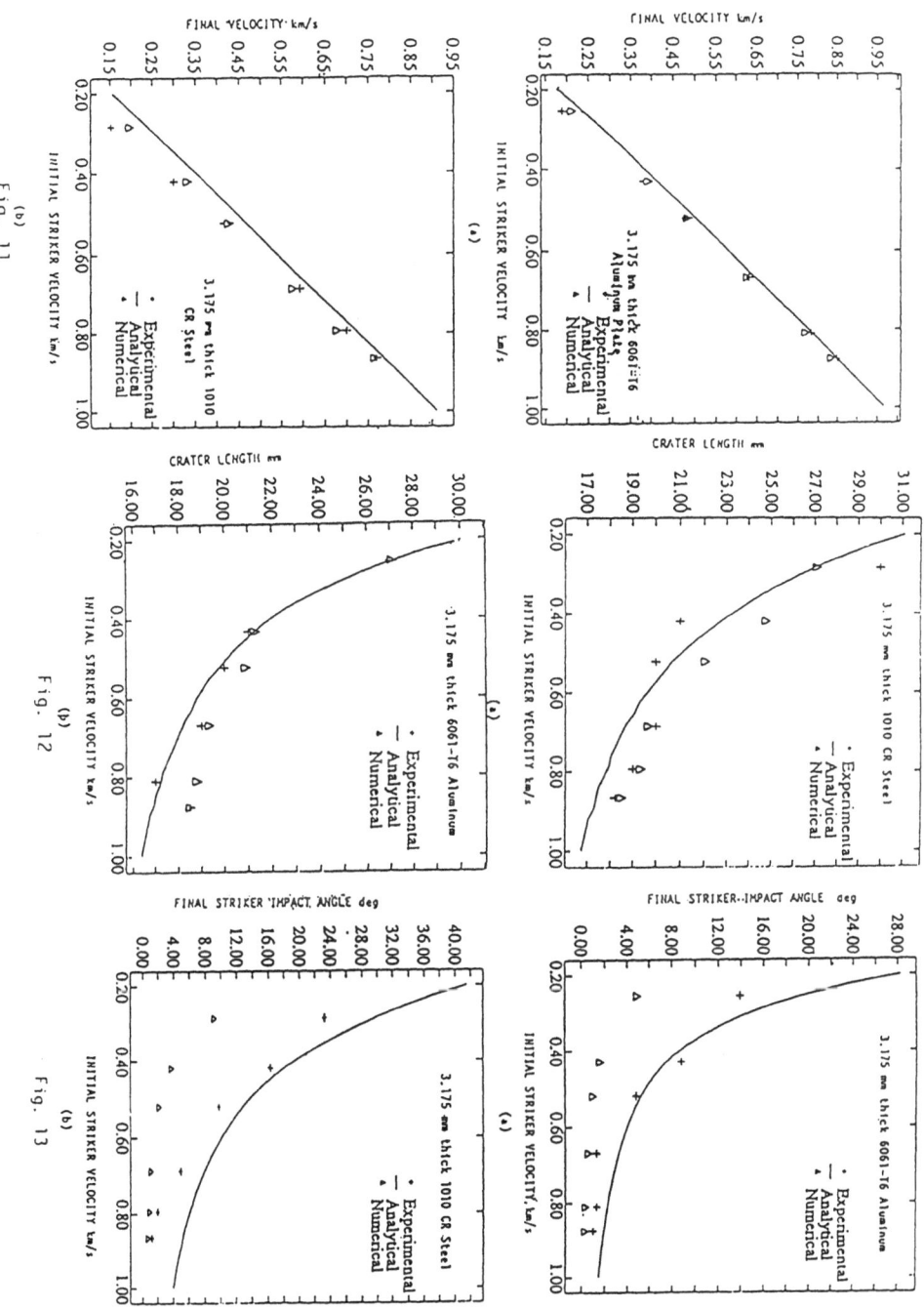

Fig. 11-13 Comparison of Blunt Striker Results for the Final Velocity, Crater Length and θ as Function of Initial Velocity: (a) 6061 T-6Al (b) 1010 CR Steel

184

Acknowledgment

This paper is based a dissertation by the second author submitted to the University of California, Berkeley. The work was supported by the Air Force Office of Scientific Research under Contract AFOSR F49620-89-C-8127.

References

[1] E. Wu and W. Goldsmith, Normal Impact of Blunt Projectiles on Moving Targets -- Experimental Study, *International Journal of Impact Engineering*, Vol. 9, 1990, pp. 389-404.

[2] E. Wu and W. Goldsmith, Normal Impact of Blunt Projectiles on Moving Target -- Analytical Considerations, *International Journal of Impact Engineering*, v. 9, 1990, pp. 405-432.

[3] P. Beynet and R. Plunkett, Plate Impact and Plastic Deformation by Projectiles, *Experimental Mechanics*, v. 11, 1971, pp. 64-70.

[4] A. Prakash, Supercomputer Simulations of Impact of Cylindrical Projectiles on Moving Plates, *Proc. of the 1991 Summer Simulation Conference*, SCSC 1991 (23rd), 1991, pp. 543-547.

[5] American Society for Metals. *Metals Handbook*, 9th ed. I. Properties and Selections: Iron and Steel, 1978.

[6] I. Kroschwitz, ed. *Concise Encyclopedia of Polymer Science and Engineering*. J. Wiley & Sons, New York, 1990.
[7] W. T. Thomsen, An Approximate Theory of Armor Penetration, *Journal of Applied Physics*, v. 26, 1955, pp. 80-82.

[8] Y. W. Mai and B. Cotterell, The Essential Work of Fracture for Tearing of Ductile Materials, *International Journal of Fracture*, v. 24, 1984, pp. 229-236.

[9] P. J. Shadbolt, R. S. J. Corran, and C. Ruiz, A Comparison of Plate Perforation Models in the Sub-ordnance Impact Velocity Range, *International Journal of Impact Engineering*, v. 1, 1983, pp. 23-49.

[10] J. H. Giovanola and I. Finnie, The Crack Opening Displacement (COD) as a Fracture Parameter and a Comparative Assessment of the COD and J Integral Concepts, *Solid Mechanics Archives*, v. 9, 1984, pp. 227-257.

[11] F. Erdogan and M. Ratwani, Fracture of Cylindrical and Spherical Shells containing a Crack, *Nuclear Engineering and Design*, v. 20, 1972, pp. 265-286.

[12] X.Hou and W. Goldsmith, Numerical Simulation of Impact on Moving Plates. In: *High Performance Computing in Computational Dynamics*. ASME, New York, CED vol. 6, 1994, pp. 121-128.

Numerical Study for Spall Fracture Conditions

Kazuo Asada [1], Hiroaki Takahashi [2]

1 Mitsubishi Heavy Industries, Ltd., Takasago R&D Center, 2-1-1 Shinhama Arai-cho Takasago 676 JAPAN
2 Mitsubishi Heavy Industries, Ltd., Sagamihara Machinery Works, Tana, Sagamihara 229 JAPAN

Summary

In the present study, the unified fracture condition was examined through comparison between the back surface velocity of flying plate impact experiment and the velocity calculated from computer code by using Ni-Cr-Mo steel (SNCM-630 steel). The computer code used here was that of one-dimensional large deformation stress wave propagation by finite-difference-method. The experimental results and computer code results were well in accord with each other, confirming the applicability of the unified fracture condition to perforation dynamics.

Key words : Spall fracture, Unified dynamic fracture condition, Flying plate impact experiment, FDM computer code, Laser interferometer

1. Introduction

When a projectile with high velocity impacts onto a target, several fracture modes (such as spall fracture with back surface separation, plug fracture with adiabatic shear band, penetration fracture with plastic flow and so forth) appear in the target. Which fracture mode will be superior to the others depends on various conditions: the structure, the material, the velocity and angle of the projectile. Under the general impact condition, only one fracture mode rarely appears in the target, and usually more than two fracture modes overlap.

Therefore, in order to identify one impact fracture strength of materials, a particular type of impact experiments should be undertaken. For example, a flying plate impact experiment by flying plate onto target plate at high velocity is to identify a spall fracture strength of target material, and a plug fracture experiment with adiabatic shear band of a thin target caused by a flat end projectile. Because different impact condition brings in different fracture strength even for the same material, it is difficult to determine one fracture strength in the general impact condition.

Then, we derive a unified fracture condition which contains three modes of spall fracture, plug fracture and penetration fracture [1]. This fracture condition is that fracture strain is affected by pressure and temperature, and that the strain is required to satisfy three fracture conditions of spall, plug and penetration. We examine the unified fracture condition through comparison between numerical analysis and experiment by using Ni-Cr-Mo steel (SNCM-630 steel). In detail, we set up the equation of the unified fracture condition from (1) the spall stress of flying plate impact experiment (for the condition corresponding to spall fracture), (2) the maximum stress and maximum strain of unaxial tensile test (for the condition corresponding to plug fracture), and (3) the maximum strain of unaxial tensile test under Bridgman's ultra-high pressure [2] (for the condition corresponding to penetration fracture). Then, we develop the code of one dimensional large deformation stress wave propagation in order to analyze flying plate impact experiment. This code is based on finite-difference-method and can be

185

used for analysis of a crack caused by spall fracture. The constitutive equation is that of elastic-plastic solid and the equation of state is Grüneisen's equation. In flying plate impact experiment, the back surface velocity of the target is measured with VISAR and compared with the velocity computed form the code. As regards the response of back surface velocity and fracture point of the target, the experimental results and computer code results are well in accordance with each other. The plastic strain which appeared in the code corresponds quantatively to the result of the fracture surface observation by SEM in the experiment.

2. Material Properties

The material properties of Ni-Cr-Mo steel (SNCM-630 steel) can be given by the constitutive equation and the equation of state. The nonlinear equation of state is important. For the constitutive equation, we adopt an elastic-plastic solid of which stress may be affected by strain rate. The value of the elastic-plastic solid is given by uniaxial tensile test, and shear split Hopkinson bar method. On assumption that the equation of state is Grüneisen's equation, the reference value of the steel is used as the value of the equation of state (3).

The constitutive equation of SNCM-630 steel (stress-strain relation) is shown in Fig.1, where E = Young's modules, ν = Poisson's ratio, Y_s = static yield stress, Eh = tangent modulus, σ_d = dynamic stress, σ_s = static stress, $\dot{\varepsilon}_p$ = equivalent plastic strain rate, n, D = dynamic material constants.

The relation between dynamic stress and static stress is given by a following equation.

$$\frac{\sigma_d}{\sigma_s} = \begin{cases} 1 & : \sigma_s \leq Y_s \\ 1 + \left(\dfrac{\dot{\varepsilon}_p}{D}\right)^{1/n} & : \sigma_s > Y_s \end{cases} \tag{1}$$

The equation of state of SNCM-630 steel is shown in Table I, where ρ_0 is initial density and C_0, S and Γ are material constants. The relation between shock wave propagation velocity C_s and particle velocity U_p is given by a following equation.

$$C_s = C_0 + S U_p \tag{2}$$

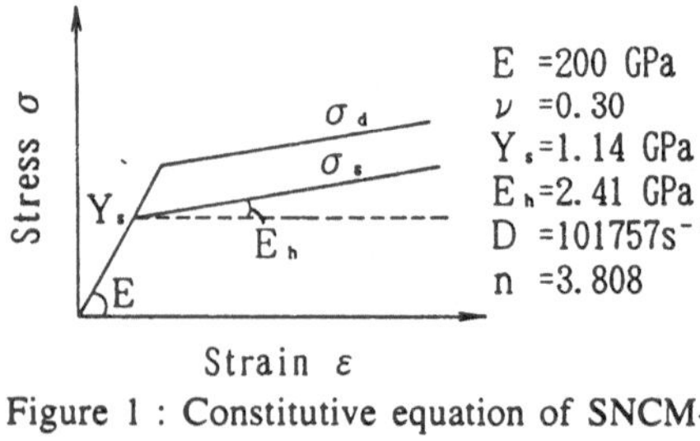

E = 200 GPa
ν = 0. 30
Y_s = 1. 14 GPa
E_h = 2. 41 GPa
D = 101757s^{-1}
n = 3. 808

Table I : Equation state of SNCM-630

Material	ρ_0 (kg/m³)	C_0 (m/s)	S (-)	Γ (-)
SNCM-630	7.82×10³	4583	1.771	1.69

Figure 1 : Constitutive equation of SNCM-630

3. Unified fracture condition

The fracture strain of material ε_f depends on pressure P, temperature T, the second invariant of stress deviations J_2', and plastic strain rate $\dot{\varepsilon}_p$. Fracture strain ε_f is given by a following equation.

$$\varepsilon_f = \varepsilon_f\left(P, T, J_2', \dot{\varepsilon}_p, \cdots\cdots\right) \tag{3}$$

Suppose that fracture strain ε_f depends on only pressure P and temperature T, and Equation (3) can be rewritten as follows.

$$\varepsilon_f = \varepsilon_f(P, T) \tag{3}'$$

Here, when σ_1, σ_2 and σ_3 are principal stress, pressure P is given by a following equation.

$$P = -\frac{1}{3}(\sigma_1 + \sigma_2 + \sigma_3) \tag{4}$$

In order to determine the function in Equation (3)' according to the three fracture conditions, we provide the assumptions on the relation between pressure and fracture strain.

    a. spall fracture is due to the conditions of spall stress and zero strain.

    b. plug fracture is due to the conditions of the minus pressure (tension) and strain corresponding to the fracture conditions of uniaxial tensile test.

    c. penetration fracture is due to the conditions of the plus pressure and strain corresponding to the fracture conditions of uniaxial tensile test under high pressure.

Concerning plug fracture and penetration fracture, however, the temperature-dependent effect should be taken into account, because the large plastic deformation of material locally causes high temperature. The conditions of fracture strain are shown in Fig.2, where ε_p is equivalent plastic strain. Assume that $\dot{\varepsilon}_1^P$, $\dot{\varepsilon}_2^P$ and $\dot{\varepsilon}_3^P$ are the components of main plastic strain rate, and ε_p is given by a following equation.

$$\varepsilon_P = \int_0^t \frac{\sqrt{2}}{3}\left[\left(\dot{\varepsilon}_1^P - \dot{\varepsilon}_2^P\right)^2 + \left(\dot{\varepsilon}_2^P - \dot{\varepsilon}_3^P\right)^2 + \left(\dot{\varepsilon}_3^P - \dot{\varepsilon}_1^P\right)^2\right]^{1/2} dt \tag{5}$$

where t is time. A, B, and C in Fig.2 are the points of spall fracture, plug fracture and penetration fracture at temperature $= T_0$, respectively. On the other hand, A', B', and C' in Fig.2 are the points of spall, plug and penetration fracture at temperature $= T_1$ (T1 > T0).

Let us now consider the conditions of the SNCM-630 steel. For Point A in Fig.2, we adopt the average of spall stresses in flying plate impact experiment. Similarly we adopt as Point B the average of the maximum stress and maximum strain in the true stress-strain relation of uniaxial tensile test at room temperature, and as Point C the maximum strain of the steel under the condition of Bridgman's high pressure. The values of pressure and strain at Point A, B and C are shown in Table II. We here use only the points A, B, C just as fracture condition, because the rise of temperature can be ignored in flying plate impact experiment.

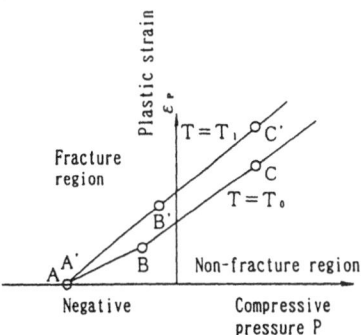

Figure 2 : Strain fracture condition

Table II : Fracture strain fracture condition for SNCM-630 (room temperature)

	Point A (spall fracture)	Point B (plug fracture)	Point C (peneter. fracture)
Pressure P (GPa)	−3.6	−1.11 *	+2.45
Strain ε_f	0	0.916	3.0

* $P = \frac{-1}{3}(\sigma_1 + \sigma_2 + \sigma_3) = \frac{-1}{3}(3.33 + 0 + 0) = -1.11$

4. Flying plate impact experiment

We made a flying plate impact experiment, using a flying plate of SNCM-630 steel (the outer diameter is 38mm, the thickness of plate is 5.05mm), a target plate of SNCM-630 steel (the outer diameter is 46mm, the thickness of plate is 10.05mm) and an powder gun (the full length is 6m, the outer diameter is 170mm, the inner diameter is 40mm), under two conditions of velocity (447m/s, 543m/s). In that experiment, we measured the back surface velocity and wave propagation velocity of the target with VISAR and pin contact gauge, and computed the spall stress of SNCM-630 steel and the point of spall fracture of the target plate from the measured data.

The VISAR data and the back surface velocity are shown in Fig.3 (when impact velocity = 447m/s) and Fig.4 (when impact velocity = 543m/s). The VISAR data indicates the intensity of interference light, and the number of the interference patterns is in proportion to the back surface velocity of the target. The zero points of time in Fig.3 and Fig.4 are not equal to impact time. Concerning spall pull-back velocity, it is shown as ΔU in each figure. The section of the target plate after experiment is shown in Fig.5 ((a) impact velocity = 447m/s, (b) impact velocity = 543m/s). The SEM photograph of spall surface is shown in Fig.6 (impact velocity = 447m/s).

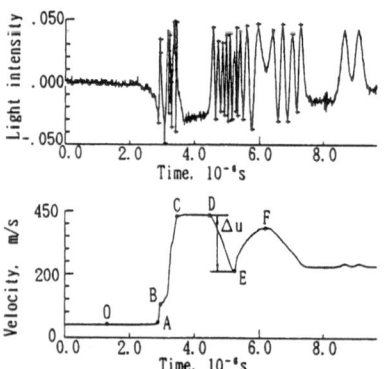

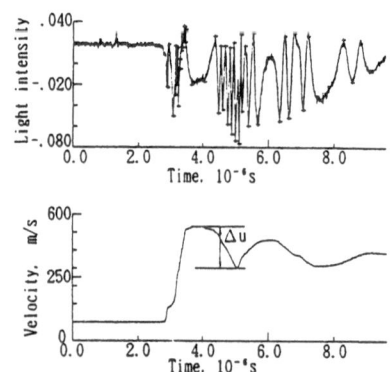

Figure 3 : VISAR data and back surface velocity of experiment (impact vel. = 447m/s)

Figure 4 : VISAR data and back surface velocity of experiment (impact vel. = 543m/s)

The thickness of the target plate is expressed as h, and the spall facture position measured from the back surface of target is expressed as X. Thus, elastic wave propagation velocity C_e, plastic wave propagation velocity C_P and spall rarefaction wave propagation velocity C_{SP} are given by a following equation.

$$C_e = h/t_A, \quad C_P = h/t_C, \quad C_{SP} = 2X/(t_E - t_D) \qquad (6\text{-}1)$$

Here, spall stress σ_s is computed from initial density ρ_0, spall rarefaction wave propagation velocity C_{SP} and spall pull-back velocity ΔU as follows:

$$\sigma_s = \frac{1}{2}\rho_0 C_{SP}\Delta U \qquad (6\text{-}2)$$

Elastic wave propagation velocity C_e, plastic wave propagation velocity C_P, spall rarefaction wave propagation velocity C_{SP}, maximum free surface velocity U_m, spall pull-back velocity ΔU, spall stress σ_s and spall fracture position measured from the back surface of the target are determined by Equations (6-1) – (6-2), the back surface velocity in Fig.3 and Fig.4, and the section of the target plate in Fig.5, as shown in Table III. Fig.5 shows that spall fracture is found at the middlw of the target plate, while Fig.6 shows that plastic deformation is found on spall fracture surface.

(a) impact velocity = 447m/s

(b) impact velocity = 543m/s

Figure 5 : Spall fracture of target plate

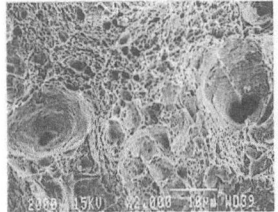

Figure 6 : Spall surface of target plate
(impact vel. = 447m/s)

Table III : Results of flying plate experiment

Impact velocity	C_s (m/s)	C_P (m/s)	C_{SP} (m/s)	U_m (m/s)	ΔU (m/s)	σ_s (GPa)	X_s (10^{-3}m)
447 (m/s)	5852	5025	3989	441	229	3.57	5
543 (m/s)	5852	5025	4016	542	233	3.66	5

5. Numerical Analysis

5.1 Theory

We here explain the equations that should be introduced into the code for analysis: basic equation, constitutive equation and equation of state.

5.1.1 Basic Equation

Conservation of mass, equation of motion and conservation of energy in Lagrangian coordinate system are, respectively following.

$$\frac{\rho}{\rho_0} = \frac{dV}{dv} \tag{7}$$

$$\rho a = -\frac{\partial\sigma}{\partial x} - \frac{\partial q}{\partial x} + (\alpha - 1)\frac{\psi}{x} \tag{8}$$

$$\rho\frac{\partial E}{\partial t} = (P+q)\frac{1}{\rho}\frac{\partial\rho}{\partial t} + P_d + \left\{-\frac{\partial h}{\partial x} + (\alpha-1)\frac{h}{x}\right\} + \rho Q \tag{9}$$

where ρ_0 – initial density, ρ – density after deformation, dV – initial volume of element, dv – volume of element after deformation, a – acceleration, σ – stress (plus compressive), q – artificial viscous stress (plus compressive), P – pressure (plus compressive), α – shape factor. x, y, z – components of three-dimensional spatial coordinate system (however, x is the direction of one-dimensional stress wave propagation). E – internal energy per unit mass, t – time, Q – calorific value per unit mass, h – heat flow, Pd – power of deviation stress, ψ – difference of stress, σ and ψ are given by following equations.

$$\sigma = -\sigma_x = P - \sigma_x^d \tag{10}$$

$$\psi = \sigma_x - \sigma_y = \sigma_x^d - \sigma_y^d = 2\sigma_x^d + \sigma_y^d \tag{11}$$

where σ_x, σ_y and σ_z are the stresses in the direction of x, y and z, respectively, and artificial viscous stress q is a sum of two viscous stresses of first-order equation (q_1) and second-order equation (q_2). q is given by a following equation.

$$q = q_1 + q_2 = \rho B_1 \Delta X^2 \left(\frac{1}{\rho}\frac{\partial \rho}{\partial t}\right)^2 + (B_2 \Delta x)C_e \frac{\partial \rho}{\partial t} \tag{12}$$

where C_e is elastic wave propagation velocity, Δx is width of space element, and B_1 and B_2 are artificial viscous constants.

5.1.2 Constitutive Equation

The constitutive equation is represented by three equations: law of flow rule for the relation between deviatoric stress rate and deviatoric strain rate, Mieses yield condition for the relation between second of deviatoric stress and static yield stress, and the equation yield stress on strain rate. They are given by the following three equations:

$$\frac{\dot{\sigma}_x^d}{d_x^d} = \frac{\dot{\sigma}_y^d}{d_y^d} = \frac{\dot{\sigma}_z^d}{d_z^d} = 2G \tag{13}$$

$$f_y = \left(\sigma_x^d\right)^2 + \left(\sigma_y^d\right)^2 + \left(\sigma_z^d\right)^2 \le \frac{2}{3}Y_s^2 \tag{14}$$

$$\frac{Y_d}{Y_s} = \begin{cases} 1 & : f_y < \frac{2}{3}Y_s^2 \\ 1 + \left(\dfrac{\dot{\varepsilon}_P}{D}\right)^{1/n} & : f_y = \frac{2}{3}Y_s^2 \end{cases} \tag{15}$$

where d_x^d, d_y^d and d_z^d are the components of deviatoric strain rate in the direction of x, y and z, respectively, f_y is yield function, and Ys and Yd are static and dynamic yield stress.

5.1.3 Equation of State

The equation of state which expresses the properties of pressure is given by Grüneisen's equation of shock-particle velocity as follows:

$$P - P_H = \Gamma\rho(E - E_H), P_H = \rho_0 C_0^2 \eta/(1 - S\eta)^2, E_H = P_H\eta/(2\rho_0) \tag{16}$$

where P_H is Hugoniot pressure, E_H is Hugoniot energy, Γ is Grüneisen's ration, and C and S are material constants providing Hugoniot pressure (refer to Equation (2)). η is volume strain and given by following equation.

$$\eta = 1 - \rho_0/\rho \tag{17}$$

5.2 Calculation Conditions

In calculation, the flying plate (thickness of plate = 5.05mm) is divided into ten equal layers. Similarly, the target plate (thickness of plate = 10.05mm) is divided into twenty. Suppose that an element satisfying fracture condition is broken, we make a model so as to separate the node at the left side of the element into two parts. (General FEM code include no functions concerning such a separation. Therefore, we believe that the capability of our code to separate nodes is unique.) Here, the values of artificial viscous constants B_1 and B_2 are 0 and 0.3, respectively. The value of spall rarefaction wave propagation velocity is the average of experimental values (= 4,000m/s).

5.3 Calculation Results

As regards results, energy relation and energy error diagrams are shown in Fig.7 (impact velocity = 447m/s) and Fig.11 (impact velocity = 543m/s), x-t diagrams are shown in Fig.8 (impact velocity = 447m/s) and Fig.12 (impact velocity = 543m/s), response diagrams of pressure and plastic strain at the middle point of target plate are shown in Fig.9 (impact velocity = 447m/s) and Fig.13 (impact velocity = 543m/s), and back surface velocity diagrams of target plate (including experimental data) are shown in Fig.10 (impact velocity = 447m/s) and Fig.14 (impact velocity = 543m/s).

For impact velocity is 447m/s, the energy relation and energy error diagram (Fig.7) shows that the kinetic energy of flying plate is transformed into the internal energy of flying plate and target plate. The diagram of Fig.7 also shows that the maximum value of energy error is 0.013, and this fact proves that a precise calculation was performed. The x-t diagram (Fig.8) shows that fracture happened in the middle point of target plate. This is well in accordance with the result of experiment (refer to Fig.5(a)). As shown in the response diagram of pressure and plastic strain at the middle point of target plate (Fig.9), fracture happens under the condition that pressure P is -3.1GPa and plastic strain ε_p is = 0.115. The values of pressure and plastic strain correspond to those of Table II (P = -3.285GPa when ε is = 0.115). From the comparison on back surface velocity between the experimental data and calculated value (refer to Fig.10), it is clear that the maximum free surface velocity of experiment corresponds to that of calculation. But, concerning spall pull-back velocity (ΔU), the calculated value (170m/s) is smaller than experimental one (229m/s), though the wave forms of both calculation and experiment are in good agreement with each other. HEL does not appear in calculation but does in experiment, probably because artificial viscous stress is introduced into calculation. For the impact velocity is 543m/s, the same results are obtained as the case for 447m/s.

The reason why the calculated value of spall pull-back velocity (ΔU) is smaller than the experimental one is probably that the pressure of calculation elements with spall fracture is smaller than experiment spall pressure (3.6GPa) due to the appearance of plastic strain.

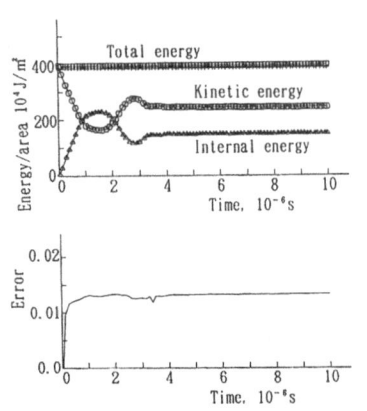

Figure 7 : Energy relation and energy error (impact vel. = 447m/s)

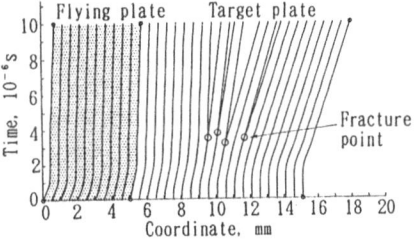

Figure 8 : x-t diagram (impact vel. = 447m/s)

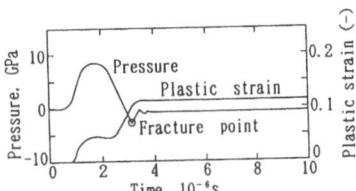

Figure 9 : Pressure and plastic strain at the middle point of target plate (impact vel. = 447m/s)

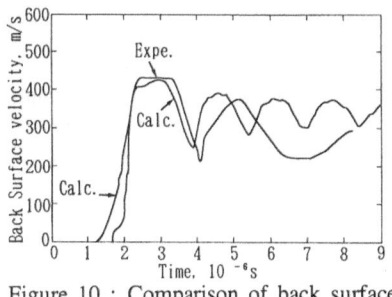

Figure 10 : Comparison of back surface
velocity (impact vel. = 447m/s)

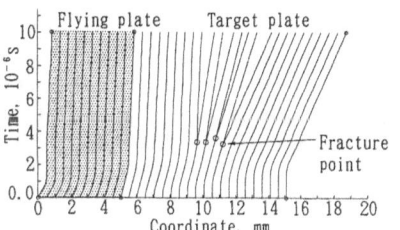

Figure 12 : x-t diagram (impact vel. = 543m/s)

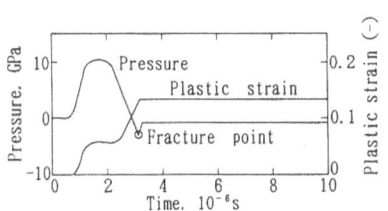

Figure 13 : Pressure and plastic strain at
the middle point of target plate
(impact vel. = 543m/s)

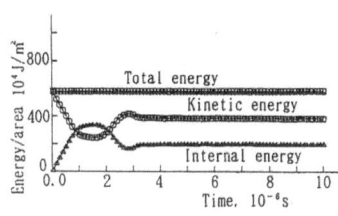

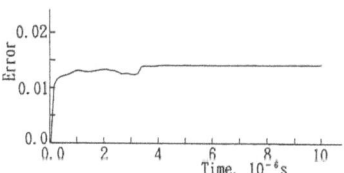

Figure 11 : Energy relation and energy error
(impact vel. = 543m/s)

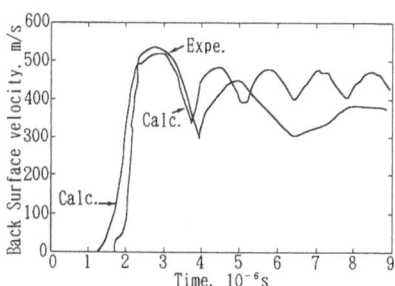

Figure 14 : Comparison of back surface
velocity (impact vel. = 543m/s)

6. Conclusions

In penetration analysis, we clarified the unified fracture condition (fracture condition on strain) which can be applied to a complicated fracture phenomenon such as overlapping modes of spall fracture, plug fracture and penetration fracture. We introduced the unified fracture condition into the code of one-dimensional large deformation impact wave propagation analysis, and by means of the code we analyzed flying plate impact experiment (spall fracture experiment) using VISAR in order to compare the wave form of back surface velocity and the spall fracture point of target plate with those of experiment. As a result, we proved that not only qualitative but quantative explanation was possible by undertaking the numerical analysis with fracture condition on strain.

References

[1] J. F. Mescall, Proc. Annu. Meet. Soc. Eng. Sci., 14th (1977).
[2] P. W. Bridgman, Jour. Appl. Phys., 17, 201 (1946).
[3] B. J. Kohn, Air Force Weapons Lab. Report, AFWL-TR-69-38 (1969).

Dynamic Fracture and Damage of Rock During Impact

H.P.Rossmanith, L.Mishnaevsky Jr., R.E. Knasmillner and K. Uenishi

Institute of Mechanics, Technical University Vienna, Austria

Abstract

Wave propagation in damageable media and the evolution of damage due to repeated impact during percussion drilling in brittle disordered materials is investigated.

Analytical, numerical and experimental work complementing each other in clarifying the complex dynamic process of damage evolution and fracture network development in polycrystalline brittle materials is presented. Upon field tests rock samples have been sectioned and a microscopic investigation of the damage and fracture zone at the bottom and the circumference of the bore hole has been performed.

Based on experimental findings a three-dimensional numerical simulation scheme has been developed which incorporates wave propagation in dissimilar granular arrangements, dynamic impact phenomena and accumulation of brittle damage of rock grains.

Keywords: Impact, strain rate, wave propagation, brittle damage, brittle rocks and brittle polycrystalline materials, percussion drilling, micro-fractures.

1 Introduction

When stress waves of appropriate intensity propagate through a brittle possibly pre-damaged and inhomogeneous material a population of micro defects will be generated and/or possibly extended. Thus, wave energy will be dissipated during the production of damage. Obviously, in periodic impact events such as encountered in percussion drilling in rock under quasi-stationary conditions a steady-state regime of micro-fractures forming the damage zone is translated through the rock. For layered configurations such as layered rock or laminated composites the formation of the damage zone becomes an instationary process with added complexity at the interfaces.

2 Percussion Drilling

In percussion drilling damage and fracture with subsequent destruction of the grain assembly is caused by the repeated impact of drill bits onto the rock surface. Because of the extreme dissimilarity in the material properties the drill bits, usually made of a sintered tungsten-carbide–cobalt alloy [1], can be considered in the numerical simulation as a rigid body. The advance of a drill bit per impact of $0.5mm$ can be decomposed into two phases: initial contact formation and the ensuing bit engagement. The typical rise-time, t_g, is about $200\mu s$ and the engagement phase lasts about $20ms$.

Figure 1 shows a percussion drilling tool with typical drill bits of conically shaped form with a spherical cap. When these drill bits indent the rock surface the local contact area increases from nearly zero to a maximum value at maximum indentation. During the entire engagement phase of

194

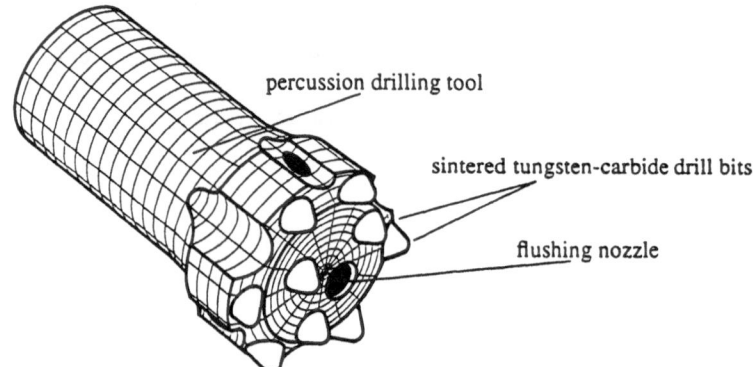

Figure 1: 2.5 inch button drill bit

the bits wave energy is transferred from the tool to the rock. For practical reasons a damage zone is formed below the drill bit from where a fracture system will start to form the chips. After impact the contact area diminishes again.

Percussion drilling is modeled as a problem where a rigid indenter impacts a half-space. This impact problem is governed by displacement boundary conditions for the drill bits. This corresponds to a prescribed time-dependent change of contact area as sketched in Fig.2.

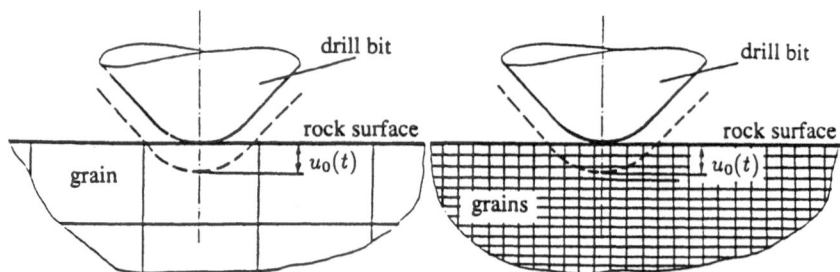

Figure 2: Indentation of a drill bit into a) coarse grained and b) fine grained granular rock

In the numerical calculations to follow the displacement, $u_0(t)$, of the center point of the drill bit region is represented by

$$u_0(t) = \begin{cases} 0 & \text{if } t < 0 \\ \hat{u}\sin^2(\pi t/2T) & \text{if } 0 \le t < t_g \\ \hat{u} & \text{if } t \ge t_g \end{cases} \tag{1}$$

where $\hat{u}$ is the maximum indentation and t_g is the rise-time according to Fig.3.

For spherically shaped indenters the variation of the radius of the area of contact, $2a(t)$, with respect to then depth of indentation $u_0(t)$ is determined by the geometrical relation

$$a(t) = \sqrt{2u_0(t)R - u^2(t)} \quad (0 \le u_0 \le R) \tag{2}$$

where R is the radius of the tip of the drill bit.

In reality, the elasticity of the rock yields a smaller contact area than obtained from a simple geometrical model. In addition, during the percussion drilling operation, the tool will not be completely

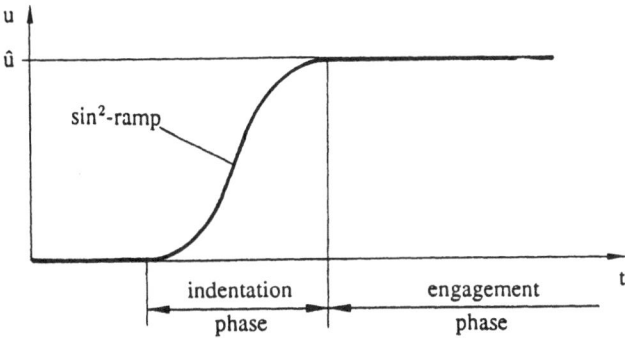

Figure 3: Drill bit advance

withdrawn between subsequent strokes and, hence, the contact conditions become extremely complex. Furthermore, the disintegration of the rock material, the formation of sub-grained and powdered material fragments and the incomplete flushing of fines will increase the complexity of the interaction process.

3 Numerical simulation

Hard rocks are composed of grains of different elastic and other properties which are randomly distributed throughout the material [2, 3]. In this numerical simulation, a simplified approach is followed where rock is modeled as a material which consists of several types of rectangular blocks with given dissimilar elastic properties. The numerical model developed allows the simulation of single and multiple impact of one or more indenters on the rock surface (Fig.4).

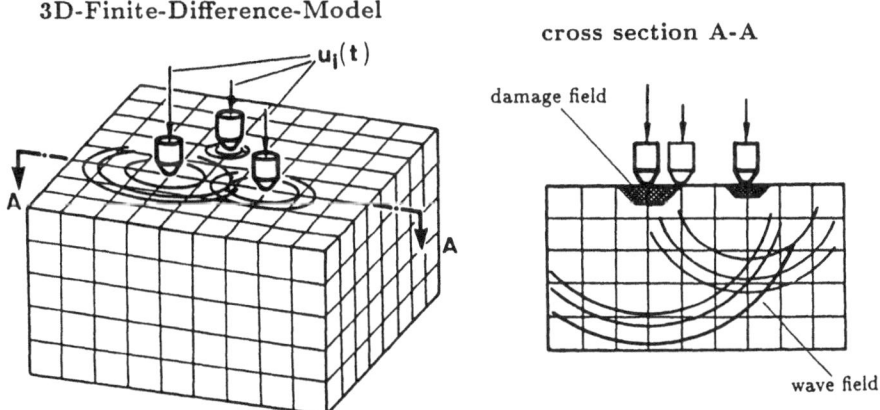

Figure 4: Numerical model of the percussion drilling process

The propagation of momentum and energy in a material modeled by an assembly of tightly connected distinctive square model blocks is assumed to occur by stepwise interaction between adjacent blocks. Each square block being squeezed by its predecessor neighbour transmits the deformation to its frontal neighbour and, by means of Poisson's effect to its lateral neighbouring elements. The process of interaction between blocks is modeled in terms of displacements and velocities.

An engineering model of wave propagation is based on the displacements of the boundaries and therefore the deformation of an arbitrary block (m, n, k) in the bulk (see Fig.5). The displacements of the boundaries of the block (m, n, k) is caused by

a) the vertical movement of the upper neighbouring block in vertical direction, block $(m-1, n, k)$;

b) the lateral expansion of four neighbouring blocks (blocks $(m, n-1, k)$, $(m, n, k-1)$, $(m, n, k+1)$ and $(m, n+1, k)$) caused by Poisson's effect.

c) the displacements caused by waves propagating in the horizontal direction (interchange the indices m and n in eqs (3-5).

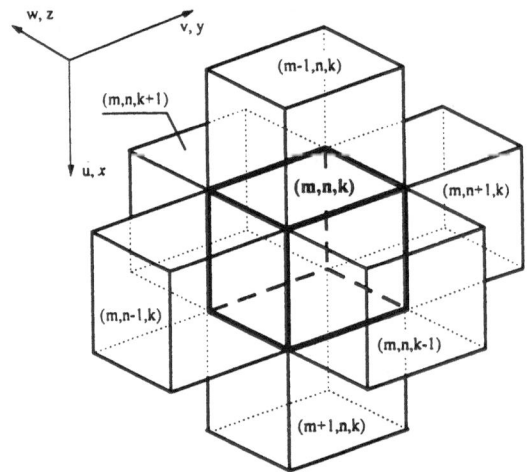

Figure 5: 3D block assembly

The deformation of the grains is progressively imparted to all neighbouring partners in a forward manner. The low mismatch of the acoustical impedances of the grain materials of granite allow to neglect the multiple reflections of the propagating waves at the interfaces of the grains.

The displacement components $[u, v, w]$ of the block (m, n, k) due to a longitudinal stress wave in terms of finite differences follow in the form

$$u^p_{m,n,k}(t + \Delta t) = u^p_{m,n,k}(t) + \left[u^p_{m-1,n,k}(t) - u^p_{m-1,n,k}(t - \Delta t)\right] T_{m-1,n,k/m,n,k} -$$
$$- \left[u^p_{m+1,n,k}(t) - u^p_{m+1,n,k}(t - \Delta t)\right] T_{m,n,k/m+1,n,k} + \tag{3}$$
$$+ \nu_{m,n,k} \left[v^p_{m,n,k}(t) - v^p_{m,n,k}(t - \Delta t)\right] /2 + \nu_{m,n,k} \left[w^p_{m,n,k}(t) - w^p_{m,n,k}(t - \Delta t)\right] /2$$

and equivalently for v and w

$$v^p_{m,n,k}(t + \Delta t) = v^p_{m,n,k}(t) + \nu_{m,n,k} \left[u^p_{m,n,k}(t) - u^p_{m,n,k}(t - \Delta t)\right] /2 + \tag{4}$$
$$+ \nu_{m,n,k} \left[w^p_{m,n,k}(t) - w^p_{m,n,k}(t - \Delta t)\right] /2 + \left[v^p_{m,n+i,k}(t) - v^p_{m,n+i,k}(t - \Delta t)\right] T_{m,n,k/m,n+i,k}$$

$$w^p_{m,n,k}(t + \Delta t) = w^p_{m,n,k}(t) + \nu_{m,n,k} \left[u^p_{m,n,k}(t) - u^p_{m,n,k}(t - \Delta t)\right] /2 + \tag{5}$$
$$+ \nu_{m,n,k} \left[v^p_{m,n,k}(t) - v^p_{m,n,k}(t - \Delta t)\right] /2 + \left[w^p_{m,n,k+j}(t) - w^p_{m,n,k+j}(t - \Delta t)\right] T_{m,n,k/m,n,k+j}$$

where $T_{a,b,e/c,d,f}$ is the transmission coefficient [4] from block (a, b, e) into block (c, d, f).

The displacement of boundaries of the blocks due to shear pulse transmission has been worked out in a similar way [5].

4 Damage evolution

When the percussion tool impacts the rock surface part of the kinetic energy of the drill bit will be transmitted to the rock. Under static conditions the force exerted by the drill bit on the rock surface causes a displacement field characterized by elastic strain energy. In the dynamic case the energy of the impactor is partly converted into elastic strain energy (deforming the surface and the grains) and partly into kinetic energy which will be radiated in the rock.

In a damageable material further damage is created by dissipation on the expenses of the wave field. Part of the energy locked in the wave field is converted into surface energy for the extension of micro-cracks.

This *damage energy* which represents the loss of wave energy due to dissipation may be obtained from an energy balance consideration for an unit volume.

The process of damage evolution in granular rock has been modeled by utilizing the approach developed by Krajcinovic and Fonseka [6, 7] where the incremental increase of vectorial damage is related to the incremental increase of strain. In analogy to plasticity theory, a damage surface will be defined in the strain space to yield a strain-based criterion for the evolution of damage. In compression the effect of pulverization due to shear stresses within the blocks onto Young's modulus has been neglected.

For modeling purpose the process of damage and fracture evolution in disordered polycrystalline rock as shown in Fig.6a is simplified. As shown in Fig.6b it is assumed that randomly oriented and randomly distributed plane penny-shaped micro-cracks are formed in each block during impact.

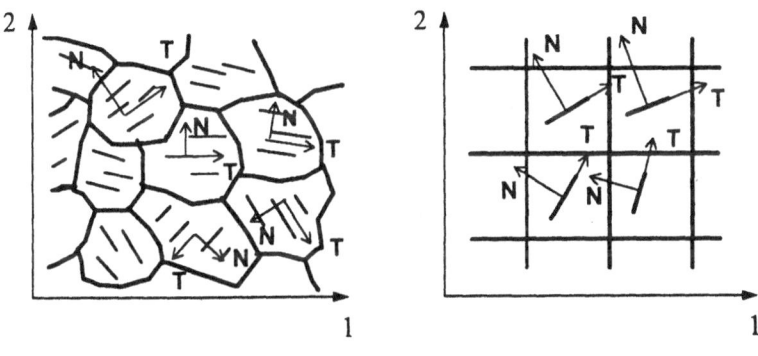

N, T ... local coordinate systems
1, 2 ... global coordinate system

Figure 6: Polycrystalline rock with grains weakened by micro-cracks: a) "real" situation and b) simplified model for numerical simulations

5 Computational results

Numerical simulations on the basis of eqs (3)-(5) have been performed using a finite difference distinctive element scheme.

Figure 7 represents the state of damage formed during a single impact obtained at 20 time-steps after initial contact. The area bounded by the innermost polygon indicates a state of damage where the material is damaged to the extent where the chips and the powder can be removed by flushing.

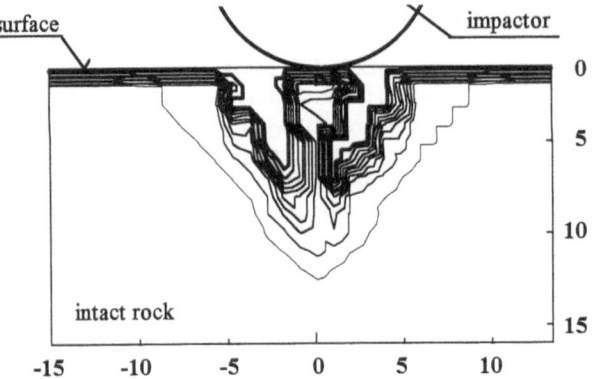

Figure 7: Evolution of damage in the center section

In these numerical simulations the blocky nature of the model material induces a kind of "mesh type" dependency of the results. The blocky structure yields a conically shaped volume of rock comminuted. Other grain structures such as polyhedrons, spheres or ellipsoids will give a different form of the damage zone. However, the form of the damage band evolved is the same.

In this model the damage zone increases with time during the indentation process. The small band characterizing the remaining damaged zone (Fig.7) which is located between the intact rock and the material to be flushed is in close agreement with the findings of the experimental investigations conducted in medium-grained granite (see Section 6).

A relation between the efficiency of energy utilization in rock fragmentation - here defined as the specific work needed to generate damage up to the flushing threshold divided by the input energy - versus depth of indentation is shown in Fig.8.

The results indicate that the efficiency is equal to 0 for small displacements which are associated to strain levels below the critical value for further increase of damage. This corresponds to the stalling of the drilling operation at low impact energy levels (mining: starving of drill). Beyond the stall region there is a strong increase with a maximum of efficiency at the time where the first spall occurs. The numerical simulations exhibit an optimal initial load level at relatively small initial displacements. In these calculations the part of energy transmitted to the rock has been considered as input energy.

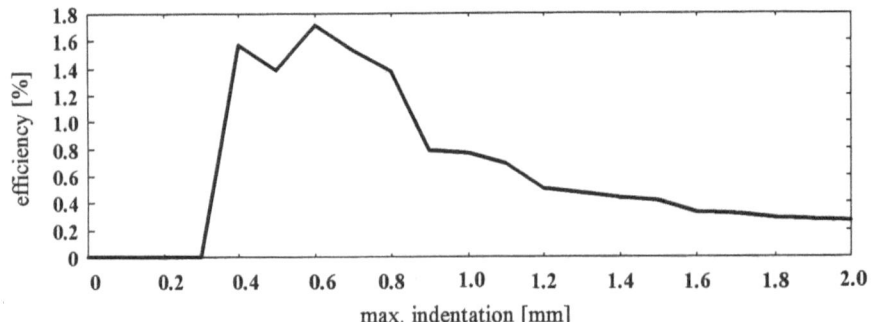

Figure 8: Efficiency of rock destruction by percussion drilling

6 Experimental investigations

To study micro-crack formation and damage evolution in rock bore-holes with a diameter of $65mm$ were percussion drilled. The hydraulic percussion drill used for the experiments operates with an impact rate of $40Hz$ and about $200J$ impact energy per stroke and in granite provides an overall steady-state drill advance of $1.2m/min$. The hydraulic hammer was stopped immediately when the center of the granite cube of dimensions of $1m$ edge length was reached. The rock samples investigated consist of a medium grained granite with crystallites of 2 to $5mm$ in size.

The large granite boulders chosen for the drilling experiments did not allow for damage due to reflected waves.

After drilling the boulders were appropriately sectioned. The carefully cleaned and oven-dried samples containing the bore-holes were then vacuum-impregnated with dyed epoxy-resin at a temperature of $80°C$. After curing of the epoxy-resin, the samples were sectioned with a diamond saw, ground and investigated with a surface microscope. The dying technique allows to distinguish between micro-cracks caused by percussion drilling and micro-cracks and crushed grains caused by the final sectioning and grinding processes [8].

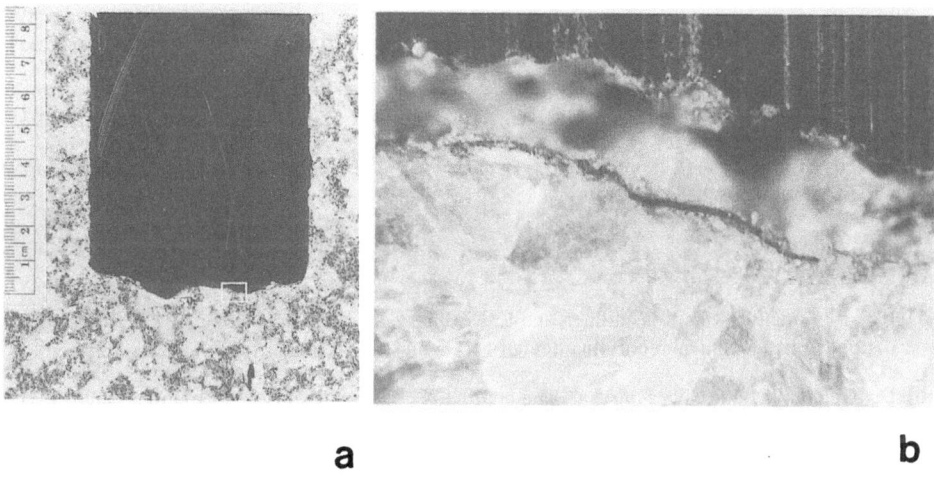

a   **b**

Figure 9: Sections of a bore-hole: a) cross section, b) enlargement

The investigation of the cross sections of the bore-hole revealed a thin layer (approx $2mm$ thick) of micro-cracks and damaged grains.

At the bore-hole bottom the thickness of this layer of micro-cracks is the same (Fig.9a). This is in agreement with numerical results. In the enlargement shown in Fig.9b some additional macro-cracks can be observed to extend parallel to the bore-hole bottom. These may be caused by shearing off the remaining asperities left behind.

This relatively small band of damage is in contrast to results of static plate edge indentation tests conducted in thin slices of rock where long cracks extending from the tip of the indenter into the material have been observed [9]. These discrepancies may be caused by the different boundary conditions and dynamic effects between the fully 3D percussion drilling experiments and the 2D experiments.

The size and orientation of the micro-cracks are mostly independent of the grain boundaries. The same behaviour is observed by investigating the cuttings: most of the chips are composed of more than one mineral and are densely micro-cracked.

It should be noted that the investigations have been carried out in medium grained granite and the results may be quite different for other types of rock.

7 Conclusions

The combination of brittle damage continuum mechanics and the theory of elastic wave propagation proved a very suitable approach for the description of impact destruction of rock and rock-like materials showing brittle behaviour.

The 3D finite block numerical code developed allows the simulation on a PC of single and multiple impact. Further development will include the impact of a sequence of bits and the implementation of the time-varying elastic properties into the wave propagation model.

The results of the experimental field tests compare well with the numerical model simulations.

References

[1] C.P. Chugh. *High Technology in Drilling and Exploration*, chapter Percussive Drills, pages 531–598. A.A. Balkema, Rotterdam, 1992.

[2] Fred G. Bell. *Engineering Geology*. Blackwell Scientific Publications, Oxford, London, 1993.

[3] Robert S. Carmichael (Editor). *Practical Handbook of Physical Properties of Rocks and Minerals*. CRC Press, Inc., Boca Raton, Florida, 1988.

[4] J.D. Achenbach. *Wave Propagation in Elastic Solids*. North-Holland Publishing Comp.-Amsterdam, London, 1973.

[5] H.P. Rossmanith, R.E. Knasmillner, and L. Mishnaevsky Jr. Report to Austrian Research Foundation (FFF). Technical report, Institute of Mechanics, TU-Vienna, 1995. (To be published).

[6] D. Krajcinovic and G.U. Fonseka. The continuous damage theory of brittle materials - Part 1: General theory. *Journal of Applied Mechanics, Trans. ASME*, 48:809–815, 1981.

[7] G.U. Fonseka and D. Krajcinovic. The continuous damage theory of brittle materials - Part 2: Uniaxial and plane response modes. *Journal of Applied Mechanics, Trans. ASME*, 48:816–824, 1981.

[8] G. Simmons and D. Richter. Microcracks in rocks. In: R.G.J. Strens (Editor), *The Physics and Chemistry of Minerals and Rocks*, pages 105–138. John Wiley & Sons, London, New York, 1974.

[9] P.-A. Lindqvist, H.-H. Lai, and O. Alm. Indentation fracture in rocks observed in-situ in a scanning electron microscope. Technical report, Division of Mining and Rock Excavation, University of Luleå, Sweden, 1982. In: Rock Fragmentation by Indentation and Disc Cutting, Doctoral Thesis.

Morphological Aspects During Dynamic Crack Propagation

Akira Kobayashi[1] and Shinji Ogihara[2]

1: Professor 2: Research Fellow

Department of Mechanical Engineering, Science University of Tokyo, Noda, Chiba 278, JAPAN

Summary

Fracture morphology in PMMA is experimentally investigated in conjunction with the dynamic crack propagation velocity profiles and the fracture surface roughness. Strain energy release rate at the transition in the fracture surface morphology is calculated to correlate with the morphological change. Several quantitative results are obtained.

Keywords: morphology, PMMA, dynamic crack propagation profiles, fracture surface roughness, strain energy release rate

1. Introduction

Fracture process in PMMA (polymethyl methacrylate) is studied over the past half century. Berry [1] points out the concerned system has only the potential energy up to the critical point, and the increase in the potential energy is equal to the work done on the system. He also states the very system has both the potential and kinetic energy beyond the critical point. At any stage the potential energy and the work done on the system can be expressed in terms of the parameters of the system, and the kinetic energy can be obtained as the difference between the work done on the system and the increase in the potential energy through a simple energy balance. From this point of view, the equations for the crack propagation velocity as a function of crack length can be derived.

For the morphology in the fracture surface, Berry [2] mentions the color and the parabolic marking are observed in the fracture surface in PMMA. The origin of the colors is attributed to the thin layer interference of the incident white light. The thin layer consists of the oriented material of lower refractive index and lower density compared with the original PMMA itself and results in the process of crazing which accompanies crack propagation in many glassy thermoplastics. The origin of the parabolic marking appears presumably due to a hesitation in the secondary fracture while it is still isolated from the primary fracture.

Döll [3] studies the dynamic crack propagation in PMMA with different molecular weight. For PMMA with the average molecular weight of 163, 000, there exists a transition in the appearance of the fracture surface, changing from smooth to coarse-structured rib-like roughness, associated with a transition in crack propagation velocity. He calculates the strain energy release rate at the transition state.

In the present report, the fracture surface morphology is experimentally investigated to try a quantitative approach associated with the crack velocity profiles, the surface roughness and the strain energy release rate, of which details are in what follows.

2. Experiment

2.1. Specimen and tensile test

PMMA (polymethyl methacrylate), Delaglass A of Asahi Chemical Co. Japan make, is used. The specimen size is as shown in Fig. 1, in which a sharp starting notch of initial crack length C_0 is introduced by a fine cutter of Fine Cut type F-31 of Heiwa Technica Japan make. The specimen size is 280mm × 100mm × 1mm or 3mm in thickness as shown in Fig.1.

A tensile tester is an Instron type AG10-TA of Shimadzu Japan make and applies a load at the cross-head speed of 0.5mm/min. ($\dot{\varepsilon}$ =1.20×10^{-5} /s) and 500mm/min. ($\dot{\varepsilon}$ =4.24×10^{-3} /s) at room temperature, as shown in Fig.2. Actually realized strain rate $\dot{\varepsilon}$ is measured by a strain gage.

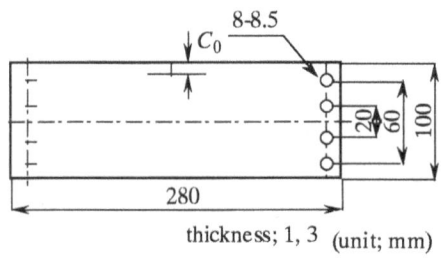

thickness; 1, 3 (unit; mm)

Figure 1. Specimen dimensions.   Figure 2. Tensile testing machine.

2.2. Crack velocity measurement

The conventional velocity gauge method is employed to measure the running crack velocity. As shown in Fig. 3, the velocity gauges consist of a series of conducting wires, duPont's conductive silver coating material of U.S. make, placed at certain intervals perpendicular to the direction of the expected path of crack propagation. Thus the average crack propagation velocity between two adjacent wires can be obtained from the electronic signal produced upon breaking of conducting wires due to the running crack propagation.

2.3. Fracture surface observation and roughness measurement

An optical microscope, BHT-MU of Olympus Japan make, is employed to observe and take photograph of fracture surfaces as shown later. The roughness in fracture surface is measured by a Surftest 402 of Mitsutoyo Co. Japan make and the maximum roughness height R_{max} is also measured as a function of crack length C.

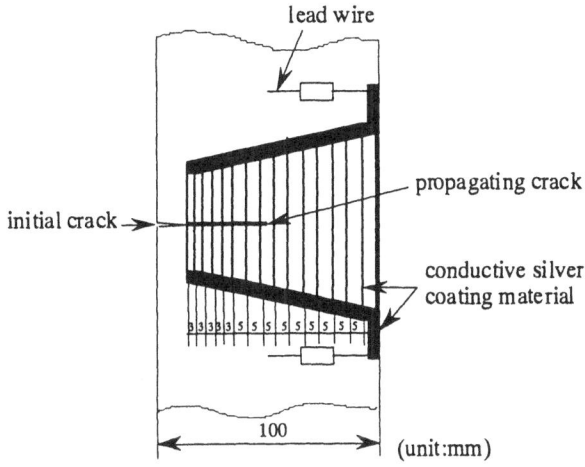

Figure 3. Velocity gauge arrangement.

3. Experimental results and discussions

3.1. Crack propagation velocity profiles

Obtained crack propagation velocity profiles are shown in Figs. 4 (a)-(d) as a function of arbitrary running crack length C. In these figures, note that two initial crack lengths are employed.

Berry [1] proposes the crack propagation velocity profile for a brittle elastic solid like PMMA as

$$v_c = \sqrt{\frac{2\pi E}{k\rho}} \sqrt{1 - \left(\frac{c_0}{c}\right)^2 - 2\left(\frac{\sigma_g}{\sigma_c}\right)\frac{c_0}{c}\left(1 - \frac{c_0}{c}\right)} \tag{1}$$

where v_c = the running crack velocity, c_0 = the initial crack length, c = arbitrary running crack length, σ_g = the Griffith critical stress, σ_c = the observed ultimate stress, k = a numerical constant, ρ = the density, and E = the Young's modulus. Equation (1) has two limiting cases, namely, an upper boundary, putting $\sigma_c = \infty$

$$v_c = \sqrt{\frac{2\pi E}{k\rho}} \sqrt{1 - \left(\frac{c_0}{c}\right)^2} \tag{2}$$

and a lower boundary, putting $\sigma_c = \sigma_g$,

$$v_c = \sqrt{\frac{2\pi E}{k\rho}} \left(1 - \frac{c_0}{c}\right) \tag{3}$$

In Figs. 4 (a)-(d), these boundaries are also shown as the dotted lines for the upper boundary and the solid lines for the lower ones. It is observed the experimental crack propagation velocity profiles are plotted within these boundaries fairly well, confirming to a brittle elastic behavior of PMMA.

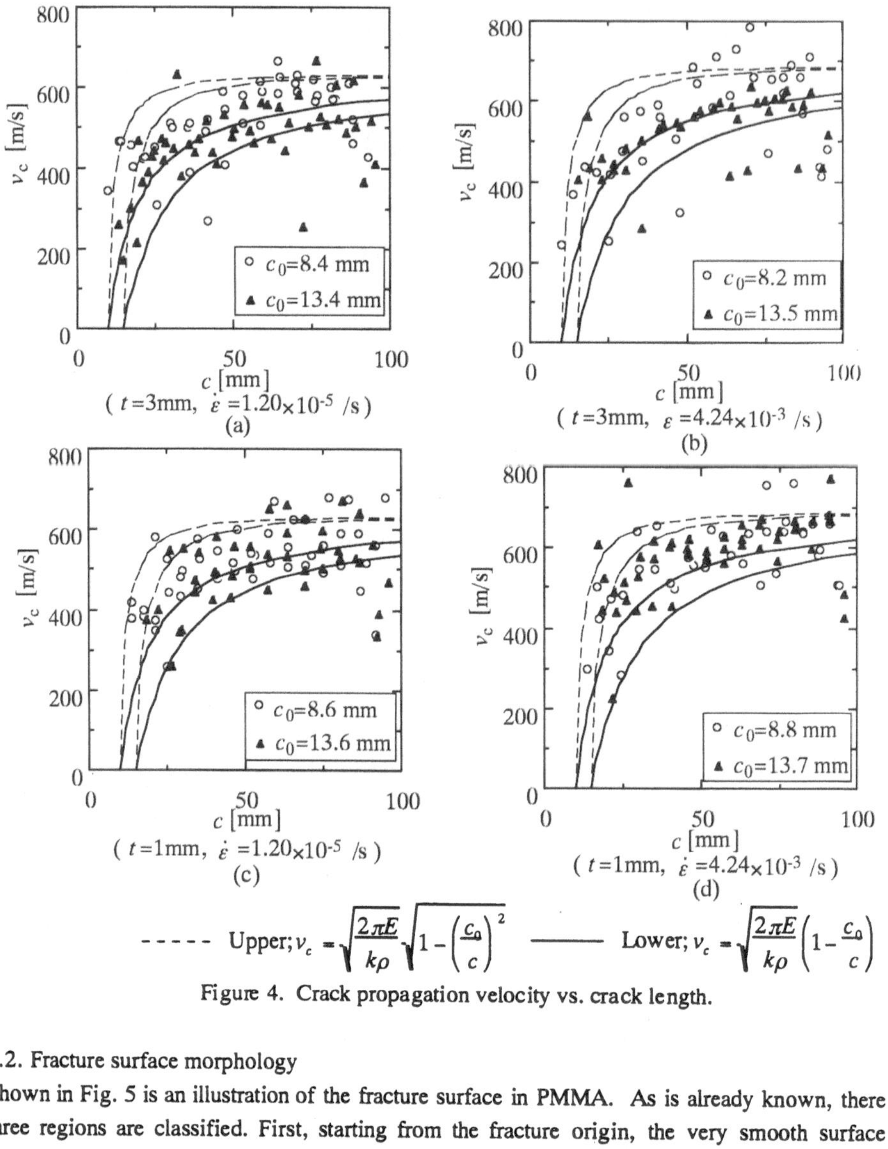

$$\text{Upper}; v_c = \sqrt{\frac{2\pi E}{k\rho}}\sqrt{1-\left(\frac{c_0}{c}\right)^2} \qquad \text{Lower}; v_c = \sqrt{\frac{2\pi E}{k\rho}}\left(1-\frac{c_0}{c}\right)$$

Figure 4. Crack propagation velocity vs. crack length.

3.2. Fracture surface morphology

Shown in Fig. 5 is an illustration of the fracture surface in PMMA. As is already known, there three regions are classified. First, starting from the fracture origin, the very smooth surface referred as the mirror area appears, next the parabolic markings are observed and are referred as the mist area, and finally rib-like surface appearances are observed referring as the hackle area.

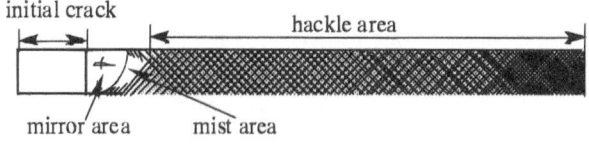

Figure 5. Schematic illustration of the fracture surface.

Figs. 6 and 7 show the fracture morphology examples. In Fig. 6(a) it is observed both the mirror and the mist area are larger compared with those in other figures. Figs. 8 (a)-(d) show the mirror and mist area as a function of the initial crack length. Note that Figs. 8 (a)-(d) correspond to Figs. 4 (a)-(d). It is found both the mirror and the mist area decrease for the longer initial crack length as seen in Fig. 8 (d), while the others are rather in wide scatter. The mirror and the mist area decrease with the increase in the initial crack length only at a higher cross-head speed of $\dot{\varepsilon}$ =4.24×10^{-3} /s, so far as the present test results are concerned.

(a) Initial crack length 8.6mm

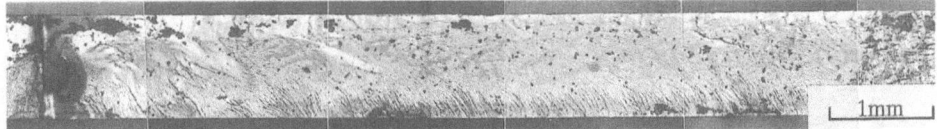

(b) Initial crack length 13.6mm
Figure 6. The transition in the appearance of the fracture surface of PMMA.
(t =1mm, $\dot{\varepsilon}$ =1.20×10^{-5} /s)
(Crack propagation direction from left to right.)

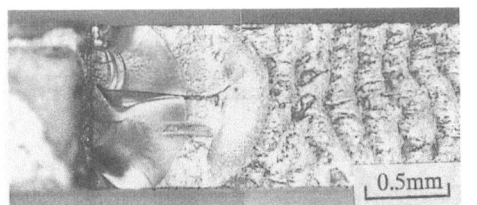 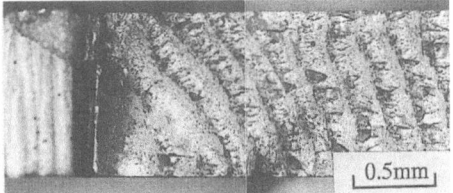

(a) Initial crack length 8.8mm   (b) Initial crack length 13.7mm
Figure 7. The transition in the appearance of the fracture surface of PMMA.
(t=1mm, $\dot{\varepsilon}$ =4.24×10^{-3}1/s)
(Crack propagation direction from left to right.)

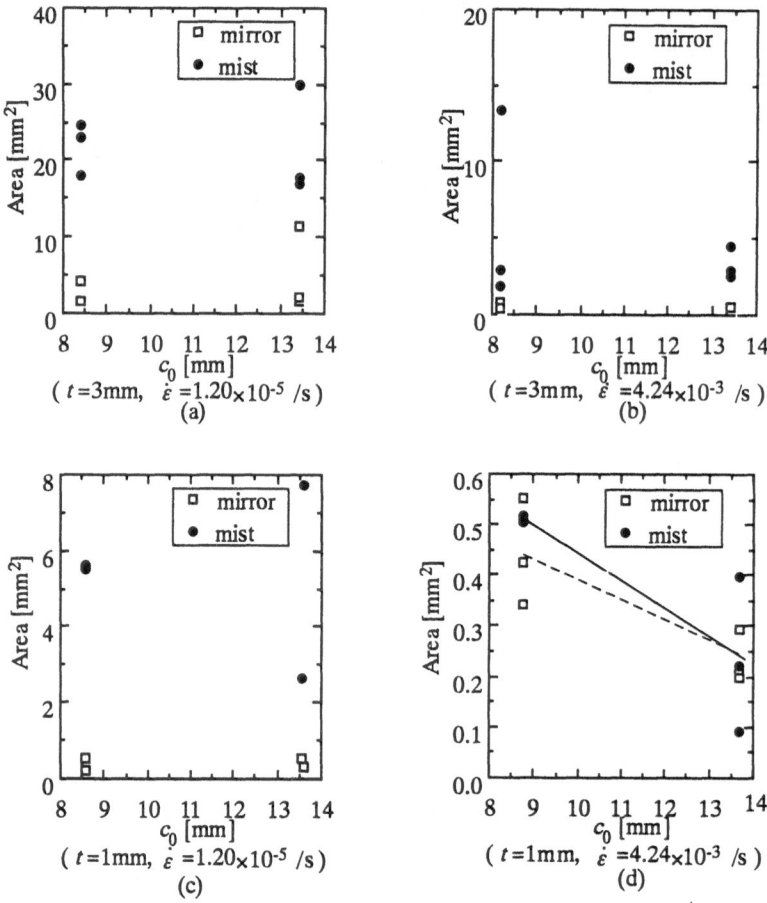

Figure 8. Area vs. initial crack length.

3.3. Roughness

Roughness R_{max} measured is shown in Figs. 9 (a)-(d). Generally speaking, the roughness in the fracture surface increases as a crack advances. Also the roughness in the fracture surface in Fig. 9(b) somewhat increases compared with other cases.

3.4. Strain energy release rate at the morphological transition

Here the strain energy release rate at the morphological transitions is considered after Döll [3].

The strain energy release rate G is given by

$$G = 1.27\pi\frac{\sigma^2 c}{E}\left(1 - v^2\right)$$

(4)

where σ = the fracture stress, c = the crack length at the transition, and v = Poisson's ratio. From 20 PMMA specimens the average strain energy release rates are obtained as shown in Table

1, in which $E = 3.32\text{GPa}$ for 0.5mm/min. ($\dot{\varepsilon} = 1.20 \times 10^{-5}$ /s), $E = 3.91\text{GPa}$ for 500mm/min. ($\dot{\varepsilon} = 4.24 \times 10^{-3}$ /s) and $\nu = 0.35$ for both cross-head speeds are employed.

Table 1 shows the average strain energy release rate G at the transition thus calculated. In Table 1, G_1 = the average strain energy release rate at the transition from the mirror to the mist area, and G_2 = the average strain energy release rate at the transition from the mist to the hackle area. As seen from Table 1, G_1 and G_2 are always greater at lower cross-head speed of 0.5mm/min. ($\dot{\varepsilon} = 1.20 \times 10^{-5}$ /s) than those at 500mm/min. ($\dot{\varepsilon} = 4.24 \times 10^{-3}$ /s). G_2 is always greater than G_1 in any case irrespective of the cross-head speed except for $G_2 = 0.347\text{kJ/m}^2$ at 500mm/min. ($\dot{\varepsilon} = 4.24 \times 10^{-3}$ /s).

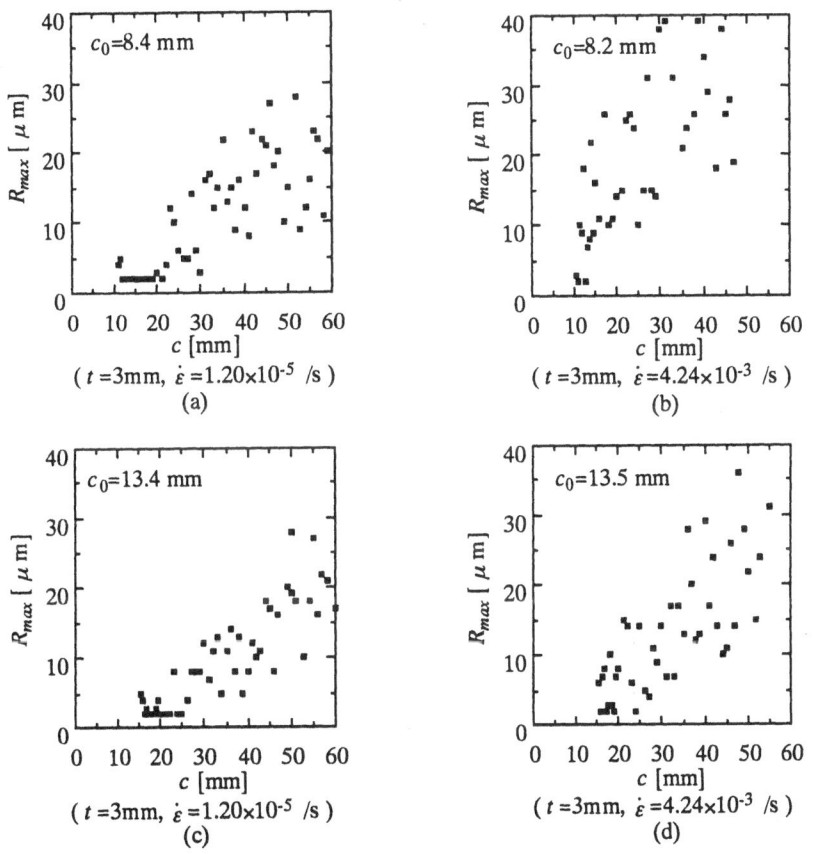

Figure 9. R_{max} vs. crack length.

Döll [3] reports the strain energy release rate from the mist to the hackle are is $0.83 \pm 0.10\text{kJ/m}^2$ at the cross-head speed of 12mm/min. The present experimental values, G_1 and G_2, rather agree with Döll's value in order of magnitude. Thus the change in fracture surface morphology in terms of G_1 and G_2 is correlated with the crack velocity profile.

Table 1. The average strain energy release rate at transition.

	cross-head speed 0.5mm/min. ($\dot{\varepsilon} =1.20\times10^{-5}$ /s)	initial crack length c_0 mm	cross-head speed 500mm/min. ($\dot{\varepsilon} =4.24\times10^{-3}$ /s)	initial crack length c_0 mm	thickness t mm
G_1	0.484kJ/m^2		0.231kJ/m^2		
		8.4		8.2	3
G_2	0.873		0.270		
G_1	0.613		0.361		
		13.4		13.5	3
G_2	0.878		0.380		
G_1	0.313		0.238		
		8.6		8.8	1
G_2	0.501		0.264		
G_1	0.630		0.447		
		13.6		13.7	1
G_2	0.706		0.347		

cf. Döll's: 0.83±0.10kJ/m^2 at 12mm/min.

4. Conclusions

Quantitative approach to investigate morphological aspects during dynamic crack propagation is made. The fracture surface roughness is measured as a function of running crack length, finding ever increasing tendency with the increase in the crack length. Average strain energy release rate is determined both at the mirror to the mist area transition, G_1 , and at the mist to the hackle area transition, G_2 . It is found both G_1 and G_2 are always greater at lower cross-head speed, while G_2 is always greater than G_1 irrespective of the cross-head speed with only one exception.

5. Acknowledgement

The authors appreciate Mr. H. Ehara for his assistance in the present report.

References

[1] J. P. Berry. *Some kinetic considerations of the Griffith criterion for fracture*, J. Mech. Phys. Solids, vol. 8, 1960, pp. 194-206.

[2] J. P. Berry. *Fracture processes in Polymeric Material*s, J. Appl. Phys. Solids, vol. 33, 1961, pp. 1741-1744.

[3] W. Döll. *A molecular weight dependent fracture transition in polymethyl methacrylate*, J. Mater. Sci., vol. 10, 1975, pp. 935-942.

Damage of Glass Subjected to Ultra-Short Pulsed Laser Shock Wave

Xin-zen Li*, Motohiro Nakano*, Yoshiaki Yamauchi *, Keizo Kishida*and Kazuo A. Tanaka**

 *Department of Precision Science and Technology
**Course of Electro Magnetic Engineering
Osaka University, 2-1 Yamada-oka, Suita, Osaka 565,   Japan

Summary

Damage of a material from the microprocess to macro loss of strength was generally analyzed by the damage variables. The statistic average in space had to be taken to use these variables. This does not complied with the fact that fracture is finally caused by only one fatal crack. Based on the combination of NGA(Nucleation, Growth and Accumulation) model and percolation theory, we propose a novel method to study microcrack evolution that does not need any space average. The application in the study of spallation in glass is presented in this paper. The strong pulsed laser has been used to generate a very short shock wave that propagate radically from center in a disk shape specimen. Damage of spallation crack clusters of different stages were   observed in the samples. The microcrack nucleation and growth were included in the computer code to simulate the damage field. The final fracture occurs when the maximum span of microcrack cluster extends to the specimen dimension. The experimental pattern of cracks can be illustrated from two types of simulations using same material parameters in our cylindrical shock wave experiments and uniaxial laser spallation tests for thin plates by other researches. The revealed microcrack percolation system is very complicated. We expect that an elegant fracture criterion be obtained by the analysis of this meso scale microcrack system.

Keywords: soda-lime glass, laser shock wave, spallation, microcrack, percolation

1. INTRODUCTION

Glass is a material that was studied to obtain fracture rule as early as A. Griffith[1]. However, the experimental phenomena in glass were quite divergent. The spallation in thin glass plates was recently studied by T.Resseguier *et al.* by ultra short shock waves generated by laser. [2]   They found a damage zone much thicker than incident wave length near the rear surface.

To understand the principle, fracture mechanics has developed to use damage concept to get insight into the mechanism of final catastrophic failure. It was Currant *et al.* who suggested the importance of the scale range concept in the study of fracture phenomena[3]. They suggested that only the damages great enough in the meso range should be include in the research system and the ones smaller than the meso concept   were organized   as microcrack nucleation. At the same time, they

pulsed laser

SP008

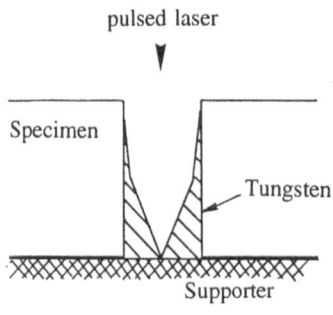
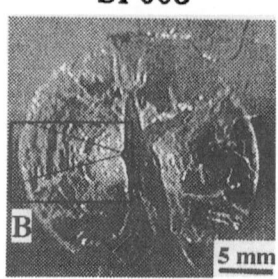

(a) specimen center designed for test using laser   (b) a specimen after experiment

Figure 1: Illustration for cylindrical shock wave experiment

point out the advantages to use ultra-short shock pulses to observe evolution of damage system. The ultra-short shock waves could provide pictures of different stages of damage evolution with clear stress history. The spallation was the   preferred process in their work. However, their theory, known as NGA(Nucleation, Growth and Accumulation) model, appeared it weakness when illustrating the last stage of microcrack accumulation. It takes space average to define damage quantities. In contrast, the practical fracture is always triggered by only one fatal crack.

There are many macro models proposed to be applied in spallation. The one established by Tuler and Butcher was a famous example[4]. A similar model was proposed in the illustration of spallation damage in the work of Resseguier *et al.*[2] mentioned before. They proposed a damage quantity following Tuler and Butcher type namely as volume percentage of voids in the glass. It fracture criterion was supposed to be the point when the void percentage exceed 20%. However, it is hard to image the 20% void percentage in a material as brittle as glass. In a summary, such models have the disadvantages of 1) The physics meaning of the damage cannot be exact corresponded to its name. 2) The growth law of the damage followed a empirical equation. and 3) The critical limit of the damage was artificially appointed.

On the other hand, it has been illustrated that the percolation process is common in all the critical phenomena. Many studies excited by the obvious similarity between percolation system and fracture phenomena[5]. Fracture surfaces were found to have fractal structure[6]. It is a pity that most of the researches using an over simplified artificial system in their simulations. The so called numerical experiment cannot be simply applied to practical applications.

In our opinion, the nucleation and growth in a specific stress field for a single crack can be well defined by the frame work of NGA theory. The microcracks field was also properly understood when the microcracks are small and their distances between each other are large so that their interaction between each other could be omitted. This is the ideal microcrack field[7]. At the later stage, the accumulated microcrack clusters, not individual ones, play the key role. The cluster size distribution and the maximum cluster decide material behavior. It is our start point here to conduct well specified experiment for a relative wide range of strain rate range and specimen shapes. Using relative reliable and simple fundamental models to produce results that could be directly compared with the experiments.

The laser induced shock waves is chosen as loading method for it can provide a nano-second shock wave with enough amplitude. We design a cylindrical shock wave experiment by using disk shape specimen. The specimen with and without a main crack were used to observe the different patterns. The laser shock experiments in thin plate in Ref.[2] provide a complement to our experiment data. The numerical simulation for spallation was reported this time.

2. EXPERIMENTS

Our experiments were performed by Gekko MII in ILE (Institute of Laser Engineering) which has $0.53 \mu m$ wave length and 0.5 nanosecond pulse duration. The tested material was chosen as soda-lime-glass. We used two groups of specimen with or without a main crack in a disk plate with a size of ϕ 20x2mm. An absorption hole of ϕ 1 mm was opened in the center of each glass disk and was coated by a layer of tungsten powder. The pulsed laser was focused to an energy density of 10^{13} W/cm^2 order. The pressure in the hole surface was deduced as high as tens of giga-pascals.

For disks without main cracks, shock induced damage were ring shaped cracks over the samples. We observed large spallation cracks several times at 3-5 mm from the side boundary. One of the disks was suffered complete fragmentation to powders of sub-millimeter size. For disks with main cracks, we observed main crack openings characterized by the crack bifurcation at the original crack tip locations. This was regarded as evidence of very high order loading rate. There were also spallation damages in the lateral regions of the main crack. Fig.1 gives the experiment configuration and a picture of one sample after shock loading.

3. NUMERICAL SIMULATIONS

3.1 Constitutive equation and stress wave propagation.

The Hugoniot data for soda-lime glass could be found in the works of Resseguier et al. for large glass plate and Dremin et al. for glass powder up to 40 GPa[8,9]. A permanent densification model for hydropressure above 3 GPa was suggested by Ref.[8]. Noticing that there was a constant stress gap of 2.4 GPa between the two experiments, it could be regarded as the difference between uniaxial pressure and the hydro-pressure. We further suggest a flow strength of $Y=3.6$ GPa. This may be understood as the shear strength level to trigger SiO_4 tetrahedral doing relative sliding in the surrounded substance.

A finite differential method was used as the main frame of stress wave calculation. Bulk stress P and volume strain ε are obtained by using Gruneisen equation of state(EOS). Partial stresses and partial strains follow Hook's law in elastic zone. When the shear stresses are over flow strength Y, we use a relationship similar to ideal plastic model. The partial stress and strain can be determined considering the deformation constraints. For example, the motion equation in cylindrical case is

$$v \left(\frac{\partial \sigma_r}{\partial r} + \frac{\sigma_r - \sigma_\theta}{r}\right) = \frac{\partial^2 u_r}{\partial t^2} \quad \text{(compression positive)} \qquad (1)$$

Where σ, u, v are stress, displacement and specific volume, respectively. Subscripts r, θ, t refer spaces and time, respectively.

For symmetric condition in our experiments, we have $u_\theta = 0$, hence

$$\varepsilon_r = \frac{\partial u_r}{\partial r} \quad \varepsilon_\theta = \frac{u_r}{r} \quad \varepsilon_z = 0 \text{ (plane strain)} \qquad (2)$$

The expression for σ_r and σ_θ beyond elastic zone can be written as

$$\sigma_r = P + \frac{1}{2G} \frac{(2\varepsilon_r - \varepsilon_\theta)}{3} \qquad \sigma_\theta = P + \frac{1}{2G} \frac{(2\varepsilon_\theta - \varepsilon_r)}{3} \qquad (3)$$

Here G is the shear modules. The results of our calculations are showed in Fig.2. It can be seen that the plane strain condition was always valid for the 2mm thickness cylindrical disks.

3.2 Microcracks field of interaction type

We assume that the microcracks can only be nucleated under elongation stress. The nucleation rate $\dot{N}$ of a specific area can be expressed by a simple model as

$$\frac{d\,N(t)}{d\,t} = \dot{N}(t) = \dot{N}_0 \exp(\frac{\sigma - \sigma_1}{\sigma_2}) \quad \text{for } \sigma > \sigma_c \qquad (4)$$

where N_0, σ_1, σ_2, σ_c are constants. For the stress field mentioned above, the stresses are only functions of variables x (or r) and t. That means the mesh represents a disk shape volume cell with thickness Δx in uniaxial strain, and a cylinder cell with a fan-shaped section. The nucleated microcracks are put into the respect mesh with random location and orientation. The length C of the crack is also a random quantity but limited in a meso scale. This crack length may be related to the diameter of a disk-shape crack in 3-D. Mott model is used to express the growth of microcracks driven by the local stresses[10]. That is

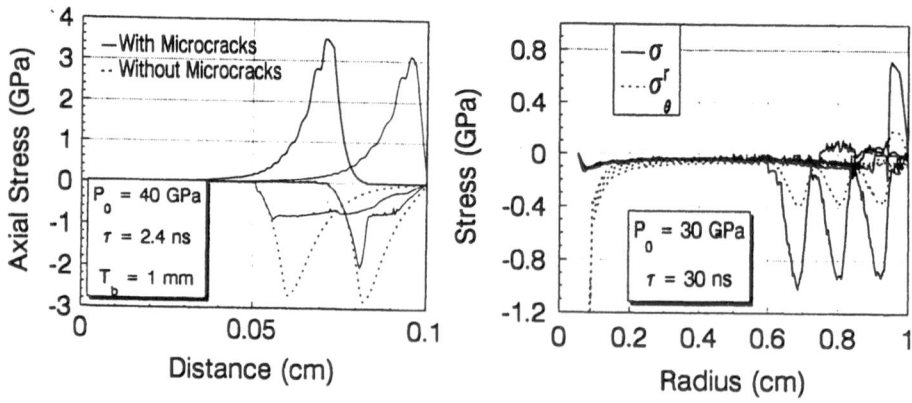

(a) uniaxial strain every 40ns from 0.12 μs   (b) cylindrical wave every 0.2 μs from 1.6 μs

Figure 2: Calculated stress profiles.

$$\frac{dC}{dt} = C_R(1 - \frac{C^*}{C}) \text{ for } C > C^* \tag{5}$$

and
$$C^* = \frac{1}{2\pi}(\frac{K_{Ic}}{S_n})^2 \tag{6}$$

where K_{Ic} is the fracture toughness under plane strain. S_n is the driven force. It can be the maximum principle stress or, as in our present program, the traction perpendicular to the crack plane. C_R is the maximum crack velocity. It is generally regarded as Raleigh wave speed.

When two crack lines cross with each other, they will connect together in the present paper. The conneeted cracks is called crack clusters. The clusters may have no limit of growth velocity.

3.3 Energy equilibrium and stress relaxation in microcrack field.

It is the interaction between the stress field and the microcracks field that plays a key role in the brittle fracture mechanics. One reliable way to count this effects of microcracks to stress field is through energy balance principle. We have energy release rate J that represents the surface energy taken by newly produced crack surfaces. The fracture toughness in Eq.(6) is related with J through relationship

$$J = \frac{(1 - v^2)}{E} K_{Ic}^2 \tag{7}$$

Here E, v are Young's modules and the Poisson's ratio. By calculating the average distance between cracks in a mesh, we can obtain the equivalent number in 3-D space. The total surface energy in the mesh could be evaluated through crack surfaces. The dilatation volume change in the local stress level could by consequently obtained. This volume change is substituted back into EOS to obtain new stress tensor. The comparison of the stress relaxation by microcracks were showed by dot lines in Fig.2(a). The quantity of stress change is controlled by fracture toughness and microcrack number. The material parameters used in the calculations are given in Table.1. To our satisfactory, although the condition are quite different, the fracture toughness values of soda-lime glass obtained by our previous work give out an good prediction for the present experiments[11].

Table 1. Material parameters used in the simulation

specific volume	v	(cc/g.)	0.3809	Raleigh wave speed	C_R	km/s.	3.06
Young's modulus	E	(GPa)	71.3	fracture toughness	K_{Ic}	MPa·m$^{1/2}$	0.7
Possion's ratio	v		0.24	energy release rate	J	J·m^{-2}	7.3

4. PERCOLATION ANALYSIS

4.1 Experiments simulation

For the transient characteristic of the stress field, the produced percolation system is a continuum percolation field of sticks with dynamic density, non-uniform distribution and non-uniform length. The available results of percolation system were too simple compared with the practical one in our study. However, it is possible to monitor the cluster size and structures directly in our method. It also easy to accept that macro fracture criteria is that when the largest cluster span extends the specimen size. Fig.3 shows our calculated damage pictures with comparison with experiments. The damage zone depth could be properly reproduced. However, the calculated cluster size **r** seems smaller than the experimental cracks **R** as showed by Fig.4. **D** and **d** in Fig.4 are average distance in experiment and calculation, respectively. We found that this was due to the crack opening velocity is too small. In another word, this suggested a higher nucleation rate in the vicinity of existing crack tip region under the Raleigh wave speed limitation,.

4.2 Fractal structure of the percolation cluster

The largest cluster in the system is called as percolation cluster. It was found that the percolation cluster has a fractal structure at the stage near percolation critical point. It is worthwhile to investigate the scaling law of the percolation cluster. Fig. 5 gives a logarithm plot of the diameter and total length of cracks in the cluster. A linear relation with slope of 1.93 was suggested. This, called as Fisher exponent, is the dimension number. The present Fisher exponent is too large. This shows that the microcracks practically connected more efficiently. The singularity in stress near crack tips should be accounted to make the result better.

It has to be stressed that the fracture surface dimension is not the same as the cluster dimension obtained here. The fracture trace is the final 'go through path' in the percolation cluster. It is a sub-space of cluster space. The whole set of the space characters may not be performed by its subspaces.

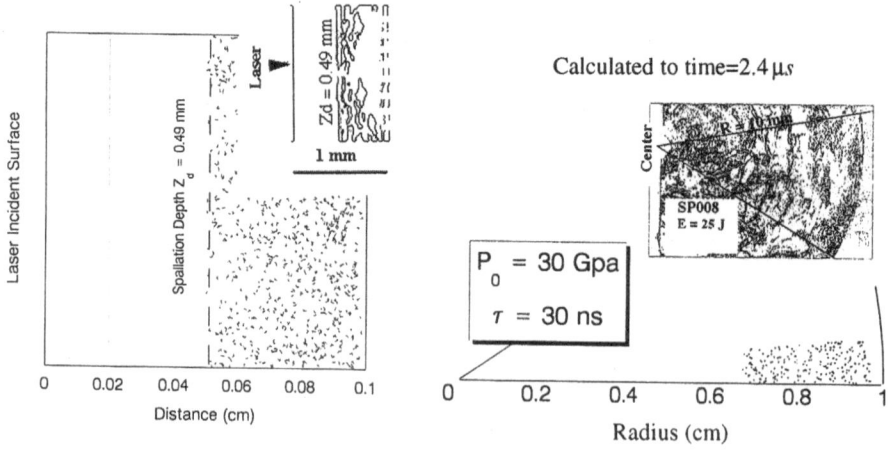

(a) uniaxial strain, upright attached is a
cluster topology of Fig.3 in Ref.[2].

(b) cylinddrical plane strain, upright attached
cluster diagram is from Fig.1(b), domain B.

Figure 3: Crack clusters in experiments and calculations.

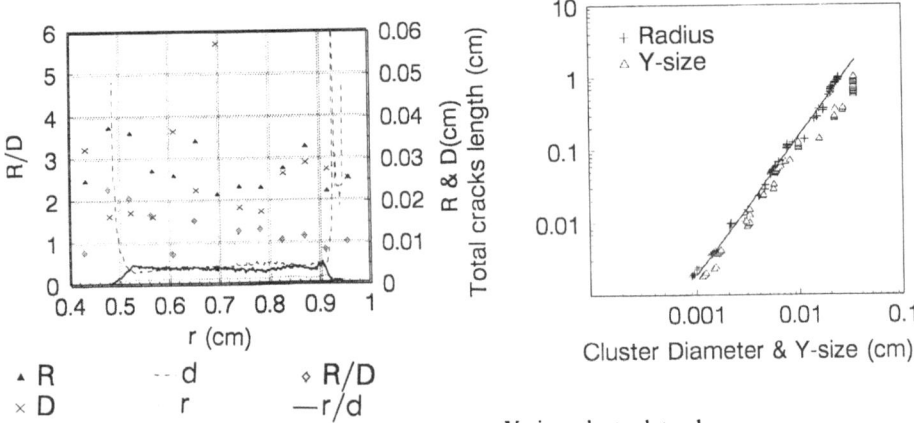

Y-size, cluster lateral span

Figure 4: Cluster size distribution   Figure 5: Logarithm plot percolation cluster of different time.

5. DISCUSSION AND CONCLUSION

It was showed in this paper a new method to obtain quantitatively fracture knowledge . The present study has taken the following advantages:

1) Using strong, ultra-short laser shocks to do fracture experiments to obtain damage pictures at its different stages of evolution.

2) Using transparent glass materials to observe the internal structures of samples. Using computer topology techniques to obtain quantitative damage structure.

3) Utilizing the symmetric of the experimental configuration. The 2-D microcrack fields of experiments could be studied based on 1-D stress calculations.

4) Comparatively wide range of experiments be simulated with unique set of constitutive and fracture parameters. There is relatively no freedom of constants arbitrariness.

This work makes us realized some first hand knowledge of the fracture process. We found that it is very important in the balance between the nucleation rate and crack tip growth velocity. This balance is mainly controlled by the loading rate. The lower loading rate is corresponding to only several cracks growth to large size; Under very high loading rate, the picture is many microcracks in small size. It is the connected crack clusters that decide the failure behavior. The failure of the samples can be reached in a very short time with a relative large damage zone. This conclusion is similar to that of the ductile fracture case[12].

In lower loading rate process, the key problem is the behavior of single crack tip. The traditional method of singularity factor may be the main parameter in the process. In contrast, the percolation system will be formed in the very high loading rate. The performance is obviously different from that predicted by averaged damages. This could be called as a kind of self organization of the initially random system. The study of high loading rate phenomena is benefit to examining many critical parameters.

In a whole speaking, the meso scale microcracks field has its own principle of motion. It is a very complicated dynamic system that is relatively less understood by human knowledge. Our present work is only a method investigation. The method can do the micro structural principle study directly and can do comparison with experiments simultaneously. Many important messages are expected to be discovered later by using this method.

The present work did not introduce a 2-D dependent stress field. We cannot consider the crack tip singularity and describe the stress in a higher precision order yet. The experiments need to be completed mainly on stress measurement . Simulations for specimen with a main crack is still on the way.

We'd like to express our thanks to Prof. M.Boustie and Dr. H. Yoneda for providing their research data. Dr. H.Shiraga and Ms. K. Watanabe are sincerely acknowledged for their technical assistants.

References

[1] A.A. Griffith, The phenomenon of rupture and flow in solids, *Phil. Trans. Roy. Soc.*, A221, 1920, pp. 163-168.

[2] T.de Ressegiur, F.Cottet, Experimental and numerical study of laser induced spallation in glass, *J. Appl. Phys.* Vol.77, 1995, pp. 3756-3761.

[3] D.R.Curran, L.Seaman, D.A.Shockey, Dynamic failure of solids, *Physics Report*, Vol.147, 1987, pp.253-388.

[4] F.R.Tuler, B.M.Butcher, A criterion for the time dependence of dynamic fracture, *Int. J. Fract.* Vol.4, 1968, pp.431-437.

[5] P.M.Duxbuxy, P.L.Leath, P.D.Beale, Breakdown properties of quenched random systems: the random-fuse network, *Phys. Rev.B*, Vol.36, 1987, pp.367-380.

[6]C.P.Cherepanov, A.S.Balankin, V.S.Ivanova, Fractal fracture mechanics-a review, *Engng. Fract. Mech.* Vol.51, 1995, pp. 997-1033.

[7] F.J.Ke, Y.L.Bai, M.F.Xia, Characteristics of ideal microcrack system evolution, *Chinese Science*, No. 6, 1990, pp.621-631.

[8]T.de Resseguier, F.Cottet, A.Migault, A 1-D model for glass dynamic behavior, *J. DYMAT*, Vol.1, 1994, pp.31-46.

[9]A.N.Dremin, C.A.Adadurov, The behavior of glass under dynamic loading, *Soviet Physics-Solid State*, Vol.6, 1964, pp1379-1383.

[10]N.F.Mott, British fracture in mild-steel plates-II, *Engineering*, Jan. 1948, pp.16-18.

[11]M. Nakano, K. Kishida, Y.Yamauchi, Y. Sogabe, Dynamic fracture initiation in brittle materials under combined mode I/II loading, *J. Physique IV*, Vol.4, C8,1994, pp. 695-700.

[12]D.L.Tonks, Percolation wave propagation, and void linkup effects in ductile fracture, *J. Physique IV*, Vol.4, C8,1994, pp. 665-670.

Dynamic Interlaminar Fracture Toughness of CFRP Composite Laminates

Tomoaki Kurokawa[1] , Takayuki Kusaka[2] , Taichi Shimazaki[3] , Yoshiaki Yamauchi[4]
and Takanori Kawashima[1]

1   Dept. of Aeronautics and Astronautics, Kyoto University, Kyoto, 606-01 Japan
2   Hyogo Prefectural Inst. of Industrial Research, Kobe, 654 Japan
3   NKK Co., 2-1 Suehiro, Tsurumi-ku, Yokohama, 230 Japan
4   Course of Prec. Engng, Osaka University, Suita, 565 Japan

Summary

New experimental techniques were devised for determining dynamic mode I, moode II and mixed mode interlaminar fracture toughness of CFRP laminates. Up to the impact velocity of 10-20m/s, rate effects on interlaminar fracture toughness of each mode were investigated and compared with the quasi static results. Test results show that critical energy release rate decreases with rate in mode II fracture and increases with rate in mixed mode, however difference must be still investigated from the view point of the difference of mode and also the difference of crack initiation or propagation.

Key words: Composite materials, interlaminar fracture toughness, impact, high strain rate

1   Introduction

Fiber reinforced composite laminates have been widely used to many industries because of its high strength-to-weight ratio, however they are sometimes pointed out to be rather weak for impact loading especially in their interlaminar strength. Impact strength of laminated composite materials have been investigated by a lot of researchers. Many were carried out by the method of Charpy test, penetration or damage test using dynatup or other firing apparatus. Since the stress states of specimens in these dynamic tests are generally very complicated, it is difficult to determine material strength very clearly. From this point of view, the test method like as SHPB is useful to realise a simple stress state in the specimen. Kawata[1] and Harding[2] were pioneers in studies to investigate uniaxial tensile strength, compressive strength and elastic modulus of many kinds of fiber reinforced composite laminates under impact loading using Hopkinson bar technique. Their results show that the impact strength are rather higher than the static strength. Concerning with the interlaminar fracture toughness under impact loading, only limited data have been reported. The rate effect was investigated for mode I by Aliyu[3], Smiley[4], Mall[5], Yaniv[6], Karger-Kocsis[7] and for mode II by Smiley[8], Friedrich[9], Maikuma[10]. However, their results are limited to comparatively low velocity up to 2-3m/s, and they do not show same tendency. The objective of this study is to investigate dynamic interlaminar fracture toughness of laminated composites for mode I, mode II and mixed mode under the impact velocity up to 20m/s and to compare with their static strength.

2   Mode II interlaminar fracture toughness

2.1   Test method

The standard quasi static test method of this type is ENF test, where an End Notched Flexural specimen is used for three point bend test as shown in fig.1. The load increases linearly with the deflection and it drops when the crack starts to extend(fig.2). Therefore it is easy to identify

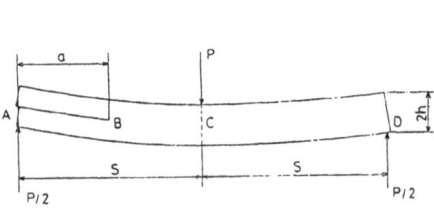

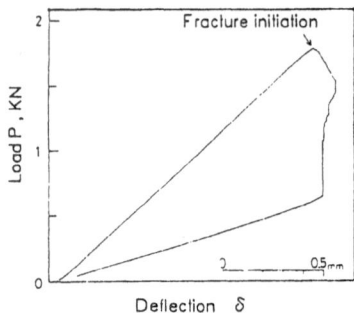

Figure 1: End Notched Flexure test

Figure 2: Load–displacement diagram of static ENF test

the critical load P_c. Then the critical energy release rate G_{IIc} can be determined by P_c and the flexural compliance of the specimen using the following fomulas,

$$G_{II}^{ST} = \frac{9a^2 P_c^2}{16 E_x w^2 h^3} \left\{ 1 + 0.2 \frac{E_x}{G_{xz}} \left(\frac{h}{a} \right)^2 \right\}$$ (1)

or

$$G_{II}^{ST} = \frac{9a^2 P_c^2}{16 E_x w^2 h^3} \left\{ 1 + 0.2 \frac{\alpha}{3} \sqrt{\frac{E_x}{G_{xz}}} \left(\frac{h}{a} \right) + \frac{\alpha^2}{30} \frac{E_x}{G_{xz}} \left(\frac{h}{a} \right)^2 \right\}$$ (2)

, where w, h, a, E_x, G_{xz} are the width, half of the thickness, crack length, Young's modulus and shear modulus of the specimen, respectively. Equation (1) was presented by Carlsson[12] using beam theory with shear deformation, and equation (2) by Whitney[13] taking account of the deformation due to singular stress field nearby crack tip expressed by the constant α, which is generally determined by finite element method.

The strain rate effects on interlaminar fracture toughness have been investigated by means of changing the rate of loading of ENF test[8]–[10]. As mentioned previously, these experimental results are limited to comparatively low velocity range because impact loading induces vibration of the specimen and makes it difficult to determine the critical energy release rate. The authors proposed the technique of combining ENF three point bend test and SHPB method with buffer material in order to suppress the flexural vibration of the specimen[11].

Schematic apparatus for dynamic ENF test is shown in fig.3. A small piece of soft metal like as lead or tin is attached on the impact end of the input bar as a buffer. Without buffer, incident wave generated has sharp rise time of about $20\mu sec$ as shown as the curve b in fig.5 and it induces vibration in the transmitted wave as shown in fig.4. The incident wave for impact with buffer of 5mm thick lead piece is shown as the curve a in fig.5, which has the rise time of about $70\mu sec$. The time histories of these incident waves can be well fitted by the following formula with rise time T, $v = v_0 \{1 - \exp(-t^2/T^2)\}$, presented as solid curves in fig.5.

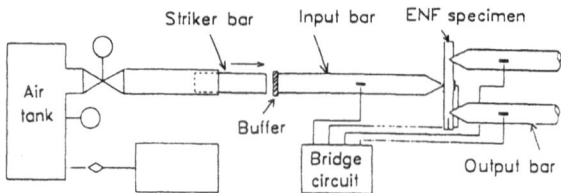

Figure 3: Dyanamic ENF experiment setup

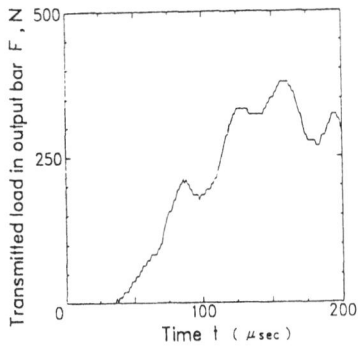

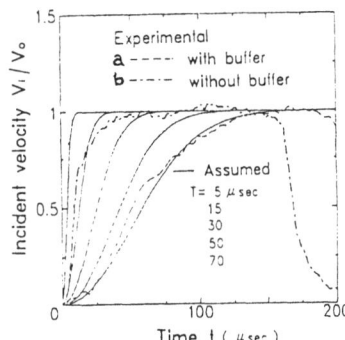

Figure 4: Transmitted wave in dynamic ENF test (impact without buffer)

Figure 5: Incident wave form with various rise times

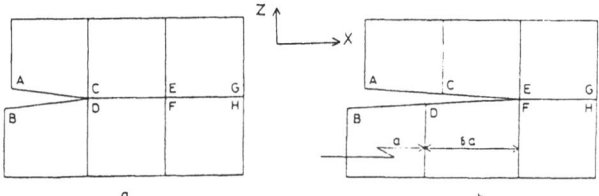

Figure 6: FEM grid model for crack closure method

Next, the time variation of the energy release rate of stress field near crack tip must be evaluated for those incident waves. The mode II energy release rate during crack extension from a to $a + \delta a$ can be evaluated by crack closure method using finite element procedure(fig.6). The formula is given by

$$G_{II}^{CC} = \frac{1}{2w\delta a}T_C^a(u_C^{a+\delta a} - u_D^{a+\delta a}) \tag{3}$$

, where T and u are the x components of nodal force and displacement, respectively. Subscripts represent nodal points and superscripts represent the position of crack tip. Scince the formula (3) requires the quantities at differnt two stages i.e. (a) and (b) in fig.6, double computation is

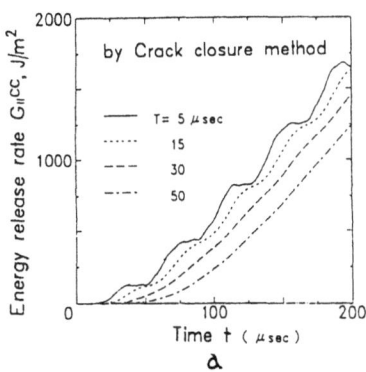

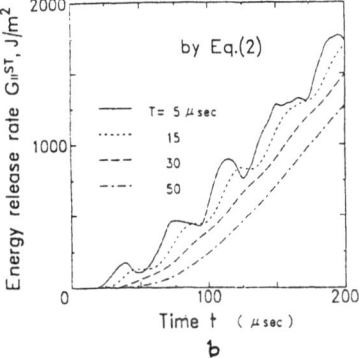

Figure 7: Time variation of energy release rate G_{II} for various incident waves

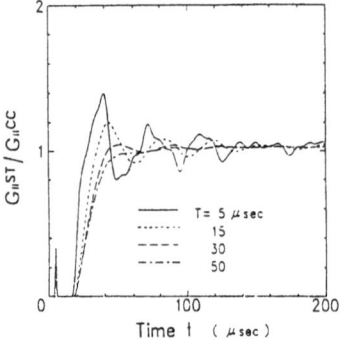

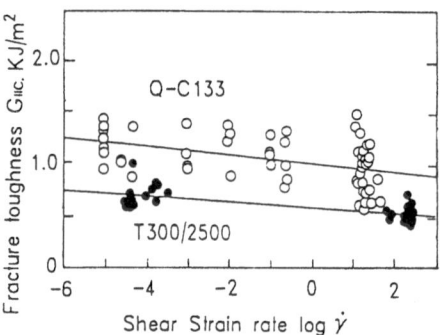

Figure 8: Time variation of the ratio between the energy release rates obtained by crack closure method and by the static formura (2)

Figure 9: Strain rate effect on mode II fracture toughness G_{IIC}

needed. The procedure can be simplified by using eq.(4) instead of (3),

$$G_{II}^{CC} = \frac{1}{2w\delta a} T_E^{a+\delta a}(u_C^{a+\delta a} - u_D^{a+\delta a}) \tag{4}$$

In order to know the time variation of dynamic energy release rate, the computation by eq.(4) must be repeated sequencially with time using the incident wave history. The results for various incident waves represented by solid curves in fig.5 are shown in fig.7(a). On the other hand, the energy release rates G_{II}^{ST}, which are obtained by applying the transmitted force histories to static eq.(2), are shown in fig.7(b). The oscilations of the curves are larger in G_{II}^{ST} than in G_{II}^{CC}, however there can be hardly seen oscillation for the incident wave with rise time T larger than 50 μsec. The ratio of G_{II}^{ST} to G_{II}^{CC} is represented in fig.8. It can be found that the both are almost equal, namely static eq. (1) or (2) can be also applied to dynamic problem.

2.2 Experimental result

Dynamic ENF tests for two kinds of CFRP materials shown in table 1 were carried out. The results are represented in fig.9. The abscissa represents the shearing strain rate given by

$$\dot{\gamma}(t) = \frac{3\dot{P}(t)}{8G_{zz}wh} \tag{5}$$

The tests were conducted using Instron type machine for the low velocity range of $\dot{\gamma} < 1\,\text{s}^{-1}$ and SHPB apparatus for the high velocity range of $\dot{\gamma} > 10\,\text{s}^{-1}$. Both materials show the tendency that critical energy release rate decreases with strain rate.

Table 1: Dimensions and mechanical properties of materials

	Dimensions(mm)			E_x	Tensile strength	v_f
	$2s$	$2h$	a	(GPa)	(MPa)	%
Q–C133	60	2.6	15	140	67	54
T300/2500	30	5.1	7.5	120	50	56

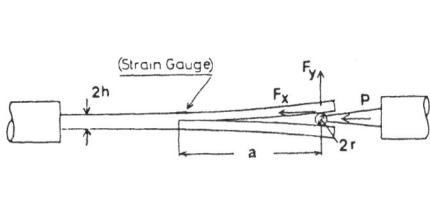

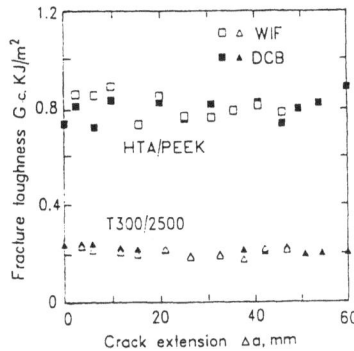

Figure 10: WIF experiment setup

Figure 11: Comparison between G_{IC} values obtained by WIF and DCB method

3 Mode I interlaminar fracture toughness

3.1 Wedge insert fracture(WIF) method

The static standard test method for mode I interlaminar fracture toughness of laminated composite material is Double Cantilever Beam(DCB) test. It is not necessarily easy to apply DCB specimen to dynamic test like as SHPB method. One idea is Wedge Loading Compact tension(WLCT) technique which was proposed by Klepaczko[12] to obtain dynamic mode I fracture toughness of metalic material. The authors devised Wedge Insert Fracture(WIF) method as a variation of WLCT technique. The schematic is shown in fig.10. A small cylinder of radius r is inserted into the crack in the mid-plane of laminated specimen of the thickness $2h$ and width b. The x,y component of the contact force between the wedge and the specimen are expressd using beam theory by

$$F_x = \frac{P}{2} = \frac{9E_x I r^2}{2A^4}\left(1 + \frac{2\mu}{3r}A\right) \qquad , \qquad F_y = \frac{3E_x I r}{A^3} \tag{6}$$

where E_x I and μ is bending rigidity of the halh thickness beam and the coefficient of the friction, respectively. And A is modified crack length which is equal to $a + \epsilon$, where a is distance between the wedge and crack tip and the constant ϵ is pre-determined by the specimen compliance in DCB test. Using eq.(6), the energy release rate G_I is reduced as

$$G_I = \frac{9E_x I r^2}{wA^4} \tag{7}$$

or

$$G_I = \frac{P}{w}\left(1 + \frac{2\mu}{3r}A\right)^{-1} \tag{8}$$

Since the eq.(8) includes the unknown coefficient μ, it cannot give an accurate value of G_I. Fig.11 shows the comparison between G_I values estimated by DCB test and WIF(eq.(7)) method for two kinds of laminated composite materials. DCB method and WIF method was repeated alternately for one specimen during 60mm crack extension. Both estimations show a good agreement.

3.2 Dynamic WIF experiment

Combining WIF method with SHPB method, dynamic mode I interlaminar fracture test was conducted. The movement of the wedge, which is connected with the end of the input bar, can

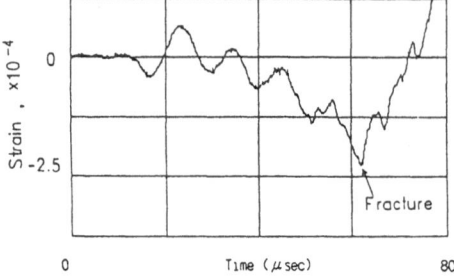

Table 2: Comparison of critical crack length at crack initiation between by static and dynamic WIF test (T300/2500)

	Critical crack length at crack initiation a_c(mm)
Static	33-36
Dynamic	34-36

Figure 12: Strain history measured on the dynamic WIF specimen (v=10.2m/s)

be known using the strain history obtained through the strain gage cemented on the input bar. Another strain gage is also cemented on the surface of the specimen nearby the crck tip. Fig.12 shows the strain history of the specimen obtained for the incident wave of impact velocity v=10.2m/s. The specimen material is CF/Epoxy(T300/2500) and the stacking sequence is $[0]_{20}$. It is found that the delamination occured at the point where the strain suddenly rose to extension side in fig.12. Combining this with the incident wave history, the critical crack length i.e. the critical energy release rate at the crack initiation can be known. Table 2 shows the comparison of the critical crack length between static and dynamic WIF test of T300/2500. Since the dynamic WIF test has been just started and only limited data have been obtained, the difference between static and dynamic G_{IC} has not been made up clear at the present stage.

4 Mixed mode interlaminar fracture toughness

4.1 Test method

The technique developed in this study for the test of this mode is oblique tension test by transverse impact.[13] The test material used is CF/Epoxy(Toray T300/2500) unidirectional eight layered laminates. The specimen is 1.1mm in thickness, 10mm in width and 150mm in length. The artificial crack was prepared inserting teflon film to the mid-plane at the end of the specimen in fanufacturing process. The specimen was fixed on the bed, and the upper half of the cracked part of the specimen was connected with a thin aluminum strip of 0.2mm thickness, 10mm width and 150mm lenth, whose end was fixed. The aluminum strip was aligned in the line of the specimen and slightly pre-tensioned. The hemi-spherical or wedge head projectile was impacted transversely to the aluminum strip near at the connecting point of the aluminum strip and the specimen (fig.13). The delamination induced by this impact is supposed to be mixed mode fracture. Fig.14 shows an example of propagation of delamination of the specimen and transverse wave along the aluminum strip. The picture was taken at four instances of 280,430,530 and 730 μs after the beginning of the strike by flashing xenon lamp.

The delamination energy can be evaluated by the following procedure. The work W done by the external force T acting to the specimen, i.e. the tensile stress of aluminum strip is given by

$$W = \int (T\sin\theta)v_y dt - \int (T\cos\theta)v_x dt \qquad (9)$$

where v_x, v_y and θ is x,y components of the velocity of the connecting point and inclination angle of the aluminum strip (fig.13). The tensile stress of the aluminum strip σ_Y is estimated as σ_Y=103Mpa using the transverse wave velocity $C_w = \sqrt{\sigma_Y/\rho}$=196m measured from fig.14.

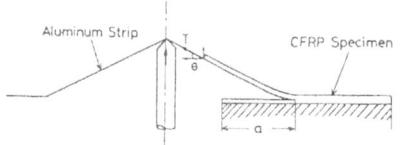

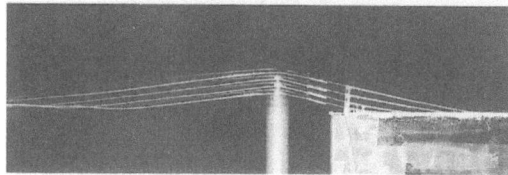

Figure 13: Schematic setup for impact oblique tension test

Figure 14: Propagation of delamination in CFRP $(v = 19.9\text{m/s}, \Delta t = 150\mu s)$

On the other hand elastic energy W_{el} in the specimen is given by

$$W_{el} = \frac{E_x I}{2} \int \left(\frac{\partial^2}{\partial x^2} w \right) dx + \frac{1}{2E_x A_S} \int T^2 dx \tag{10}$$

where w, E_x, A_S, I is deflection, longitudinal Young's modulus, cross sectional area and the moment inertia of the cross section, respectively. The kinetic energy of the specimen is calculated by

$$K = \frac{\rho_c A_S}{2} \int v_y^2 dx \tag{11}$$

where ρ_c is the density of the specimen and the contribution of v_x is neglected. In total, the delamination energy W_{del} is evaluated by

$$W_{del} = W - W_{el} - K \tag{12}$$

Therefore energy release rate G_c can be obtained as the derivative of W_{del} with respect to delamination area A_{del}

$$G_C = \frac{dW_{del}}{dA_{del}} \tag{13}$$

4.2 Experimental result

The dynamic experiment was carried out for two impact velocity ranges of 20m/s and 1m/s. Using Instron type machine, the quasi-static test was conducted, where the kinetic energy K in eq.(8) was neglected. Fig.15 shows the relation between W_{del} and A_{del} for the impact of velocity $V=19.9$m/s. The inclination of this line represents the energy release rate G_c of eq.(13). G_c values obtained by these procedures for different velocity ranges are illustrated in fig.16. In contrast with fig.9, the result of fig.16 shows the different tendency that the critical energy release rate increases with the loading rate. The difference of this tendency might be not due to the difference of fracture mode but due to the difference between crack initiation and crack propagation.

5 Conclusions

Based on the experimental results obtained, the following conclusions may be drawn:
(1) G_{IIC} obtained by dynamic ENF method shows decreasing tendemcy with loading rate.
(2) Mixed mode interlaminar fracture toughness G_C by dynamic oblique tension test shows increasing tendency with crack propagation velocity.
(3) This different tendency might be due to the difference between crack initiation and crack propagation.
(4) The differnce between dynamic and static G_{IC} has not been made up clear at the moment.

224

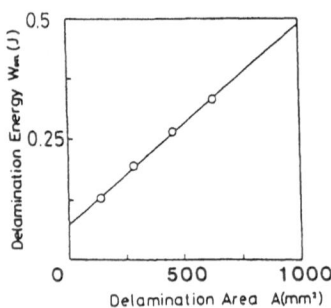

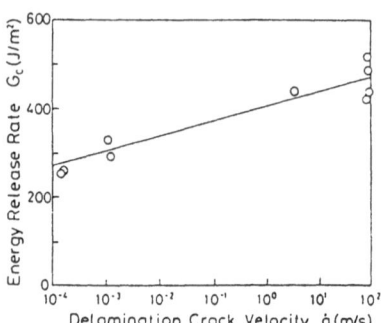

Figure 15: Delamination energy vs delamination area of T300/2500 in mixed mode
impact delamination test ($v = 19.9$m/s)

Figure 16: Rate effect on mixed mode interlaminar fracture toughness G_C

References

[1] K.Kawata, S.Hashimoto and N.Takeda, Mechanical Behaviorurs in High Velocity Tension
of Composites, *Proceedings of ICCM IV*, 1982, pp829-836

[2] J.Harding and L.M.Welsh ,*Journal of Materials Science*, 1983, Vol.18, pp1810-1826.

[3] A.A.Aliyu and I.M.Daniel, Effects of Strain Rate on Delamination Fracture Toughness of
Graphite/Epoxy, *ASTM STP*876, 1985, pp336-348.

[4] A.J.Smiley and R.B.Pipes, Rate Effects on Mode I Interlaminar Fracture Toughness in
Composite Materials, *Journal of Composite Materials*, Vol.21, 1987, pp670-687.

[5] S.Mall, G.E.Law and M.Katouzian, Loading Rate Effects on Interlaminar Fracture Toughness of a Thermoplastic Composite, *ibid.* , Vol.21,1 987. pp569-579.

[6] G.Yaniv and I.M.Daniel, Height-tapered Double Cantilever Beam Specimen for Study of
Rate Effects on Fracture Toughness of Composites, *ASTM STP*972, 1988, pp241-258.·

[7] J.Karger-Kocsis and K.Friedrich, Fracture Behavior of Injection-Molded Short and Long
Glass Fiber-Polymide 6.6 Composite, *Composites Science and Technology*, Vol.32, 1988,
pp293-325.

[8] A.J.Smiley and R.B.Pipes, Rate Sensitivity of Mode II Interlaminar Fracture Toughness
in Graphite/Epoxy and Graphite/PEEK Composite Materials, *Composite Science and
Technology*, Vol.29, 1987, pp1-15.

[9] K.Friedrich, R.Walter, L.A.Carlsson, A.J.Smiley and J.W.Gillespie,Jr. , Mechanisms for
Rate Effects on Interlaminar Fracture Toughness of Carbon/Epoxy and Carbon/Peek
Composites, *Journal of Materials Science*, Vol.24, 1989, pp3387-3398.

[10] H.Maikuma, J.W.Gillespie,Jr. and D.J.Wilkins, Mode II Interlaminar Fracture of the
Center Notch Flexural Specimen under Impact Loading, *Journal of Composite Materials*,
Vol.24, 1990, pp124-149.

[11] T.Kusaka, Y.Yamauchi and T.Kurokawa, Effects of Strain Rate on Mode II Interlaminar
Fracture Toughness in Carbon-Fibre/Epoxy Laminated Composites, *Journal de Physique
(Colloque C8)*, Vol.4, 1994, pp.C8-671-676.

[12] J.R.Klepaczko, Application of the Split Hopkinson Pressure Bar to Fracture Dynamics,
Mechanical Properties at High Rate of Strain, Ed. J.Harding, The Institute of Physics,
1979, pp201-208.

[13] T.Kurokawa, Y.Yamauchi and T.Shimazaki, Propagation of Delamination and Impact
Strength of CFRP, *Proceedings of The JSASS/JSME Structures Conference*, Vol.33, 1991,
pp190-193. *(in Japanese)*

The Problems of High Rate Loading: The New Criterion of Fracture, The Erosion, The Asymmetric Impact Loading

Nikita F. Morozov[1] and Yuri V. Petrov[2]

1 Math. & Mech. Faculty, St.-Petersburg State University, Bibliotechnaya pl., 2
  Petrodvorets, St.-Petersburg, 198904, RUSSIA
2 Strength & Fracture Mech. Group, IPTRAN, Russian Academy of Sciences
  12 Linia, 13, St.-Petersburg, 199178, RUSSIA

Summary

The new incubation time criterion in connection with the problem of modeling of some of the specific effects of dynamic fracture is discussed. An approach based on a system of fixed material constants describing macro strength properties of the material is considered. The corresponding approach allows one to manage without the a priori given rate dependencies of dynamic strength and fracture toughness. The problem of erosion and the problem of crack growth initiation due to asymmetric loading on the basis of the incubation time conception are analyzed.

Keywords: Incubation time, impact loading, crack dynamics, fracture criterion, solid particle erosion

1 Incubation time criterion

The progress into experimental mechanics that took place during the last years has led to the understanding of the fact that dynamic fracture in brittle solids is remarkable for its specific nature. The corresponding experiments discovered some principal effects that have no interpretation within the framework of conventional models of brittle failure. Among them are the dynamic branch of the time dependence of strength in spalling (Zlatin et al. [1], 1975; Nikiforofsky and Shemjakin [2], 1979); the dependence of critical stress intensity factor of crack growth initiation on loading history (Ravi-Chandar and Knauss [3], 1984; Kalthoff [4], 1986; etc.); the behavior of the short pulse load threshold amplitudes causing the failure at the crack tip (Kalthoff and Shockey [5], 1977; Shockey et al. [6], 1986); etc.

The experiments have shown that the main contradictions of the traditional models reveal themselves when failure occurs during the short time intervals after the start of the loading process. Morozov and Petrov [7] (1990) proposed an approach to the analysis of brittle failure in dynamic based on the incubation time criterion:

$$\frac{1}{\tau}\int_{t-\tau}^{t}\frac{1}{d}\int_{0}^{d}\sigma(r,\theta,t')drdt' \leq \sigma_{c} \qquad (1)$$

Here d and τ are material structure size and material structure time of failure, σ_c is static strength of the material, $\sigma(r,\theta,\tau)$ is the tension stress at the crack tip $(r=0)$. The material structure size d is to be determined in accordance with the data of quasi-static tests of crack containing specimens and in the case of plane deformation state it may be expressed by the simple formula:

$$d = \frac{2}{\pi} \frac{K_{Ic}^2}{\sigma_c^2} \qquad (2)$$

The material structure time τ is responsible for the dynamic peculiarities of the macro fracture process and for each material it should be found from experiments. In accordance with this approach σ_c, K_{Ic} and τ make up the system of fixed material constants describing macro strength properties of the material. Petrov [8] (1991) has shown that the criterion (1) reflects the discrete nature of dynamic fracture of brittle solids. In the special case of slow deformation the quasi-statical criterion by Neiber-Novozhilov (Neiber [9], 1937; Novozhilov [10], 1969) from (1) comes out.

In the case of intact materials the criterion (1) reduces to the form:

$$\frac{1}{\tau} \int_{t-\tau}^{t} \sigma(t') dt' \le \sigma_c \qquad (3)$$

The analysis of the particular problems of dynamic fracture mechanics is associated with the appropriate choice of the parameter τ. Here we shall consider two basic opportunities:

a) The incubation time τ is defined through the material structure size of fracture:

$$\tau = \frac{d}{c} = \frac{d\sqrt{\rho}}{k}, \qquad (4)$$

where c is the maximum wave velocity, ρ is the density of continuum, k is the constant depending on the deformation material properties. According to this definition the incubation time has a physical meaning of the minimum time period required for the interaction between two neighboring material structure cells.

The incubation-time criterion with the parameter τ selected through the formula (4) allows one to describe effectively the temporal dependence of strength and the fracture zone geometry in conditions of spalling (Morozov et al. [11], 1990; Petrov and Utkin [12], 1992). Thus, the definition (4) provides good corroboration between incubation time criterion and well-known experiments in the case of "defectless" materials.

b) The incubation time τ does not directly depend on the material structure size of failure. This takes place when a problem of initiation of the macro crack growth is considered. Nucleation, growth and coalescence of micro defects in the special process zone region at the crack tip precedes the growth of the macro crack. These processes are accompanied by a local stress relaxation and do change the effective material properties. The incubation time τ is to be considered as the principal integral characteristic of the processes in the corresponding process zone region. Petrov and Morozov [13] (1994) proved that in the case of macro cracks the material structure time τ can be interpreted as an incubation time in well-known minimum time criterion proposed and explored by

Kalthoff and Shockey [5] (1977), Homma et al. [14] (1983), and Shockey et al. [6] (1986):

$$\tau = t_{inc} \tag{5}$$

The aforementioned dependence of the fracture toughness on loading history (Ravi-Chandar and Knauss [3], 1984) and the specific behavior of the short loading pulse threshold amplitudes (Kalthoff and Shockey [5], 1977) can be explained and effectively analyzed by means of the incubation time criterion under the condition (5). The corresponding results were obtained by Petrov and Morozov [13] (1994).

2 Incubation time and the problem of erosion

The solid particle impact velocity corresponding to the beginning of target material loss during the steady state erosion can be considered as a critical or threshold velocity (Urbanovich et al. [15], 1994). It is a principal characteristic that bears an information about dynamic strength properties of materials subjected to the impact loading. In this section the connection between the threshold velocity W and the incubation time τ is investigated.

One of the principal features of the erosion process is that the material surface is subjected to the extremely short impact actions. The evaluation of failure in these conditions may be done only on the basis of special criteria reflecting the specific of the fast fracture mechanics. The incubation time criterion (3) is an effective means for this analysis. On the other hand, the erosion data can be used in the assessment of dynamic strength properties of the material. We shall consider the simplest way of determination of the incubation time on the basis of erosion data.

Let a spherical particle of the radius R falls with the velocity v onto the surface of an elastic half-space. Using the classic Hertz impact theory approximation (Kolesnikov and Morozov [16], 1989), we describe the motion of the particle by the following equation:

$$m\frac{d^2h}{dt^2} = -P, \tag{6}$$

where:

$$P(t) = k(R)h^{3/2}(t), \qquad k(R) = \frac{4}{3}\sqrt{R}\,\frac{E}{(1-v^2)}, \tag{7}$$

At the beginning of the impact event we have: $dh/dt=v$. The maximum penetration h_0 takes place when $dh/dt=0$. Solving the equation (6), we have:

$$h_0(v,R) = \left(\frac{5mv^2}{4k}\right)^{2/5}, \qquad t_0(v,R) = \frac{2h_0}{v}\int_0^1 \frac{d\gamma}{\sqrt{1-\gamma^{5/2}}} = 2,94\frac{h_0}{v} \tag{8}$$

where t_0 is the duration of the impact event. The penetration function $h(t)$ can be approximated by the simple formula (Kolesnikov and Morozov [16], 1989):

$$h(t) = h_0 \sin(\pi\, t / t_0) \tag{9}$$

The maximum tension stress at the contact area edge is given by the expression (Lawn and Wilshaw [17], 1975):

$$\sigma(t,v,R) = \frac{1-2v}{2}\frac{P(t,v,R)}{\pi\, a^2(t,v,R)}, \quad a(t,v,R) = \left[3P(t,v,R)(1-v^2)\frac{R}{4E}\right]^{1/3} \tag{10}$$

where the contact force $P(t,v,R)$ is to be found by (7)-(9).

Let $v=W$ is the threshold velocity corresponding to the beginning of failure. We consider the function:

$$f(\tau,v,R) = \max_t \int_{t-\tau}^{t} \sigma(s,v,R)ds - \sigma_c \tau \tag{11}$$

According to (3), we determine the incubation time as a positive root of the equation:

$$f(\tau,W,R)=0 \tag{12}$$

for the given W and R.

These formulae can be used in determining of the incubation time through the erosion data. The corresponding calculations were performed for the aluminum alloy. The threshold velocity W was determined experimentally and it turned out to be equal to 33m/s for $R=150\mu$m (Urbanovich et al., 1994). Calculations were conducted for the material parameters: $E = 73$GPa, $v = 0.3$ $\sigma_c = 460$MPa. It was found out that the function $f(\tau,W,R)$ had a single positive root. The calculated value of the incubation time turned out to be equal to 0.5μs. The temporal dependence of strength in conditions of spalling, calculated through the incubation time criterion (3) with the value $\tau=0.5\mu$s, is presented on the Fig.1. It is seen that the dependence is remarkable by the presence of both static and dynamic branches. The experimental data by Zlatin et al. [1] (1975), presented on the Fig.1, show a good corroboration between the spalling fracture experiment and the incubation time criterion in which τ is taken from the erosion data. It is noteworthy that approximately the same value of the incubation time for this particular material can also be obtained by the formulae (2) and (4):

$$\sigma_c = 460MPa, \quad K_{Ic} = 37MPa\sqrt{m}, \quad c = 6500m/s, \quad 2K_{Ic}^2 / \left(\pi\sigma_c^2 c\right) \approx 0.6\,\mu s$$

3 On the fracture at the crack tip due to asymmetric impact loading

The incubation time criterion (1) provides an effective way of determination of the direction of failure in crack containing materials. The corresponding example with respect to the shear band paradox (Kalthoff and Winkler [18], 1987; Lee and Freund [19], 1990) was published by Petrov and Utkin [20] (1994). Here we shall outline their results.

The particular boundary value problem to be analyzed is shown schematically in Fig.2. It can be described analytically by the following equations:

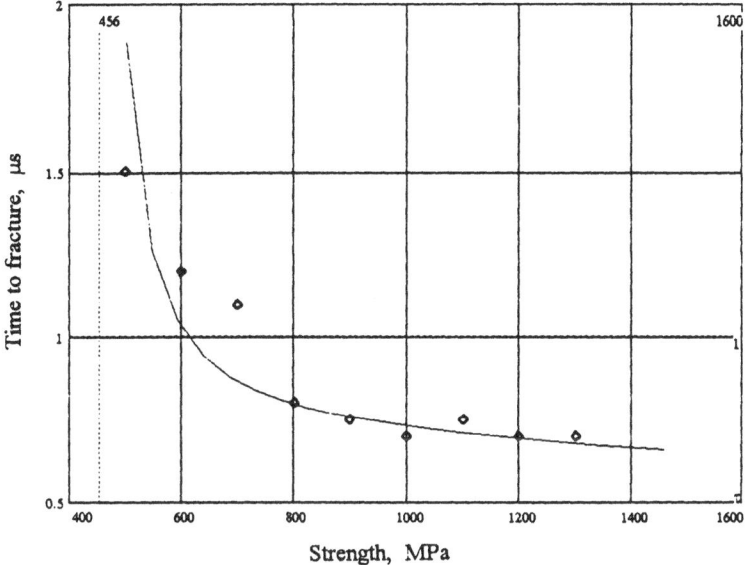

Figure 1. The temporal dependence of strength for the aluminum alloy B95 calculated through the incubation time criterion (3). The material constants are: $E = 73$GPa, $v = 0.3$ $\sigma_c = 460$MPa, $\tau = 0.5\mu$s. The experimental data are taken from the paper by Zlatin et al. (1975).

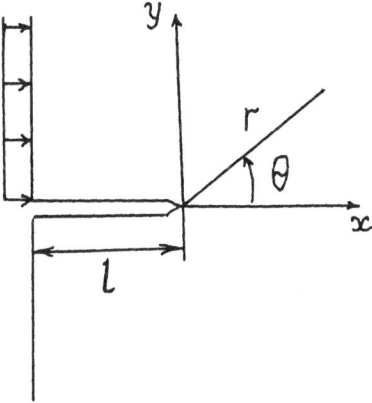

Figure 2. The edge crack asymmetric impact loading problem.

$$u_x = \frac{\partial \varphi}{\partial x} + \frac{\partial \psi}{\partial y}, \quad u_y = \frac{\partial \varphi}{\partial y} - \frac{\partial \psi}{\partial x}$$

$$\frac{\partial^2 \varphi}{\partial x^2} + \frac{\partial^2 \varphi}{\partial y^2} - a^2 \frac{\partial^2 \varphi}{\partial t^2} = 0, \quad \frac{\partial^2 \psi}{\partial x^2} + \frac{\partial^2 \psi}{\partial y^2} - b^2 \frac{\partial^2 \psi}{\partial t^2} = 0$$

$$a = 1/c_d = \sqrt{\rho/(\lambda + 2\mu)}, \quad b = 1/c_s = \sqrt{\rho/\mu} \tag{13}$$

$$\sigma_x(-l, y, t) = 0, \quad \sigma_{xy}(-l, y, t) = 0, \quad y < 0$$

$$u_x(-l, y, t) = \int_0^t v(t')dt', \quad \sigma_{xy}(-l, y, t) = 0, \quad y > 0$$

$$\sigma_y(x, \pm 0, t) = 0, \quad \sigma_{xy}(x, \pm 0, t) = 0, \quad x < 0$$

Here ϕ, ψ and a, b are longitudinal and shear displacement potentials and inverse waves speeds respectively; λ, μ and ρ are the Lame constants and the mass density; $v(t)$ is the loading velocity prescribed for non-negative values of its argument t. In our particular case we shall consider $v(t)=VH(t)$, where $H(t)$ is a Heaviside step function. The boundary value conditions are expressed in terms of the stress components.

The corresponding asymptotic solution for the stress components at the crack tip is given by the expression:

$$\sigma_{ij} = \frac{K_I(t)}{\sqrt{2\pi r}} \cdot f^{(I)}_{ij}(\theta) + \frac{K_{II}(t)}{\sqrt{2\pi r}} \cdot f^{(II)}_{ij}(\theta) + \sigma^{(R)}_{ij}(r, \vartheta, t) + O(r^{3/2}), r \to 0 \tag{14}$$

Here (r, θ) are polar coordinates at the crack tip.

The asymptotic solution we are going to use for the analysis consists of both singular and regular parts. To determine the direction of crack propagation, we have to apply a criterion of fracture that is irrespective to the presence of singularity. The incubation time criterion provides us such opportunity.

To determine the direction of crack propagation we shall calculate time to fracture t_* on each line extended at the angle $\theta(-\pi/2 \le \vartheta \le \pi/2)$ from the crack tip. We assume that the crack grows in a direction θ for which time to fracture is minimum.

The singular part of the asymptotic solution (was analyzed by Lee and Freund [19] (1990). The corresponding expressions for the stress intensity factors in can be taken from their paper. Combining it with the regular terms of the asymptotic (14) and using the criterion (1) we receive the following:

Calculations according to (14) has shown that propagating waves bear tensile stresses providing a tension near the line of symmetry: $x>-l$, $y=0$. This tension is not big in magnitude but it appears at the crack tip much faster than a tension in the singular stress field. Therefore, during a short time interval after the coming of the longitudinal stress wave to the crack tip a mean tension in the regular stress field prevails over a mean tension in the singular stress field. This entails a quite different behavior of the crack during a high rate impact event in comparison with what only the singular term analysis gives. Calculations were conducted for the material constants similar with what steels have. The incubation time was taken to be equal to $5\mu s$. The analysis of failure based on a combination of both singular and regular parts of the asymptotic solution and incubation time fracture criterion (1) allows one to predict three specific manners of the crack response. The results of the calculations can be summarized as following:

(a) At low loading velocity the crack does not propagate.

(b) When loading velocity exceeds the certain value, a failure is observed that in general corresponds to the fracture in quasi static pure mode-II conditions. In this instance, the crack catastrophically propagates at the angle around -78 deg with respect to the ligament, and its instability can be effectively predicted by the classical fracture mechanics (Lee and Freund, 1990).

(c) The further increase of loading velocity leads to the specific dynamic effect. The regular part of the asymptotic solution is very important for the early time range of the impact event, as the corresponding transient wave, produced by the impact, bears a tension that arises faster than the stress intensity factors. Thus, when loading velocity is big enough the crack propagates in a direction that is controlled by both parts of the asymptotic solution. It turned out to be approximately equal to +4 deg. This was observed in the experiments by Kalthoff and Winkler [18] (1987).

Therefore, the analysis has shown that the regular terms of the stress field asymptotic at the crack tip may be essential for a prediction of the specific failure behavior in dynamic.

Acknowledgments

The research described in this publication was made possible in part by Grant No. NW3300 from the International Science Foundation and Grant No. 94-01-01393-a from the Russian Fundamental Research Foundation.

References

[1] Zlatin, N.A., Pugavhev, G.S., Mochalov, S.M., and Bragov, A.M., 1975, Time Dependence of Strength of Metals, *Izv. AN SSSR, FTT,* Vol.17, No.9, pp.2599-2602 (in Russian).

[2] Nikiforofsky, V.S. and Shemjakin, E.I., 1979, *Dynamic Fracture of Solids,* Nauka, Novosibirsk (in Russian).

[3] Ravi-Chandar, K., and Knauss, W.G., 1984, An Experimental Investigation into Dynamic Fracture: 1. Crack Initiation and Arrest, *Int. J. Fract.,* Vol.25, pp.247-262

[4] Kalthoff, J.F., 1986, Fracture Behavior Under High Rates of Loading, *Engineering Fracture Mechanics,* Vol.23, pp. 289-298.

[5] Kalthoff, J.F., and Shockey, D.A., 1977, Instability of Cracks Under Impulse Loads, *Journal of Applied Physics,* Vol.48, pp.986-993.

[6] Shockey, D.A., Erlich, D.C., Kalthoff, J.F., and Homma, H., 1986, Short Pulse Fracture Mechanics, *Eng. Fract. Mechanics,* Vol.23, pp.311-319.

[7] Morozov, N.F., and Petrov, Y.V., 1990, Dynamic Fracture Toughness in Crack Growth Initiation Problems, *Izv. AN SSSR. MTT (Solid Mechanics),* No.6, pp.108-111 (in Russian).

[8] Petrov, Y.V., 1991, On "Quantum" Nature of Dynamic Fracture of Brittle Media, *Dokl. AN SSSR*, Vol.321, No.1, pp.66-68 (in Russian).

[9] Neuber, H., 1937, Kerbspannungslehre (Notch Stresses), Julius Verlag, Berlin.

[10] Novozhilov, V.V., 1969, Necessary and Sufficient Criterion of Brittle Strength, *Prikl. Matematika I Mechanika*, Vol.33, No.2, pp.212-222 (in Russian).

[11] Morozov, N.F., Petrov, Y.V., and Utkin, A.A., 1990, On the Analysis of Spalling in the Frameworks of Structural Fracture Mechanics, *Dokl. AN SSSR*, Vol.313, No.2, pp.276-279 (in Russian).

[12] Petrov, Y.V., and Utkin, A.A., 1992, Structure-Time Peculiarities of Dynamic Fracture of Materials, In *Macro- and Micro Mechanical Aspects of Fracture*, Abstracts, Proc. of the EUROMECH-291 (June 22-27, 1992, St. Petersburg, Russia).

[13] Petrov, Y.V., Morozov, N.F., 1994, On the Modeling of Fracture of Brittle Solids, *ASME Journal of Applied Mechanics*, Vol. 61, pp.710-712.

[14] Homma, H., Shockey, D.A., and Murayama, Y., 1983, Response of Cracks in Structural Materials to Short Pulse Loads, *J. Mech. Phys. Solids*, Vol.31, pp.261-279.

[15] Urbanovich L.I. Kramchenkov E.M., Chunosov Y.N., Morozov N.F., Petrov Y.V., Utkin A.A., 1994, Influence of Mechanical and Physical Properties of Both Solid Particles and Target Materials on the Critical Impact Velocity, *Proc. of the Int. Conf. (Arzamas-16)*.

[16] Kolesnikov, Y.V., and Morozov, E.M., 1989, *Contact Fracture Mechanics*, Moscow, Nauka (in Russian).

[17] Lawn, B.R., and Wilshaw, T.R., 1975, Indentation fracture: principles and application, *J. Mater. Sci.*, Vol.10, No.6, pp.1049-1081.

[18] Kalthoff, J.F., and Winkler, S., 1987, Failure Mode Transition at High Rates of Shear Loading, *Impact '87 (International Conference on Impact Loading and Dynamic Behavior of Materials)*, May, Bremen, West Germany.

[19] Lee, Y.J., and Freund, L.B., 1990, Fracture Initiation Due to Asymmetric Impact Loading of an Edge Cracked Plate, *ASME Journal of Applied Mechanics*, Vol.57, pp.104-111.

[20] Petrov, Y.V., and Utkin, A.A., 1994, On Fracture Initiation Due to Asymmetric Impact Loading, In *Experiment and Macroscopic Theory in Crack Propagation*, Abstracts, Proc. of the EUROMECH-326 (September 25-28, 1994, Kielce, Poland).

A Study of Fracture Behaviours in Ceramic Plates by Impact Load and Its Application to a Development of Hole-Punching Technique

Taketoshi Nojima[1] and Ken-ichi Sakaguchi[2]

1 Division of Aeronautics and Astronautics, Kyoto Univ., Kyoto,606-01, Japan
2 Division of Mechanical Engineering, Fukui National College of Technology, Sabae, 916,Japan

Summary

Sphere impact tests and static indentation tests were carried out in various kinds of boundary conditions to characterize fracture behaviours by point loading in brittle materials (a soda lime glass and an alumina) and also to investigate a possibility of hole punching in these plates. It was quite difficult to form a hole in these plates without serious radial cracks when the plates were not satisfactorily clamped. By precompressing the plates, punching works were performed in the glass and some engineering ceramics, and it has become clear that these plates can be well punched. The developed punching technique realized a wide range of punching from normal to quite small size holes less than 1mm as well as punching many holes by a single operation.

Key words: Impact, damage, fracture, punching, glass, ceramics, brittle material

1. Introduction

Extensive experimental works have been reported on the damage and fracture behaviours in some glasses and ceramic plates by impact load [1-4]. These works are mainly concerned with characterization for production of Hertzian ring crack and its growth as well as for the formation mechanism of severer damages such as median and lateral cracks. However, no systematic works have been reported to investigate a possibility of punching in these materials by impact load, and it seems that sufficient efforts have not been made in developing punching techniques. In the first part of this paper, sphere impact data are reported, and damage and fracture behaviours are characterized. In the second part a punching method in these brittle materials is presented.

2. Experimental method and specimens

A conventional air gun shown in Fig.1 was used for impact point loading tests. A toughened zirconia sphere (the diameter d=2 and 5mm) set in a duralumin sabot was propelled to impact onto glass or alumina ceramic plates. The target plates were put on steel block with a 30mm diameter hole (supported; named SU) or without a hole (rigid back; RB). Clamping tests (daimeter D=10 and 30mm; CL-D10, CL-D30) were also performed by installing a target plate in specially designed clamping apparatus (Fig.1 (b)) made of a steel frame and a compact jack. The clamping apparatus is able to precompress the plate up to 98KN around a punched hole. Static sphere indentation tests were also performed to clear and characterize the difference in damage behaviours between impact and static tests.

Impact punching tests were performed by experimental system shown in Fig.2, which consists of a loading (dropping) steel bar and a clamping device (the same as Fig.1 (b)). The bar is dropped (maximum fall height; 0.6m) for punching, and a load necessary for punching is directly measured by strain gauges attached to the bar. The top of the bar was machined to 3 kinds of shapes shown in Fig.3(a)~(c) (the top shapes of (a)~(c) are flat, spherical and blunted conical, and named F punch, S punch and C punch, respectively). Static punching tests were also carried out by using the same shaped punches.

Test samples for point impact (or indentation) tests are square cut (45×45mm) commercial soda-lime glass of 1.4~11.5mm in thickness (annealed at 500 ℃ for 30min.), and 1 and 2mm thick alumina ceramic (96% alumina, Kyocera). For punching tests, not only the alumina and the glass but toughened zirconia and silicon-nitride (1 and 2mm thickness, Kyocera and Nikato) were used.

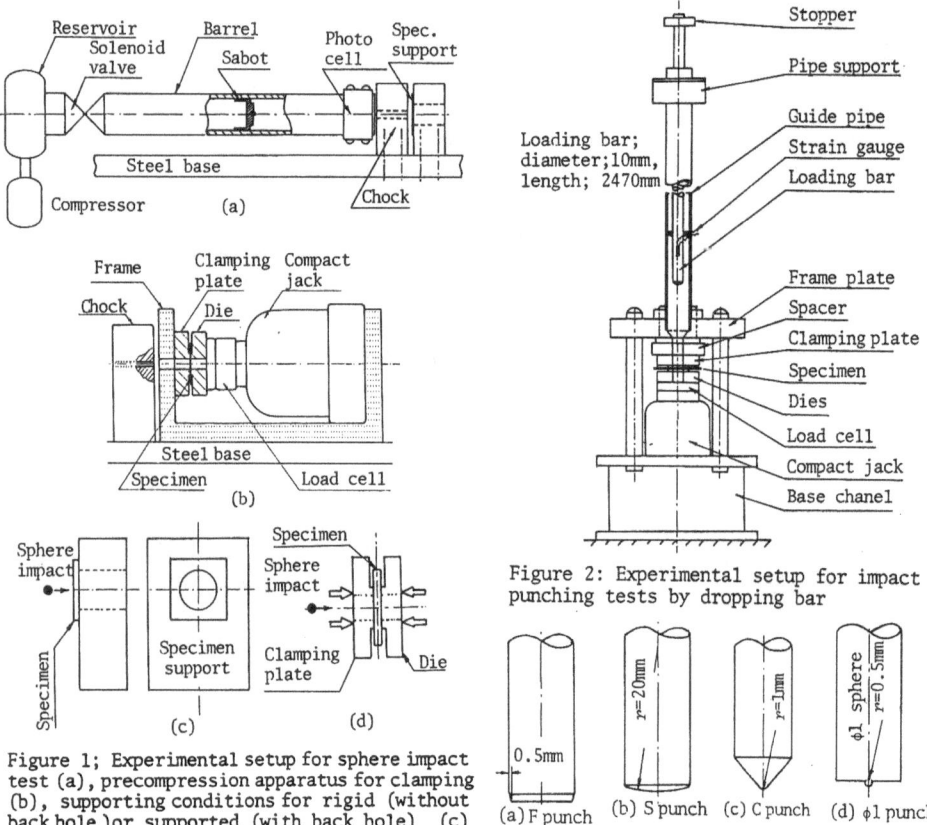

Figure 1; Experimental setup for sphere impact test (a), precompression apparatus for clamping (b), supporting conditions for rigid (without back hole)or supported (with back hole) (c) and clamping condition in (b), (d)

Figure 2: Experimental setup for impact punching tests by dropping bar

Figure 3; Top shapes of dropping bars (a)~(c) and punch tops for static test (a)~(d)

3. Experimental results and discussions

3.1 Sphere impact and static indentation tests

In order to get fundamental experimental data for damage and fracture behaviours in ceramic plates by impact load, toughened Zr sphere (2mm in diameter) was propelled onto the glass or the alumina ceramic plates by the apparatus shown in Fig.1. Impact velocities tested were in the range of 10~120m/s. Systematic impact tests were carried out in 1.4~11.5mm thick glass plates, because not only produced damages, cracks and their growth can be easily observed, but also a possibility of punching in these materials seems to be fairly judged by the data analysis. Damage and fracture patterns observed by the series of impact tests (SU-D30) are illustrated on impact velocity V_0-plate thickness diagram (Fig.4), and corresponding typical damages and cracks are shown in Fig.5. In Fig.4, the damage and fracture patterns are devided into 5 categories;

Region 0 ; no damage region,
Region I ; growth of Hertzian ring crack (HRC) to conical crack,
Region II ; clear chipping at impact site and median cracks,
Region III; heavy chipping, conspicuous median and lateral cracks (plate is substantially fractured) and
Region IV; formation of conical fracture (conical crack reaches rear surface).

Region II will be furthermore devided into II-(a) and II-(b), which are separated according to whether distinct median cracks have grown. Region IV is also devided into 2 regions; in IV-(a) (hatched in Fig.4), a conical fracture appears with quite smooth fracture surface (t=2.8mm) or with distinguished hackles (t=1.4~1.9mm), and IV-(b) is for high impact velocities where radial cracks appear around conical fracture. Fig.5 (a), (b) and (c) correspond to Region I, II-(a) and II-(b), respectively, and Fig.(d) and (e) correspond to Regions IV-(a) and IV-(b).

In Fig.4 for SU-D30, conical fracture appears at V_0=30~50m/s. In a series of the tests in RB condition, the conical fracture also appears at this velocity range and fracture patterns in thicker plates do not differ so much from the map shown in Fig.4.

Static indentation tests (d=2mm) of the glass were performed in the condition of RB, CL-D10 and CL-D30 (clamping stress σ_C=50MPa, V_0=0.5mm/min.). Characterized damage and

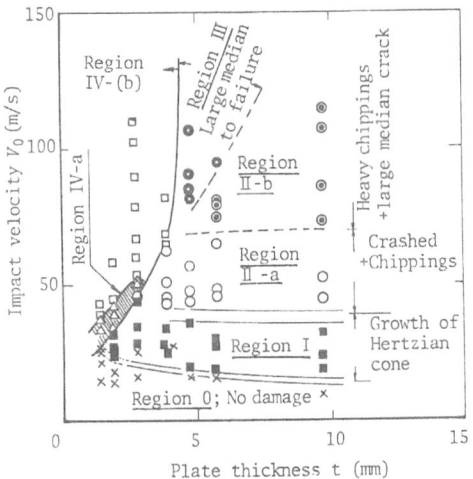

Figure 4: Damages and fracture pattern map in soda-lime glass impacted by 2mm diameter toughened Zr sphere (D=30mm, simply supported; SU-D30) (patterns are devided into 5 categories on impact velocity - plate thickness diagram).

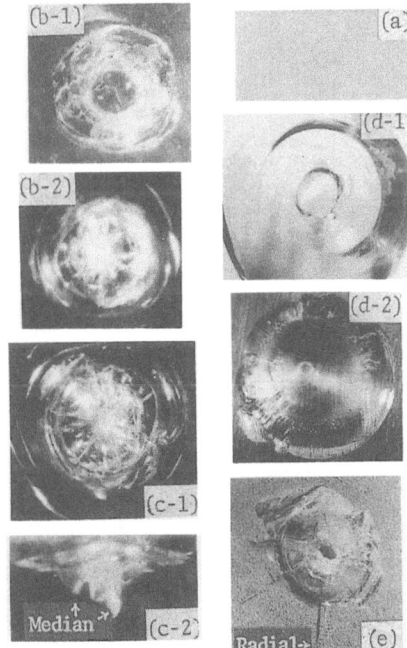

Figure 5: Photos of typical damage changes in thick glass plates (t>5mm) ((a)→(b)→(c)) and in thin plate ((a)→(d)→(e)) with increasing impact velocity (top views except for (c-2). (a); Hertzian ring and conical crack, (b); crashed at impact site (with small radial cracks), (c); median cracks (c-2; side view), note that conical crack turns to horizontal direction, (d-1); perfect conical fracture with smooth cracked surface (penetrated from center, t=2.8mm, V_0=50m /s) and (d-2); conical fracture with hackle (t=1.9mm. V_0=35m/s), and (e); radial cracks from conical crack at higher velocity (t=3.9mm, V_0=75m/s)

fracture patterns are illustrated on load P_I-t diagram in Fig.6 (a) and (b) for RB and CL-D30. Typical $P_I - \delta$ relations in these tests are shown in Fig.7. From the relations, P_{max} value and pop-in load (point A in Fig.7) are determined. They are shown in Fig.6 by double and hollow circles, respectively. Fig.6 shows that HRC is produced by a quite low load. Critical load for forming HRC, P_H, slightly increases as the thickness becomes thinner (the same trend is seen in impact data of Fig.4). The reason for the increase will be due to the increase of contact area by elastic deflection of the plate. In thick plate (t>5~6mm), the HRC grows to conical crack, it becomes multiple, subsequently loaded site is crashed, and finally median cracks start to grow. The thin plate (t<3mm) for CL-D30 is penetrated by star-like cracks producing a small hole at loaded point. In RB condition, the plate fractures by radial cracks (Fig.6 (a)). By comparing Fig.6 (b) with Fig.4, it can be noticed that damage and fracture behaviours do not fundamentally differ so much, excepting that in the thin plates, conical fracture is produced by the growth of HRC in impact tests, while in static test, the plate is penetrated with star-like cracks.

By the series of these tests, it becomes clear that a plate can not be penetrated without radial cracks, if clamping force σ_C is not enough. Therefore in order to study a possibility of punching work, the glass and alumina ceramic plates were clamped up to 200MPa, and a 5mm diameter Zr sphere was propelled by using the apparatus shown in Fig.1 (b). Typical examples for penetrated holes are shown in Fig.8. The photos show that the glass plate is well punched and the inside of clamped site is clearly crashed (Fig.(a), white region). In the alumina plate, radial cracks are completely arrested at clamped position (arrowed in Fig.(b)). These facts indicate that if the σ_C is enough, ideal punching work will be possible in the brittle materials.

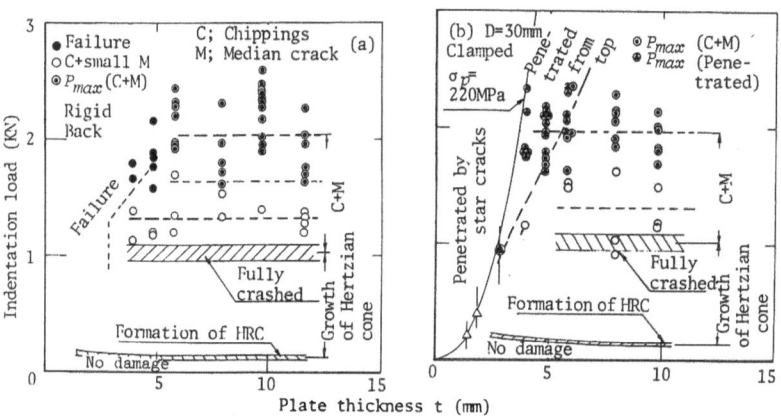

Figure 6; Damage pattern on indentation load-plate thickness diagram in soda-lime glass (V_0= 0.5mm/min., d=2mm), (a)rigid back (RB) condition, (b): supported (D=30mm) condition (SU-D30)

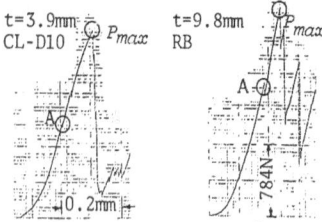

Figure 7; Indentation load P_I-displacement δ curves (V_0=0.5mm/min.)

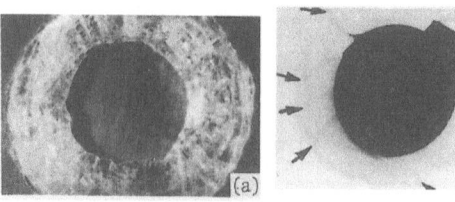

Figure 8; Photos of impacted glass (a) and alumina (b) plates (impacted by ϕ5mm Zr sphere, V=95m/s)

3.2 Punching tests

Impact punching tests were carried out on the glass and alumina ceramic plates by using the loading bars with 3 kinds of top shapes F, C and S punches in Fig.3 (a)~(c). Typical load-time traces obtained by the tests (V_0=2.4m/s) are shown in Fig.9 (a)~(c). In the shear type punching tests by F punch, P_{max}-value becomes quite large. Although thin plates (e.g.,t=1mm, D=10mm) can be well punched without radial cracks around punched hole under quite high clamping stress (σ_C=250~300MPa), the crack always appears in the punching works of thicker plates. On the other hand, in the tests by C and S punches, the P_{max} is quite low, and necessary clamping stress for ideal punching works becomes quite small; e.g., in the tests by S punch, it is about 60MPa (Fig.10). By these experimental facts, C and S punches are proved to be more available in the punching works of brittle materials. Therefore in the following punching works, these punches are used to investigate a practicable range of punching works.

Although the plates are loaded at the center in both punching works, punching process (crack formation mechanism) is different. The C and S punches have the top radii of 1 and 20mm, respectively. It has been well known that the necessary load to produce HRC, P_H, depends upon the diameter d of indenter [1]. Sphere indentation tests were performed to evaluate the P_H-values. The obtained results for the glass and alumina ceramic are shown in Fig.11 (a) and (b) (RB). The figure shows that the P_H values increase almost linearly with the diameter in both materials ($P_H \simeq$0.065d in the glass and $P_H \simeq$0.125d in the alumina, P_H in KN and d in mm); e.g., in the alumina, the P_H value in C punch (R=1mm) is 0.25KN, while the P_H value in S punch (R=20mm) is estimated to be 5.0KN. This implies that in the tests by C punch, HRC is easily formed at the impact site, while in S punch it is not easy to produce the HRC. In order to know the process of hole formation, it is evaluated in the following that either the stress for HRC or bending stress at the rear surface is predominant.

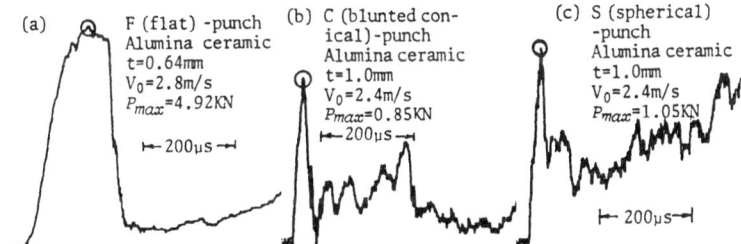

Figure 9; Load-time traces measured in loading bar

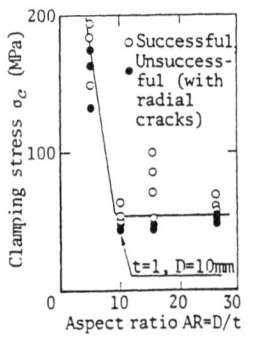

Figure 10; σ_C-AR relation (alumina)

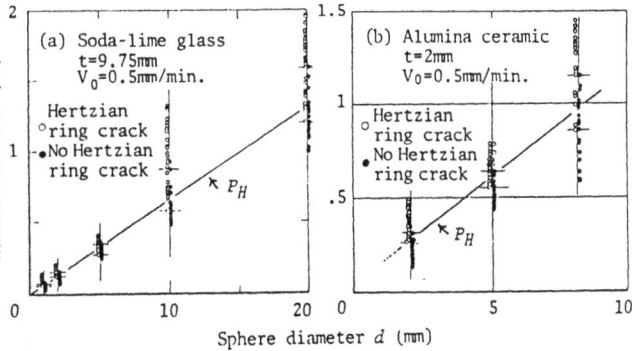

Figure 11; Critical load P_H for producing Hertzian ring crack

When a thin plate is clamped (diameter; D) and loaded at its center, the maximum bending stress σ^* (radial stress at the center of the rear surface) is given as,

$$\sigma^* = (1+\nu)P/t^2[0.485\ln(D/2t) + 0.52]. \tag{1}$$

Eq.(1) gives $P = (\sigma^* t^2)[(1+\nu)(0.485\ln(D/2t)+0.52]^{-1}$. Defining $\sigma_a = P/(\pi D^2/4)$ and putting $\sigma^* = $ constant (the bending strength), $\tilde{\sigma}_b = \sigma_a/\sigma^*$ is expressed by

$$\tilde{\sigma}_b = 4/[\pi(1+\nu)](AR)^{-2}[0.485\ln(AR/2) + 0.52]^{-1}, \tag{2}$$

where AR=D/t. $\tilde{\sigma}_b$-AR relation calculated by Eq.(2) is illustrated by thick solid line in Fig.12. On the other hand, by introducing an average indentation stress for forming HRC, $\sigma_H = P_H/(\pi D^2/4)$, and defining $\tilde{\sigma}_H = \sigma_H/\sigma^*$, $\tilde{\sigma}_H$ is expressed by

$$\tilde{\sigma}_H = P_H/(\pi\sigma^* D^2/4). \tag{3}$$

The P_H values in the alumina ceramic are about 0.25 and 5KN for C (R=1mm) and S (R=20mm) punches, respectively. By putting σ^*=350MPa, $\tilde{\sigma}_H$ values are obtained. The $\tilde{\sigma}_H$ values are shown by (horizontal) dotted lines in Fig.12. The figure shows that for AR=5, the stress $\tilde{\sigma}_H$ for forming HRC in S punch (point A) is larger than the bending stress $\tilde{\sigma}_b$ (point B), while in C punch, the $\tilde{\sigma}_H$ (point C) is smaller than the $\tilde{\sigma}_b$. In the case of AR=10, the bending stress $\tilde{\sigma}_b$ (point E) is smaller than the stress for HRC in both punches. This suggests that in a thin plate (AR; large), star-like radial cracks produced at rear surface will control the formation of a hole, while in a thick plate (AR; small), the formation of HRC and its growth are predominant in the punching works.

Crack formation mechanismin in the punching works will be summarized as follows; in the punching works by C punch (Fig.13 (a)), star-like cracks are produced at the rear surface and a thin plate is penetrated by the cracks showing a quite small hole at impact site (the diameter D^* at the rear surface is smaller than D). As the plate becomes thicker (Fig.(b)), a small conical crack is formed and grows to the rear surface with simultanious star-like radial cracks. In thick plate (Fig.(c)), a conical crack grows downwards with simultanious heavy chipping and lateral cracks at impact site. In the punching work by S punch (Fig.(d)), star-like cracks are always produced and a small hole is formed at impact site. Top diameter of the hole (D_u) is enlarged by the progression of the punch. Therefore the shape of the punched region is trancated conical. As the D_u reaches D, punching load increases rapidly [5]. The enlarging process

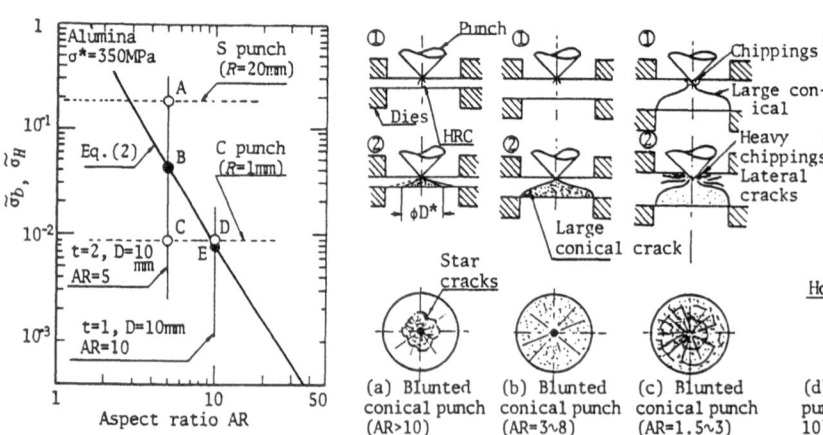

Figure 12; $\tilde{\sigma}_b$-AR and $\tilde{\sigma}_H$ - AR relation ($\tilde{\sigma}_b$; bending, $\tilde{\sigma}_H$; HRC)

Figure 13; Schematic illustration for punching processes

of the D_u in the C punch is fundamentally similar to this process. Corresponding photographs of the punching processes are shown in Fig.14.

3.3 Rate effect on Penetration load

A series of impact and static punching tests were performed in the glass (t=1.9 and 2.8mm) and the alumina ceramic (t=1 and 2mm)(D=10mm) by using C and S punches (the punch (d) in Fig.3 is also used in static tests of 2.8mm thick glass). Obtained loads for penetrating the plates (the P_{max} value in Fig.9) are illustrated in Fig.15 as a function of V_0. It can be seen from the figure that the P_{max}-value does not depend so much on the top shapes of used punches in thin plates (1.8mm thick glass and 1mm thick almina), while in thick plates (2.8mm glass and 2mm alumina), the P_{max}-value depends very much on the punches; the P_{max}-value increases considerably with the top radious. The trends of the dependences can be explained very well by the formation of HRC and its growth mentioned in Section 3.2. In both materials, the penetration load increases with loading rate. The n value in a power relation between P_{max} and V_0, $V_0 \propto (P_{max})^n$, is about 25~80 in static range. The rate effect seems to be larger in the impact region. However, in the series of the slow impact tests, characteristic features of impact effect have not been found by the observations of punched plate.

3.4 Punching of Engineering Ceramics and Their Products

Various kinds of ceramics were punched by impact loading tests by using spherical and blunted conical punches. Fig.16 (a) and (b) show punched holes in a glass plate (t=1.1mm) and alumina ceramic plate (t=1.0mm, D=10mm). Fig.16 (c) and (d) shows punched holes in silicon-nitride plate (t=1.0mm , D=10mm, bending strength 700MPa) and toughened zirconia ceramic plates (t=2mm, D=10mm, bending strength 1100MPa), respectively. In these ceramics, the P_{max}

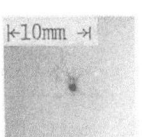   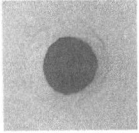

(a) Alumina, t=1mm, D=10mm (rear surface) (b) Soda-lime glass, t=2.8mm, D=10mm (rear) (c) Soda-lime glass, t= 3.9mm. D=10mm (side view) (d) Alumina, t=2mm D=10mm (rear surface)

Figure 14; Appearances of cracked ceramic plates (photos correspond to each processes in Fig.13)

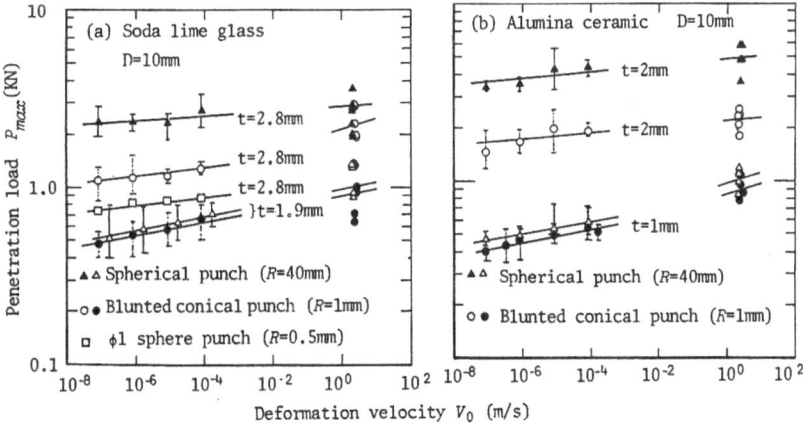

Figure 15; Penetration load (P_{max} in Fig.9) as a function of loading velocity

240

required to punch is about 5-6KN, and necessary clamping stress is about 230-270MPa for preventing radial cracks. Fig.16 (d) and (e) show quite small holes punched in 0.31mm thick alumina ceramic. Fig.16 (f) shows a bifoliation-shaped hole. These products show that the present punching method is quite stable in its processing. All plates are checked by fluorescent penetrant dye to check radial cracks around punched hole.

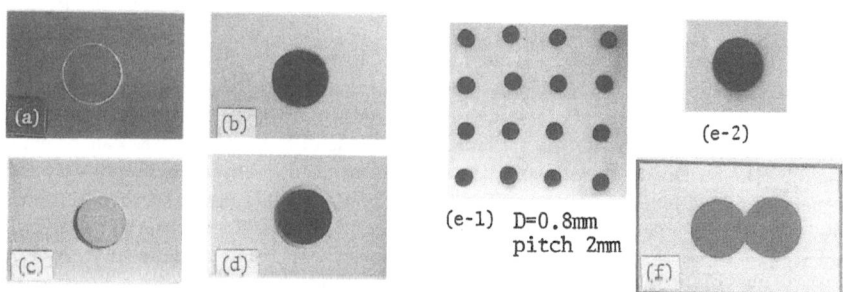

Figure 16; Appearances of punched ceramic plates, (a) slide glass (t=1.1mm), (b) alumina (t=1 mm), (c) silicon-nitride (t=1mm), (d) toughened Zr (t=2mm) and (e-1)alumina (t=0.31mm,D=0.8 mm, by single operation, hammered), (e-2) magnification of (e-1), and (f) bifoliation-shaped hole in 1mm thick alumina plate,(D=10mm except for (e-1))

4. Conclusions

A series of impact tests in brittle materials were performed not only to characterize damage and fracture behaviours but to study a possibility of punching works by impact load. Obtained experimental results are summarized as follows;

(1) crack formation and its growth due to a sphere impact onto ceramic plates were observed in a soda lime glass and alumina ceramic. In a thin plate, although conical fracture is formed and well spalled, sphere size hole can not be formed without severe radial cracks. In a thick plate, median cracks finally grow at impact site, and the plate also fractures.

(2) Experimental data for fracture behaviours indicate that if a thin ceramic plate is satisfactorily clamped, punching work will be well performed. In order to develop a punching technique for engineering use (without any radial cracks), a series of punching tests were carried out by using a blunted conical and spherical punches by clamping the plate. Thin glass or alumina plate (t=~1mm, D=10mm) can be well punched when clamping force (clamping stress around the punched region) is about 60MPa, and it is about 200~270MPa for tough engineering ceramics (toughened Zr and silicon-nitride, D=10mm, t=2mm). This punching technique has realized a wide range of punching including a formation of quite small hole less than 1mm and also punching many holes by a single operation. If the aspect ratio of the punched hole, AR=D/t, is larger than about 3, it seems that punching work is successful.

References

[1] J.R.A. Tillett, Fracture of Glass by Spherical indenters, *Proc. Roy. Soc.*, Vol.B69, 1956, pp.47-54

[2] C.G.K. Night, M.V. Swain, M.M. Chaudhri, Impact of small spheres on glass surface, *Journal of Materials Science*, Vol.12, 1977, pp.1573-1586

[3] Y.M. Tsai, H. Kolsky, A study of the Fractures produced in Glass Blocks by impact, *Journal of Mech. phys. Solids.*, Vol.15, 1967,pp.263-278

[4] J.E. Fields, Q. Sun, D. Townsend, Ballistic impact of ceramics, *Inst of Physics, Conf. Series No.102*, 1989, pp.387-393

[5] T. Nojima, K. Sakaguchi, F. Sugiyama, Formation of Holes in Engineering Ceramic plates by Press-Working, *Transactions of the JEME*, Vol.C-61, 1995,pp.1697-1702